K. K. Dompere

Cost-Benefit Analysis and the Theory of Fuzzy Decisions

Studies in Fuzziness and Soft Computing, Volume 158
Editor-in-chief
Prof. Janusz Kacprzyk
Systems Research Institute
Polish Academy of Sciences
ul. Newelska 6
01-447 Warsaw
Poland
E-mail: kacprzyk@ibspan.waw.pl

Further volumes of this series can be found on our homepage: springeronline.com

Vol. 142. M. Nikravesh, L.A. Zadeh, V. Korotkikh (Eds.)
Fuzzy Partial Differential Equations and Relational Equations, 2004
ISBN 3-540-20322-2

Vol. 143. L. Rutkowski
New Soft Computing Techniques for System Modelling, Pattern Classification and Image Processing, 2004
ISBN 3-540-20584-5

Vol. 144. Z. Sun, G.R. Finnie
Intelligent Techniques in E-Commerce, 2004
ISBN 3-540-20518-7

Vol. 145. J. Gil-Aluja
Fuzzy Sets in the Management of Uncertainty, 2004
ISBN 3-540-20341-9

Vol. 146. J.A. Gámez, S. Moral, A. Salmerón (Eds.)
Advances in Bayesian Networks, 2004
ISBN 3-540-20876-3

Vol. 147. K. Watanabe, M.M.A. Hashem
New Algorithms and their Applications to Evolutionary Robots, 2004
ISBN 3-540-20901-8

Vol. 148. C. Martin-Vide, V. Mitrana, G. Păun (Eds.)
Formal Languages and Applications, 2004
ISBN 3-540-20907-7

Vol. 149. J.J. Buckley
Fuzzy Statistics, 2004
ISBN 3-540-21084-9

Vol. 150. L. Bull (Ed.)
Applications of Learning Classifier Systems, 2004
ISBN 3-540-21109-8

Vol. 151. T. Kowalczyk, E. Pleszczyńska, F. Ruland (Eds.)
Grade Models and Methods for Data Analysis, 2004
ISBN 3-540-21120-9

Vol. 152. J. Rajapakse, L. Wang (Eds.)
Neural Information Processing: Research and Development, 2004
ISBN 3-540-21123-3

Vol. 153. J. Fulcher, L.C. Jain (Eds.)
Applied Intelligent Systems, 2004
ISBN 3-540-21153-5

Vol. 154. B. Liu
Uncertainty Theory, 2004
ISBN 3-540-21333-3

Vol. 155. G. Resconi, J.L. Jain
Intelligent Agents, 2004
ISBN 3-540-22003-8

Vol. 156. R. Tadeusiewicz, M.R. Ogiela
Medical Image Understanding Technology, 2004
ISBN 3-540-21985-4

Vol. 157. R.A. Aliev, F. Fazlollahi, R.R. Aliev
Soft Computing and its Applications in Business and Economics, 2004
ISBN 3-540-22138-7

Kofi K. Dompere

Cost-Benefit Analysis and the Theory of Fuzzy Decisions

Identification and Measurement Theory

Springer

Prof. Dr. Kofi Kissi Dompere
Howard University
Department of Economics
2400 Sixth Street, N.W.
20059 Washington, D.C.
USA
E-mail: kdompere@howard.edu

ISSN 1434-9922
ISBN 3-540-22154-9 Springer Berlin Heidelberg New York

Library of Congress Control Number: 2004106808

Springer is a part of Springer Science+Business Media
springeronline.com

Printed in Germany

Typesetting: camera-ready by author
Cover design: E. Kirchner, Springer-Verlag, Heidelberg
Printed on acid free paper 62/3020/M - 5 4 3 2 1 0

Dedications

To Joan Robinson,
Who always had the courage and talent
To take to task
The logical paradigm of established economic theory
And the corresponding ideological twists.

To Kwame Nkrumah
Who had the courage and vision
To work to decolonize Africa
And to democratize education in Ghana
Without which my scholarly works would not have
Been possible.

To Anton Wilhelm Amo Afer
The eminent Ghanaian philosopher scientist of 18th century
For his contribution to philosophy
And global intellectual heritage.

Acknowledgements

I wish to express my gratitude to all my friends, especially Professors Mbaye Cham of Howard University, Peter Akim of the University of Maryland at Baltimore, Kofi O. Nti, Director of the School of Administration, University of Ghana; Owura Mahama Bawa, John Kofi Christian, Alhaji Benson and Kwabena Nkrumah who have supported me in spirit and perseverance. Without their encouragements this monograph would have been difficult to finish. All of them have influenced my personality, productivity and vision in different ways. I also thank the initial members of the North American Fuzzy Information Processing Society (NAFIPS) for their vision of research on fuzzy logic, mathematics and technologies leading to paradigm shifting in information processing as a general theory of decision. I would like to express my gratitude to the following professors and scientists: J. Kacprzyk of Systems Research Institute of Polish Academy of Sciences for his initial encouragement when I presented the essential structure of my research of integrating cost-benefit analysis and theory of fuzzy decisions into a unified theory of decision making in a conference organized by NAFIPS; Lofti Zadeh for reading the initial draft with comments and encouragements; and also H. J. Zimmermann for his comments and suggestions. I also would like to thank the following people at Howard University: Dean Orlando Taylor and Associate Dean Emanuel Glakpe of The Graduate School of Arts and Sciences. Dr. Taylor for his vision of a research university and Dr. Glakpe for his vision of an excellent system of graduate programs and Ms. Santima Merchant Huff, the Administrative Assistant of the Department of Economics for her secretarial assistance. I have benefited from all comments of referees and participants in various conferences on decision sciences, operation research and economics.

All errors are my responsibility. Controversial ideas are intentional that are demanded by an attempt to shift the paradigm of cost-benefit analysis and broaden the fields of application in a manner that will hopefully enhance decision-choice rationality where fuzzy rationality encompasses classical and bounded rationalities.

Preface

> The genus of definitions for the theoretical sciences is (the province of) the habitus of the intellective intention, for the practical sciences, however, that of the effective intention; the objects and ends constitute the specific difference
> There is nothing in the intellect that has not already been in the senses, that is, in the sensory organs, that has not already been in sensible things from which are distinguished things not perceptible to the senses. Nothing can be inferred therefrom except the operation of the mind, sensation and the thing itself.
> Real learning is cognition of things in themselves. It thus has the basis of its certainty in the known thing. This is established in two ways: by demonstration in the case of contemplative things, and by induction in the case of things perceptible to the senses. In contrast with real learning there is possible, probable and fictive learning.
>
> Antonius Gvilielmus Amo Afer (1827)

This research has been long in the making. Its conception began in my last years in the doctoral program at Temple University, Philadelphia, Pa. It was simultaneously conceived with my two books on the Neo-Keynesian Theory of Optimal aggregate investment and output dynamics [201] [202] as well as reflections on the methodology of decision-choice rationality and development economics [440] [441]. Economic theories and social policies were viewed to have, among other things, one important thing in common in that they relate to decision making under different types of information and information processing regime. I have always viewed economic development and social policies from the viewpoint of prescriptive science rather than explanatory science as they relate to human activities in production-distribution-consumption decisions. As a matter of logical process socio-economic development must be viewed as on enveloping of past decisions guided by their own decision-choice rationality. The decision-choice space of human socioeconomic activities is vague and uncertain. The theories of socioeconomic decisions based on utility, for which economists have made impressive contribution, are riddled with problems of subjectivity, vagueness, limited information, soft data and the like. Vagueness and limited information lead to uncertainty of outcomes of the decision making process. Vagueness leads to fuzzy uncertainty while limited information leads to probabilistic uncertainty. A number of these problems of decision making are assumed away in the classical logical system of decision making. Those that are treated are modeled in terms of probabilistic uncertainty. The classical logic and its extension to probabilistic logic assume a mechanistic characterization of human linguistic communication in decision in terms of purity and exactness. The logic operates on a binary process of truth verification which is really satisfied in human communications and decisions.

Given these difficulties I invested resource in studying the utility foundation of decision theories. I was convinced that while the utility theory provides a powerful logical tool it is not sufficient to deal with the problem of general vagueness and hence different logical channels must be sought. I first went to the theory of latent variables to see whether there was some logical relieve. I was not satisfied or at least I did not find a way out. By accident I came across a flyer of International Journal of Fuzzy Sets and System at Temple University. My initial impression was that the knowledge of fuzzy sets and its logic may help the modeling of subjective phenomenon and latent characteristics in the theories of socioeconomic decisions and their extensions. From this initial impression I sought the knowledge of fuzzy sets and systems, its logical and mathematical foundations. This drive led to the writing of my essay on the theory of fuzzy decisions [694] that is directly related to the classic essay by Professors Bellman and Zadeh [661].

My understanding of classical logic its mathematics as applied to decision analysis to arrive at classical rationality, the critique of bounded rationality, and fuzzy logic and its mathematical foundation as applied to decision analysis led me to write an essay on the epistemics of decision-choice rationality [440]. In this essay I suggested that the framework of modeling *bounded rationality* or the boundedness in *classical decision rationality* is through the theory of fuzzy decisions that leads to *fuzzy rationality*. The point I was making in a nutshell is that the classical rationality is supported by its logical and mathematical paradigm while the fuzzy rationality is supported by its logical and mathematical paradigm. By shifting the paradigm of classical rationality to that of fuzzy rationality new insides may be obtained of old decision problems as well as solutions may be found for new problems in the decision analysis process.

The characteristics of bounded rationality are subjective phenomena in terms of limitations of human information-processing capability broadly defined to include linguistic vagueness, approximations and imprecise concept formations, all of which originate from perceptions and influenced by the ideology of the socioeconomic system given the information collection and recording structure. These characteristics are exactly part of the general elements that shape the environment of decision-making. Through the spectacle of fuzzy rationality these characteristics define fuzzy variables whose outcomes are conceived as a spectrum of grades or shades of quantity and quality from which exactness is abstracted by fuzzy logic and its mathematical processes. It may be pointed out that in general scientific modeling, theoretical construct and analysis we have to deal with quantity, quality and time no matter how poorly they are conceived and measured. The classical scientific work considered mostly quantity and time. The Marxian and Schumpeterian dynamics added quality. The Nkrumaist transformation dynamics added category and categorial conversion.

The implication of all these together is that the universe must be conceived in linguistic categories (sets) where elements slowly fade from one category to another through a logical process of categorial conversion. In nature, such categorial conversion takes place through natural processes. In social setting such categorial conversion takes place through human decisions with social institutions acting as the channels of transmission where lower levels of development fade into higher levels in terms of gradation. The gradation process and categorial conversion seem to fit into the fuzzy process that is guided by fuzzy rationality. Through this process, the fuzzy logic and its mathematics become powerful tools for dealing with subjective phenomenon such as utility and disutility. Utility is benefit and disutility is cost at any decision time point exhibiting the principle of duality in grades. With this conception, I was convinced that the logic and mathematics of fuzzy decisions can be extended to cost-benefit analysis where costs and benefits are viewed as subjective phenomena that are characterized by vagueness and imprecision requiring "soft computing".

Each decision alternative is viewed as cost-benefit duality. The selection of any alternative must thus be guided by cost-benefit rationality. Such cost-benefit rationality may be fuzzy or non-fuzzy depending on the set of assumptions imposed on the decision system by the analyst. The non-fuzzy characterization of the decision process simply equates humanistic system with perfectly mechanistic system. By so doing we assume away all the linguistic characteristics of human communication by simply reducing the admissible structure of the information set that supports the decision making. Without going into an epistemic discussion on fuzzy rationality, it can be shown that the information support of fuzzy rationality is greater than the information support of the classical rationality whether we are dealing with random sets or fuzzy sets. For each random set there is a corresponding fuzzy set but not the other way around. Random sets are subsets of fuzzy random sets.

The task of connecting cost-benefit analysis to fuzzy decision began when I stared teaching a graduate course in cost-benefit analysis. I saw cost-benefit analysis as a general theory of decision. As such I rejected the case-study method as an effective way of presenting its framework for decision analysis. Cost-benefit analysis must be considered as a general theory of decision-making in both fuzzy and non-fuzzy environment. The epistemic position puts all problems of optimization in cost-benefit footing. In this respect, there are two broad classes of optimization where the objective function is viewed as either benefit or cost while the constraint set is viewed as either cost or benefit. All decision problems are either to maximize benefit subject to cost or minimize cost subject to benefit. The units of measurement of costs and benefits will vary across decision problems. The general objective, however, is to decide on the alternative that optimizes the welfare of the system whether conceived as humanistic or non-humanistic.

Complementing the view of cost-benefit analysis as a general theory of decision making is the view that within the epistemics of science, cost-benefit analysis belongs to prescriptive science but not explanatory science. Cost-Benefit analysis as a prescriptive science has as its primary goal the improvement of "what there is". It is not directed toward the explanation of "what there is", the known reality. It is neither directed toward predicting the occurrences of events based on the current reality. The main objective of cost-benefit analysis is to construct action-based rules for selecting optimal act in other to bring about desired outcomes into an environment where the outcomes were previously not in existence. In other words, cost-benefit analysis is to construct decision algorithms in other to select the potential among the set of conceivable potentials for actualization. The driving force is the improvement of reality through cost-benefit balances in humanistic and non-humanistic systems.

In this respect cost-benefit analysis is a rational construct of model representation of human decision and choice behavior. It simply translates into construction of human decision rationality on the basis of costs and benefits to guide optimal selection process by recommending the best alternative. The intelligent process in recommending the best alternative on basis of cost and benefit is *cost-benefit rationality*. The cost-benefit rationality is supported by a cost-benefit paradigm that simply states that every decision element may be conceived as a control system composed by its welfare and the alternatives as potential controllers that may be selected to maintain or improve the system's welfare. The development of cost benefit rationality requires information on cost and benefit that are associated with each alternative in the choice set given the goal or objective of the decision.

The process of cost-benefit analysis requires two broad steps:
1. the development of real cost-benefit databases
2. the construction of cost-benefit criterion

Step one may be divided into sub-steps that may be specified as
   a. real micro-data construct through identification and measurement of real cost and benefit characteristics;
   b. real and nominal macro-data construct given identification and measurement of cost and benefit characteristics through the use of efficient prices.

The real micro data construct requires the identification of real cost and benefit characteristics and the manner in which they are distributed over entities. The real and nominal macro-data construct requires overcoming cross-sectional aggregation of heterogeneous real characteristics of costs and benefits.

Step two may be sub-divided into:
   a. aggregations of intertemporal cost and benefit cross-sectional aggregates into present value equivalences through a choice of efficient social rate of interest and a discounting process;
   b. the construction of criterion index from the present values in order to induce ranking over all the elements in the choice set.

The two steps are accomplished with a number of theoretical constructs that provide sets of self-contained algorithms for application. This volume is devoted to identification and measurement theory of costs and benefits in an environment where the information is fuzzy, broadly defined. The objective is to first identify costs and benefits, and the units for their real accounts.

The book is organized into six chapters. Chapter 1 discusses the relationships among decision, cost and benefit in order to introduce the subject to all interested decision agents. Chapter 2 presents the logical structure of the theory of computable cost-benefit identification matrices as a framework for identifying and measuring real cost and benefit characteristics for any decision alternative. Chapter 3 presents a fuzzy-decision theory of social cost and costing within the logical framework of the theory of computable cost-benefit identification matrices. Chapter 4 is devoted to the development of fuzzy decision theory of social benefits and benefit accounting in order to complete the development of the real cost-benefit databases. The cost and benefit accounting through the logic of computable identification matrices is presented as soft computing where subjectivity, inexactness and vague characterizations in the identified cost and benefit are accounted for.

Chapter 5 presents the essential mathematics for fuzzy decision processes while Chapter 6 is devoted to the basic theory of fuzzy decisions in order to make Chapter 3 and 4 easily assessable to those familiar with cost-benefit analysis but unfamiliar with research and literature on fuzzy decision and its mathematics. Chapter 2 is neutral that may be useful in all areas of decision making. The book is completed with references that are grouped under subject areas. This grouping of references will help to expose the multi-disciplinary approach for the development of cost-benefit decision making in the framework of soft and fuzzy computing. The theories of efficient prices, social discount rate, aggregation of real costs and benefits, social discounting and cost-benefit criteria are developed in a separate volume entitled *Cost-Benefit Analysis and the Theory of Fuzzy Decision: The Fuzzy value theory* [931b].

Table of Contents

1 Decision, Cost and Benefit

Social development translates into a dynamic sequence of success-failure experiences of human socioeconomic transformation. The success-failure experiences rest on the foundations of human decisions and choices whose outcomes establish the path of social changes. All decisions have outcomes. The outcomes are judged to be good (desirable) or bad (undesirable) depending on whether they satisfy the objectives and goals of the decision agent. The set of rules that leads to realization of good decisions and choices constitutes the *decision-choice rationality* [440]. The sequence of steps for implementing these rules of good decisions is the decision algorithm. The rules of good decisions obtained from explanations and predictions may form the basis to prescribe an optimal decision behavior.

1.1 Decision and Choice

The systematic development of the system of explanations and predictions of observed regularities in good and bad decisions relative to goals and objectives is viewed as decision theory. The scope of interest of decision theory may be seen as the critical examination of goals and objectives, the meaning of selection rules and the systematic development of consistent algorithms to realize those rules that must be prescribed and followed to realize successful decisions and choices. The set of selection rules for successful decisions and the constructed algorithms for realizing good decisions forms the *calculus of decision* on the basis of which wise choices are made.

Generally, decision theory may be either explanatory or prescriptive or combination of both. A decision theory is said to be *positive* when it is restricted to the domain of explanation and prediction on the basis of data and the explanatory principles that it implies. It is, however, said to be *normative* when it is restricted to the domain of prescriptions that maximize the probability of successful decision outcome through the practice of prescriptive rules. Prescriptive theory is not only normative but *ex-*

ante in practice. Explanatory theory on the other hand is not only positive but *ex-post* in logical practice.

Implied in the normative theory of decision is a presupposition that the decision agent is knowledgeable about algorithms and has full understanding of successful decisions. Cost-benefit analysis and all its ramifications, to which we shall attend, fall under the umbrella of prescriptive theory of decision and choice. In other words, it is principally ex-ante decision analysis. The objective here is to decide on the highest-welfare benefit project to which resources are to be committed before practice. Ex-post decision analysis through cost-benefit framework can also be done for the purpose of learning and decision adjustments.

1.2 A Reflection on Cost-Benefit Analysis

Cost-benefit analysis may be considered as an input-output process in decision making in general. In this respect cost-benefit analysis projects a general ex-ante algorithm of decision and choice. Every decision has associated costs and benefits. The measurements in which costs and benefits are registered for different types of decision may vary. The relative standing of the overall costs and benefits defines the health and efficiency of the decision. In other words, it defines the betterment of the system on which the results of the decision impact.

Considered either as prescriptive theory of decision and choice or as a system of applied calculi of decision making, cost-benefit analysis draws its analytical foundations form the theory of welfare economics and more generally from *normative science*. Epistemically, cost-benefit analysis belongs to prescriptive science [441]. The concept of welfare must be broadly interpreted if cost-benefit analysis is to provide general theoretical and applied principles of decision making over a broad spectrum of social activities. For example, the general concept of optimization, composing of objective functions, constraint set, choice variables and algorithms to arrive at optimum, defined in specific sense is nothing more than cost-benefit analysis.

The whole decision construct is to optimize the systems welfare irrespective of whether it is humanistic or non-humanistic or mechanistic. Thus whether we are thinking in terms of time, distance, goods, health, risk, education, military engagements, enactment or repeal of a statute, alternative social policies and many others, we are thinking about relative composition of costs and benefits and how they affect the general health of the system. It is this fundamental framework of relative position of costs and benefits that provides the methods and techniques of cost-benefit analysis their analytical power, acceptance and broad applications.

Every decision is a judgment. Such a judgment may or may not be information supported in the sense of positive or normative science. Cost-benefit analysis constitutes a normative paradigm for decision making and feasibility studies. Even though cost-benefit analytic framework is generally accepted by all sectors of different societies there are some important criticisms that analyst must address in the application and suggested decision. Some of these criticisms demand special analytic attention. The framework allows decision agents to demonstrate the intelligence and soundness of their decision and recommended choices before they are implemented. In other words it meets conditions of some form of rationality. Some normative aspects of the intelligence and soundness may be sensitive to some of the criticisms.

The areas of decision that the methods and techniques of cost-benefit analysis have been applied to are numerous and vary in structure and form. Among the areas of application, one may mention the valuation of life and injuries in insurance and litigation, disease control, environmental problems, water resource, information technology, social infrastructure problems, social regulations and safety, and many others. Since the areas of application are many, the practitioners and users of cost-benefit analysis must be aware of the strength and limitations of the sets of tools and methods of analysis that come with the framework. This awareness will allow the development of an ability to appreciate and resolve theoretical difficulties and new problems which invariably will arise in the practice of decision and hence choice in many areas of human life as the terrain of application is broadened to include decisions in many areas of human life and activities involving humanistic and non-humanistic systems.

These difficulties of applying cost-benefit methods may be traced to the problem of theoretical understanding of normative science to the extent that it implies prescriptive propositions that are ultimately derived from a set of acceptable norms of the society. To compare is to evaluate and to choose is to judge. All decisions and policies require some norm of judgment regarding their "relative good" to the decision agent or to the person on behalf of which the decision is made. Implied in this statement is the concept of decision-choice rationality [440] [405a]. As it will be argued, a decision problem is said to exist if more than one alternative are present for selection. This is also true of choice problems. Anytime that there is a decision and choice to be made, the problem of establishing a rank order over available alternatives emerges. The calculus of cost-benefit analysis takes claim to providing us with such a rank order. The rank order is only meaningful relative to goals and objectives irrespective of whether the decision is social or individual, humanistic or non-humanistic, or mechanistic. For normative science to be useful as a guide to decision and choice, it must not only lead to the establishment of ranking of alternatives but also must lead to the establishment of desirable social goals within the social norms. Furthermore, it must lead to a critical examination of the acceptable social norms against which desirable social goals and objectives are developed. Normative science rests on ethical foundations which are also determined by it. In other words, they mutually create and determine each other. The goals and objectives indicate in some form the potential welfare of the system.

For normative science to be complete and accepted as prescriptive science it must also examine the norms of society and evaluate them without appeal to divine justification of the social norms. It can of course appeal to the imagination of man where such imagination might have evolved through long periods of human experience. In this respect, subjectivity underlies normative science which draws its analytic power from people's imagination where such imagination leads people to set the potential against the actual by creating new conditions in replacement of the old. For human imagination to guide future through setting the *potential* against the *actual* it must impel a sense of right or good based on either a general consensus or the consensus of a decisive class who determines and sets social goals as well as decides the path of the evolving social history. If such a social consensus is assumed then, it is hoped that social decisions should not run counter to such a social consensus. For the methods and techniques of cost-benefit analysis to be useful to all kinds of social decisions, the cost benefit analysis must provide a way of arriving at social consensus within the norms of society. It must also help to determine the net social welfare implied in the potential relative to the actual.

The traditional approach to the problems of social decision and choice is to base cost-benefit analysis on welfare economics within the market setting. The approach is such that social and individual decisions are put in the same analytical footing. The analytical framework is marginalism where net social welfare is at maximum when marginal social cost is equal to marginal social benefit on the aggregate. This becomes an optimal decision rule if the social objective is maximum welfare. The question that arises is: what is the meaning of social welfare and how does one measure and compute it?

Within the traditional welfare economics, social welfare is equated with allocative efficiency based on market setting. In the light of this, two allocative maxims have been developed. The first allocative maxim is that the decision analyst accepts as his or her objective data the subjective evaluations by the individual members of society regarding their welfare at any given social state. The second allocative maxim is that social states or choices can be ranked by aggregate value of individual net gains. The first and second maxims are supported by a distributional maxim which states that social welfare is "higher" if a distribution of social output is "better" in some specified sense. In cost-benefit analysis net gains and net welfare are viewed in terms of net benefit of the system.

A number of theoretical and practical difficulties emerge when cost-benefit analysis is abstracted from the underlying foundations of welfare economics that are based on market imputations. First we encounter the problem as to how do individuals compute their personal net gains in any given social state that may be potentially consequent of a social decision. Second, we are confronted with the problem of the units of measurement of the net gains and whether such net gains can provide consistent ranking of individual preferences over social states and whether such ranking can be maintained over intertemporal states. Third, there is the problem of aggregation of individual net gains that would allow the social net gains to be established over social states in such a way as to produce social (collective) rank order over social states or alternatives for social choice.

In order to provide solutions or answers to these problems or questions, one may adopt the utilitarian view that individuals derive some enjoyment from each possible alternative. The individuals seek to maximize such enjoyment where such enjoyment is summarized by utility and represented by utility indices. But the maintenance of such an enjoyment in each possible state has a cost. This cost, from the utilitarian view, can only be seen as disutility. Thus the individual net gains in any social state are represented by his net utility as subjectively valued.

Since utility is not a directly measurable concept at least in terms of our current scientific knowledge or, is an ordinal measure, the problem is to find a cardinal measure that will not only preserve individual ranking over states but will allow for stable aggregation of individual net gains for each possible state. In a social organization based on market arrangements such measures are taken to be produced by market in terms of monetary prices if certain market configuration and information structure are assumed. Thus, market prices, costs and revenue, on the basis of which social costs and benefits are computed, are measured in the utility space. If the required market configuration does not exist then the uses of these prices may be important distortions in evaluation of social states in terms of net gains.

In actual practice of individual and social decisions, such market configuration does not exist. There are numerous kinds of individual and social decisions that have cost and benefit characteristics that markets do not exist for their evaluations. Even if markets exist for all benefit and cost characteristics for any possible social state we are confronted with the problem as to whether equal weights must be assigned to individual votes (in terms of their monetary endowment) in the market place. In other words, should the evaluation include the concept of fairness which of course is neglected by the allocative maxims? This concept of fairness forms the underlying conflict in the subjective evaluation of social states. In many cases it is an overriding factor in ranking social states. Ironically, market prices cannot be found for benefit-cost characteristics associated with alternative distributional configurations.

Finally, the process whereby the social net gains are based on individual gains presupposes that individuals know "what is best" for them at all times and their subjective evaluations represent "what is best". There are two important problems if such presupposition is accepted as a foundation for cost-benefit analysis. Should the proposition that "individuals know what is best for them" be taken as an ideology or as a scientific reality? Three possible answers may be provided. 1) The proposition may be taken as a factual judgment which is generally true but difficult to prove. 2) The proposition may be taken as a *fundamental ethical postulate* in the sense that a "good society" is one that ought to organize and act as if the individuals know what is best for them than any other person. 3) The proposition may also be taken as a *postulate of political_expediency* in the sense that it is suitable for a particular type of political organization for the individuals to act as if they know what is best for them.

The proposition as a postulate of factual judgment suffers from the problem of empirical verification and hence cannot be demonstrated neither to unquestionably exist nor can it be taken as a logical truth. The proposition as an ethical postulate suffers from the problem of subjective judgment and cannot be taken as a universal ethical truth. The proposition taken as a political expediency suffers from the same problems as when it is taken as an ethical postulate. If the proposition is taken as an ethical postulate then social decisions are guided by ethical norms. If it is taken as a political postulate, then social decisions and outcomes are guided by the prevailing political norms. On the other hand if the proposition is of a factual postulate then social decisions are guided according to the norms of nature.

Another important problem is that, even if one of these postulates is accepted and that individuals know what is best for them, does it mean that on the aggregate what is best for the individual may be best for the society as a whole? Furthermore, should all individual best interest be counted in the process of arriving at the general collective good? Another important question is whether the individual ranking of social states according to his best interest will coincide with his ranking according to what he sees as social good if he could isolate himself? Finally, how do we integrate the concept of fairness into the individual evaluative process of social states?

The utilitarian welfare-economics approach as a foundation for cost-benefit analysis suffers from the assumed conditions of existence of a well-behaved utility space for the computation of social welfare optimum even if most of these questions can meaningfully be answered. Furthermore, there is no reason for cost-benefit analysis of social decision to derive its foundation from methodological individualism as has been emphasized in most literature. In particular, there is no reason why cost-benefit analysis of social decision should be based on market mechanism. Any method that allows social costs and benefits to be computed in such a way that it meets logical consistency within the structure of science in acceptable. The critical problem is to design algorithms that allow social costs and benefits to be computed within some measurable domain. The basic objective is to establish rank order either on social states or alternatives for choice. The process and the algorithms should be free from any institutional configuration since the area of applications may be very broad to include mechanistic systems. However, they must be flexible so as to be practically relevant to social decisions for any given institutional configuration.

Another important difficulty in the use of the traditional welfare economics as foundations of cost-benefit analysis is that it is very difficult to construct a theory of social rate of discount in an attempt to link either the present to the future or the past to the present. The difficulty increases in intensity when one must establish a time process where each time has an assigned weight that connects intertemporal values of costs and benefits to the present. These weights from the point of view of intertemporal social decisions cannot be established by appeal to the utility of intertemporal consumption in a fashion of Fisher [367]. To date commodities is to accept a particular time order. To connect intertemporal commodity types to equivalent values, commodity weights are needed at each time point. In traditional cost-benefit analysis these intertemporal weights are taken as given outside the domain of decision. To define these weights in terms of utility of consumption is to assume the problem away. There is no pressing reason why the social discount rate must be taken as a market outcome. Even from the utilitarian tradition, it is easily shown by following Fisher [367] that the individual rates of time preference are measured in the utility space just as all prices are measured in utility space.

When one considers time (given that time has been properly constructed) in a decision domain, one may speak of *benefit time* and *cost time* as seen from the utility space. From the decision point of view the same time point appears as dual as such to view the rate of discount in terms of intertemporal consumption without explicit qualification amounts to viewing it in terms of benefit time alone. Of course intertemporal consumption tradeoff may be viewed simultaneously as benefit and opportunity cost. This approach which may be attributed to the Fisherian school neglects the problem of cost time as valued differently from benefit time. The rate of discount either private or social must be computed from the underlying conflict between cost time and benefit time as seen in the utility space. In this respect, there is no theory of discount rate.

1.3 A General Analytical Structure of Cost-Benefit Analysis

The basic logical framework for performing cost-benefit analysis involves a social decision space, $\mathbb{D}$, composed of a space of objectives, Ω, a space of alternatives, $\mathbb{A}$, and a cost-benefit space, Θ. The space of objectives defines the subjective elements around which selections are made for their realization. The space of alternatives defines a set of competing elements from which a selection is to be made to realize a chosen objective. The cost-benefit space defines the criterion for and constraints on the selection. The decision space defines the social vision and the possible paths of the future history. Given these spaces, the social decision space may be formally defined.

Definition 1.3.1

A social decision space, $\mathbb{D}$ is composed of triplet of spaces of social objectives, Ω alternatives, $\mathbb{A}$ and cost-benefit space, Θ whose generic elements are, δ, ω, α and θ respectively. Thus

$$\mathbb{D}=\Omega \otimes \mathbb{A} \otimes \Theta = \left\{ \delta = \left(\omega, \alpha, \theta \right) \mid \omega \in \Omega, \alpha \in \mathbb{A}, \theta \in \Theta \right\}$$

where $\otimes$, is a Cartesian product for sets.

In order to operationalize the definition 1.3.1 we need an explicit understanding of the nature and form of Ω, $\mathbb{A}$, and Θ.

Definition 1.3.2

The space of objectives, $\Omega(\beta,t)$ with generic element $\omega(\beta,t)$, is a collection of all possible elements that may be followed by an individual organization or a society at any given time, *t*, and a social parametric norm β.

The space of objectives is said to be social if it pertains to the society. It is said to be organizational if it pertains to an organized collective that is a sub-set of the society. It is said to be personal if the objectives are held by an individual. The space $\Omega(\bullet)$ will also be referred to as an objective set. The space of objectives may be partitioned into attainable set, $\mathbb{V}$ and non-attainable set, $\mathbb{V}^c$ with generic element ν. Thus

$$\Omega = \left\{ \mathbb{V}^c \cup \mathbb{V} \mid \nu \in \mathbb{V} \Rightarrow \nu \notin \mathbb{V}^c \right\}$$

The elements in sets $\mathbb{V}$ and $\mathbb{V}^c$ have relationship to each other. A decision problem is said to be *multiobjective* if $\#\mathbb{V} > 1$. It is said to be *no-objective* decision problem if $\#\mathbb{V}=0$ and it is said to be single-objective decision problem if $\#\mathbb{V} = 1$.

Definition 1.3.3

The space of decision alternatives, $\mathbb{A}(\omega)$ with generic, $\alpha(\omega)$ is a set of programs for a choice action relative to $\omega(\bullet) \in \Omega(\bullet)$. It is said to be non-trivial if $\#\mathbb{A}>1$, null if $\#\mathbb{A} = 0$ and trivial if $\#\mathbb{A}=1$ given the space of decision objectives, $\Omega(\bullet)$.

The definition 1.3.3 conceptually establishes a set of alternative programs whose choice and implementation have a potential in bringing into attainment the elements in the objective set, $\Omega(\bullet)$, that are attainable. In order to select any of these programs we need to examine and analyze for choice the set of programs that may lead to the accomplishment of an objective $\omega(\bullet) \in \Omega$. Each element $\omega(\alpha) \in \Omega(\bullet)$ may have many different characteristics that may be complementary, supplementary or competitive (conflicting) in the choice process. These characteristics allow each $\alpha \in \mathbb{A}$ to be identified and distinguished. Thus for any given $\alpha(\bullet) \in \mathbb{A}(\omega)$, there would be a set of characteristics.

Definition 1.3.4

The general set of attributes from which each $\alpha(\bullet) \in \mathbb{A}$ may be uniquely identified for selection is called a characteristic set, $\mathbb{X}$.

The decision problem is said to be multi-attribute if $\#\mathbb{X} > 1$, a single attribute if $\#\mathbb{X} = 1$. There is no element $\alpha \in \mathbb{A}$ if $\#\mathbb{X} = 0$ for any α. The symbol $\#\mathbb{X}$ denotes the number of elements in the set $\mathbb{X}$ or may be viewed conceptually as the number of characteristics or attributes. If the characteristics are fuzzy, we may elect to call the set as a *fuzzy bag*. Each attribute $\mathrm{x} \in \mathbb{X}$ may either enhance or hinder an alternative $\alpha \in \mathbb{A}$ in helping in the attainment of $\omega \in \Omega$. Such enhancement and hindrance are cast as technical benefits and costs respectively. In order to exercise choice action on $\alpha(\bullet) \in \mathbb{A}(\omega)$ to effect the outcome of $\omega \in \mathbb{V} \subset \Omega(\bullet)$ we need to examine the real costs and benefits of the characteristic set that applies to $\alpha \in \mathbb{A}$ for any given objective space, $\Omega(\beta, t)$. The logical process involved in the analysis of costs and benefits associated with the characteristic set, $\mathbb{X}$, leading to choice action on $\alpha \in \mathbb{A}$ given $\omega \in \Omega$ is called cost-benefit analysis. Let us look at the general analytical structure.

Definition 1.3.5

The cost-benefit space is composed of real cost and benefit attributes that contribute negatively and positively to the general preference of an element $\mathrm{x} \in \mathbb{X}$ leading to the selection of $\alpha(\bullet) \in \mathbb{A}(\omega)$ in the accomplishment of objective, $\omega \in \Omega(\beta, t)$.

Definition 1.3.6

If the characteristic $\mathrm{x} \in \mathbb{X}$ enhances the preference of $\alpha \in \mathbb{A}$ then it is said to be a positive contribution and hence *real benefit characteristic*. If it hinders the preference over $\alpha \in \mathbb{A}$ then it is said to be a negative contribution and hence a *real cost characteristic*. Thus the general preference of $\alpha \in \mathbb{A}$ is defined by specific preferences of $x \in \mathbb{X}$ associated with $\alpha \in \mathbb{A}$.

1.4 The Decision Structure and Cost-Benefit Analysis

From the previous discussions and the above definitions, we can state in a more or less general terms the logical framework of cost-benefit analysis given the social objectives. The general framework requires that we:

1. Construct technically and socially attainable objective set, $\mathbb{V}$, and check whether it is multiobjective or single objective given the spaces of decision and social objectives;
2. Construct the space of program alternatives, $\mathbb{A}(\alpha)$ relative to the attainable objective set, $\mathbb{V} \subset \Omega(\beta, t)$.
3. Identify the real attributes of $\alpha(\bullet) \in \mathbb{A}$ that distinguish each member in set $\mathbb{A}$. In other words, construct the characteristic set, $\mathbb{X}(\alpha)$, associated with $\alpha \in \mathbb{A}(\bullet)$.
4. Identify whether $x \in \mathbb{X}(\alpha)$ is a real cost or benefit in order to construct the cost-benefit space as well as determine whose benefits and costs must be included in the accounting process;
5. Develop a process that allows different real benefit and cost characteristics to be reduced into a common unit of measure at any time point. That is construct cross sectional aggregation over different cost and benefit characteristics in order to obtain cross-sectional cost and benefit data.
6. Estimate the time flows of real costs and benefits. In other words, construct dynamic real cost-benefit data.
7. Develop a process that will allow the future real cost-benefit estimates to be linked to the present in order to construct aggregation of intertemporal values.
8. Conduct sensitivity analysis on cost-benefit data due to uncertainty and fuzziness.
9. Develop a process and criterion index for ranking the choice-decision alternatives in accordance with the constructed cost-benefit data. In other words, construct choice criterion on the basis of cost-benefit data.

10. Select the best alternative that would allow the chosen objective to be attained.

These ten steps provide us with a basic and conceptual algorithm for cost-benefit analysis as a theory and application of decision making on the basis of cost and benefit values. The first step involves examining the elements of the objective set, $\Omega(\beta,t)$ to obtain those elements that are technically attainable in order to construct the attainable set, $\mathbb{V} \subset \Omega$. The elements in the space of objectives are either given or they may be constructed. They may be economic or non-economic in their orientation. The elements in $\mathbb{V}$ are considered technically attainable. They may, however, be unattainable economically. The second step requires that we put together a set of program alternatives, $\mathbb{A}(\omega)$, that will lead to the actualization of the elements in the attainable set $\mathbb{V} \subset \Omega(\beta,t)$. The elements in the space of decision alternatives, $\mathbb{A}(\omega)$, for any given $\omega \in \mathbb{V} \subset \Omega(\beta,t)$ must be selected on the basis of their technical feasibility and potential to achieve the element in the attainable objective set.

Given the set of decision alternatives, $\mathbb{A}(\omega)$ relative to a decision object, $\omega \in \Omega(\beta,t)$ the step three above requires us to identify the attributes of each $\alpha \in \mathbb{A}$ and enlist them to construct the characteristic set, $\mathbb{X}(\alpha)$ relative to α. This is followed by a fourth step where the elements of the characteristic set $\mathbb{X}$ are examined and separated into cost characteristics and benefit characteristics. This allows for the construction of a *cost-benefit space*, Θ. Steps three and four constitute *characteristic identification process* that demands a comprehensive accounting. Step five involves *static aggregation* of heterogeneous cost and benefit entities. The static aggregation requires the construction of *prices* or unit values of the real cost and benefit characteristics at each point of time. The prices provide us with measures of relative values of different cost and benefit characteristics. The problem of obtaining the relative values belongs to the *theory of optimal* and *efficient prices.*

The sixth step requires us to estimate or forecast the benefits and costs at each time point over the estimated life of the project or program or the decision alternative. This is the problem of *cost-benefit forecasting.* The forecasting process allows the intertemporal benefit-cost data to be constructed. The seventh step involves *intertemporal aggregation* where time-point benefit and cost values are reduced to their present value equivalencies for summability. It requires the development of a framework to compute *social discount rate.* The framework belongs to the construction of the *theory of social discount rate.* Step eight allows one to account for uncertainty, vagueness and fuzziness through sensitivity analysis as to the stability of the aggregate cost-benefit data. The ninth step requires one to develop a ranking function that allows the elements in the set of decision alternatives to be ranked for choice action. This involves the development of *cost-benefit criterion* on the basis of which the step ten is implemented.

The cost-benefit analysis may be viewed as a general process of decision theory and application. It defines and establishes a logical sequence of steps that allows a decision to be made with reasonable probability of success. From the defined sequence of problem solving the framework of cost-benefit analysis as a decision aid, given the sets of social objectives and decision alternatives, is nothing more than:

1. A collection of information on costs and benefits about decision alternatives,
2. Processing the cost-benefit information; and
3. Using the results to decide and choose the decision alternative that will "best" accomplish the objectives.

The theories of consumer and producer behaviors in neoclassical economical analysis may be seen in such a cost-benefit framework. As such cost-benefit analysis viewed from the classical logical framework is one of many approaches that may be brought to develop the applicational framework of decision making based on costs and benefits. We shall turn our attention to such a framework.

1.5 The Basic Social Welfare Foundation of Cost-Benefit Analysis

The neoclassical microeconomic decision framework provides a way to justify the use of cost-benefit technique and methods as decision tools. The thinking here is that every decision has costs and benefits. These costs and benefits may be measured in different units that may be related to input-output process and the health of the decision system. The costs and benefits implied in the economic theory of production and consumption may be illustrated with two persons, A and B, two inputs of capital, K and labor, L and two commodity output, x and y of the economy. This 2^3 (or 2 x 2 x 2) model of the general economy provides essential elements and computational implications of costs and benefits, implied in any decision alternative. The classification of implied costs and benefits will vary from one decision situation to another. The 2^3 model of the economy may be summarized in cost and benefit conditions as:

COST CONDITIONS

1. Two individuals, A and B, two different inputs, and L, and two different outputs, x and y
2. Each input and output are homogeneous and perfectly divisible
3. Each output is produced by using some quantities of K and L fixed at $\bar{K}$ and $\bar{L}$ respectively such that $\bar{K}=K^x+K^y$ and $\bar{L}=L^x+L^y$ where superscripts indicate the quantity of input used in production of the outputs x and y respectively.
4. The outputs x and y are produced according to the following input-output relations

$$x = F(K^x, L^x) \text{ and } y = G(K^y, L^y)$$

where K and L are the *real social costs* measured in input values as well as computable in output units.

BENEFIT CONDITIONS

5. The quantities of outputs x and y are distributed to A and B individuals for consumption such that: $x = x^A + x^B$ and $y = y^A + y^B$
6. The consumption of quantities, x and y, given to individuals A and B, provides them with *real benefits* measured in terms of some units of satisfaction or happiness where:
   a) The satisfaction that A and B get from consumption of x and y are represented by their respective utility indexes of, u^A and u^B. These utility indexes depend on the quantities of x and y consumed and are measured in ordinal units
   b) These utility indexes are not interdependent as such they may be represented as

$$u^A = u^A\left(x^A, y^A\right) \text{ and}$$

$$u^B = u^B\left(x^B, y^B\right)$$

In the above specification, numbers 1-4 define real cost conditions while numbers 5 and 6 define the benefit conditions of decision on choice of alternative social welfare states.

Any project or decision alternative when selected and implemented alters the social state and the welfare of the society by affecting the relative costs and benefits of the members of the society. The implementation of any project alternative uses resources of the society and hence deprives the implementation of other projects under conditions of fixed and limited resource. In general, the overall social welfare of the community may be represented by and computed in terms of an index of the form

$$W{=}W\left(u^A, u^B\right), \ W_i > 0, \ W_{ii} < 0, \ W_{ij} = 0$$

where $i = A, B$. In this respect, any computed value of the social welfare index, W, represents the net benefit level of the society involved in resource allocation (cost), output production (benefit) and output distribution (benefit distribution). The general behavioral assumption is the postulate of non-satiation where the individuals as well as the community are never satisfied in the sense that more net benefit is always preferred.

The problem of resource allocation, output production and distribution is such that every alternative $\alpha \in \mathbb{A}(\omega)$ selected must increase the benefit of the society until the social welfare, $W(\bullet)$ is at a maximum. In other words every decision alternative (composed of socioeconomic regulations, controls, projects, and law) if selected must improve the overall social benefit or welfare of the decision system. In the problem that we are setting up for allocation, production, and distribution the alternatives are defined in terms of input combinations that may be allocated for optimal production and output combinations to be distributed for consumption such that social welfare is optimized. The choice problem given the institutional arrangements of law, politics and economics may be set up in a manner that brings to focus the implied costs and benefits of various decision alternatives. The problem is simply:

The Problem

$$\text{DECISION} \left\{ \max W = W\left(u^A, u^B\right) \right.$$

$$\text{S.T} \quad \text{COST} \left\{ \begin{array}{ll} \text{Initial endowment:} & \left\{ \begin{array}{l} \bar{K} = K^x + K^y \\ \bar{L} = L^x + L^y \end{array} \right. \\ \text{Technology:} & \left\{ \begin{array}{l} x = F\left(K^x, L^x\right) \\ y = G\left(K^y, L^y\right) \end{array} \right. \end{array} \right.$$

$$\text{BENEFIT:} \left\{ \begin{array}{ll} \text{Distribution:} & \left\{ \begin{array}{l} x = x^A + x^B \\ y = y^A + y^B \end{array} \right. \\ \text{Taste and Preferences:} & \left\{ \begin{array}{l} u^A = u^A\left(x^A, y^A\right) \\ u^B = u^B\left(x^B, y^B\right) \end{array} \right. \end{array} \right.$$

(1.5.1)

The welfare optimization problem involves:

1) real cost allocation decision that is concerned with the choice of a combination of K and L among the set of all alternative combinations of K and L such that real productions of x and y are maximized and cost is minimized

between the use of K and L given the technical know-how;

2) real benefit allocation that involves a choice from a set of distribution alternatives of outputs x and y combinations for individuals A and B such that not only are both utilities of A and B maximized but that the social welfare (benefit) is maximized given the social cost conditions and social preferences. The social welfare optimization may be viewed as a trade-off of costs and benefits in the social welfare index where cost is a disutility and benefit is a utility.

The solution to the above optimization problem provides us with a logical system, computing algorithms and justification to use cost-benefit analysis to resolve certain class of decision problems under defined conditions and information structure. The framework is substantially based on but not limited to market institutions of particular type as well as individual costs and benefits that may be separated from those of society at least conceptually. Strict application of the framework thus limits the logical power of cost-benefit analysis to be extended to deal with other classes of decision alternatives in assessing the relative merits of different regimes of resource commitment, especially those involved in social decisions and cases where markets may not exist. The limitations associated with the framework will be discussed in order to introduce some methodological approaches that will enhance the theoretical and applicational framework of computable and non-computable cost-benefit analysis.

Interpreted from economic theory of decision making, the *cost* of a choice of a decision alternative to both the individual and society is specified in terms of real resource usage and measured in terms of output of the best decision alternative. The benefit of the same decision alternative to both the individual and society is defined in terms of product satisfaction based on the individual subjective preferences that collectively define the social welfare of any socioeconomic state. The ranking of the decision alternatives are examined in terms of relative marginal resource costs and benefits. The selection process of decision and choice alternatives is based on either *cost efficiency* or *benefit_efficiency* with respect to production and utilization of the results of alternatives. These entail paying attention to allocative efficiency, product-mix efficiency (that is portfolio efficiency) and distributive efficiency that address social equity.

The concepts and measurements of cost and benefit efficiencies are intimately tied to the allocative and product-mix efficiencies. Together with distributive efficiency, the social welfare optimum is determined.

The maximum social welfare optimum is such that the positive distance between total benefit and total cost, reduced to a common unit of measure, is the greatest at a given data and distributive configuration. This is attained when the marginal social benefit and social cost are not only equalized but equalized with those of individual members of the society.

One may abstract from the optimization problem of eqn. (1.5.1) by considering a decision to change the prevailing social state by undertaking a social project or enactment of a law. Let the project have decision alternatives that vary with an aggregate output, Q. Furthermore, let the associated benefit and cost be increasing function of output, Q, of the form $B(Q)$ and $C(Q)$ with marginal benefit and cost-functions, $b(Q)$ and $c(Q)$ respectively where $(db/dQ) < 0$ and $(dc/dQ) > 0$. The decision problem may be specified as:

$$\min_{Q \in A} b(Q)$$

$$\text{S.T. } A = \{Q \mid [c(Q) - b(Q) \leq 0]\}$$

(1.5.2)

The decision requires one to choose a decision alternative with scale of output that minimizes the marginal benefit such that the output level belongs to the rage of output where marginal benefit is equal to or greater than marginal cost. The solution may be graphically represented as in Figure 1.5.1 where at Q*, c(Q*) = b(Q*) and B(Q) – C(Q) is the maximum.

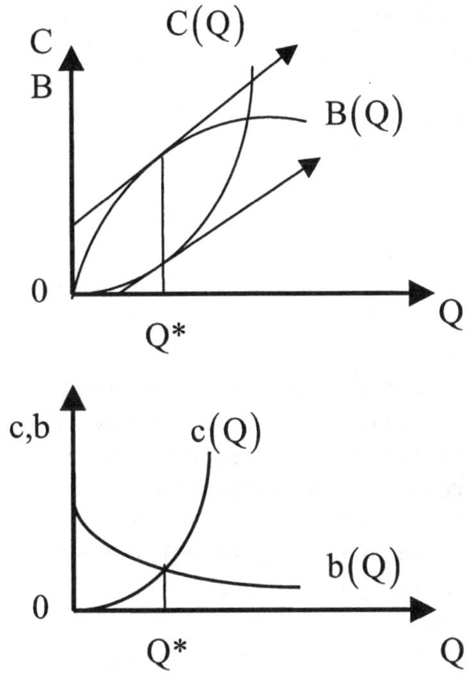

Figure 1.5.1 A Graphical Representation of the Solution to Net-Benefit Maximization

In a dual form, the problem may be written as:

$$\max_{Q \in B} c(Q)$$
$$\text{S.T. } B = \{Q \mid b(Q) - c(Q) \geq 0\}$$

(1.5.3)

In all decision alternatives there are private and social evaluations of costs and benefits at the margin of resource usage. These evaluations originate from the welfare perceptions of the individuals and the manner in which costs and benefits affect the relative standing of the individuals in the society. It is thus possible for the individual marginal benefit and cost to be equal in order to attain individual welfare optimum and yet both of them may diverge from marginal social benefit and cost due to the nature of institutional arrangements of production, distribution and consumption. A currently dominant economic theory projects a position of cost-benefit evaluative principle where on the private side of social configuration an individual welfare is at the maximum when the individual marginal resource cost equals marginal resource benefit. On the side of the societies (general public) all individual marginal resource costs and benefits have the same value as well as equal to the social marginal resource cost which must also be equal to social marginal product benefit with no discrepancies. Thus for the social organization to attain a welfare maximum the institutional arrangements and information structure that support decision evaluations must be right and that all individual benefit and cost evaluations must be accounted for in the social state.

The social organization that will, given the initial resource endowment, lead to such a social welfare optimum is called Paretian society and its economy is called the Paretian economy. The Paretian economy is *cost effective* and *fair* in that given the initial resource endowment the marginal resource cost for the production of any type of output is the same irrespective of who produces it. It is *benefit effective* and fair with respect to product usage if the marginal benefit from each product produced is the same for all commodities irrespective of its user. It is *product-mix efficient* if the marginal resource cost is equal to the marginal resource benefit irrespective of the product produced or the producer and user. It is also *distributive efficient* and *fair* if the ratio of marginal benefit of a commodity for any two individuals is the same as the ratio of marginal social benefits that these individuals contribute to the general social welfare. Any state in the Paretian economy is Paretian optimal if allocation, production and distribution (that is costs and benefits) cannot be reorganized to increase the net benefit of at least one individual without decreasing the net benefit of at least another.

1.6 Costs-Benefit Analysis and other Decision Tools

Human decision and choice activities are exercised over a wide range of natural and social elements that involve *good* and *bad* outcomes. The decision and choice activities thus lead to desirable and undesirable outcomes. The accumulated information on these outcomes constitutes the available social information set, $\mathbb{I}$. This social information set forms the foundation of common knowledge of humanity. The human cognitive process is such that we can learn from the outcomes of the individual and collective successes and failures that are contained in the social information set. The learning involves the processing of the social information set to abstract rules of good decision. Such rules of good decision are referred to as decision-choice rationality [440] that may be used as a guide to future decisions. The rules and the constructed guides are called decision aids and tools or prescriptive models of human decision activities. These rules are classified under the taxonomy of decision tools.

1.6.1 Taxonomy of Decision Tools

There are various decision aids and tools that may be brought to bear on the problem of choice. The usefulness of these decision aids and tools will depend on the class of decision problems that require solutions. The decision aids include *informed subjective judgment* and *proper concept formation.* The decision tools are made up of all kinds of mathematics and logic that can be brought to bear on the decision problem. The nature of the tool will depend on the complexities of the problem of choice. Probably, the most important decision aid is the subjective judgment.

The logical tools that my be brought to aid decisions my be classified into one of these categories of,

1) statistical and probabilistic tools
2) classical optimization tools
3) Non-classical optimization tools, and
4) Hybrids (i.e.: a mixture of any of elements of any of the 3 above).

These tools may be broken down to *static* and *dynamic.* They may also be grouped into *deterministic* and *stochastic.* Among the hybrids are:

1) cost-benefit analysis
2) cost-effectiveness analysis (see Chapter 3) and
3) benefit-effectiveness analysis (see Chapter 4).

For the purpose of *social choice* some of these tools may be inappropriate. We shall highlight the inappropriateness of some of these tools by simply looking at the possible similarities and differences that tend to characterize them.

It is interesting to note that all these tools may come under the giant umbrella of classical and non-classical operations research or optimization tools. The analysis may be conducted from the *organic view-point* or from the viewpoint of individual. In other words, we may analyze the decision from either *systems* stand point or from the consideration of *subsystems.* In economic analysis, general equilibrium analysis is an example of the former while the partial equilibrium analysis is an example of the latter. Whether the analysis for a decision is considered from the system standpoint or from the subsystem viewpoint will depend on the *linkage effects* and interrelationships of the outcomes.

An example of linkage effects in economic decision is when we have interdependent consumption or production of commodities or some individual's derived costs and benefits in consumption or production of commodities or some individual's derived cost and benefits in consumption are dependent on the costs and benefits of others. Such linkages are referred to as externalities in decision and choice. The degree and complexities of the system's analysis required in decision and choice will depend on the linkages of benefit-cost effects that are derived from such a decision. The more extended are the linkages, the more organic and comprehensive will the analysis of choice be.

When a system's decision is being considered, the optimization of its net efficiency presents some problems and difficulties. From the view point of analysis for choice, some of the tools of decision aids may be inappropriate. As such, it is apt to be cognizance of the usefulness and short comings of these tools and what they can and are usually used for. This will allow the job of the decision analysis to be related to the appropriate tool.

1.6.2 The Classical Operations Research for Decision-Making

The classical operations research as a decision aid is directed towards decision and choice problems where the variables lend themselves to *easy_qualification*. It involves optimization of effectiveness (in a specific sense) where such effectiveness can directly or indirectly be represented by a quantitative measure. It can, therefore, be used as a decision aid to physical systems' decisions or other types of decisions where the variables are measurable. The tools of the classical optimization are useful for those classes of problems where subjectivity and soft data may be considered significantly small. When the analysis involves trade-offs that maybe interdependent and the quantification of the important decision variables are vary difficult to achieve then we are in a situation where one must resort to either non-classical tools of decision analysis or a hybrid of classical and non-classical tools. Furthermore, there are a class of social decisions and choices where the presence of *lumpiness, interdependence and linkages* prevents the attainment of optimal decision through the classical techniques of choice. When such occurs one may depend on non-classical tools. The classical optimization process may also be viewed as a process of cost-benefit analysis. The nature of the costs and benefits must be appropriately interpreted. In fact all available measures for humanistic and non-humanistic systems have cost and benefit implications that must be carefully interpreted.

1.6.3 Some Non-Classical Operations Research for Decision-Making

There are many facets of non-classical operations research analysis. The decision tools that come under this are those that may be brought to help the analysis of problems of choice that do not easily lend themselves to the classical tools of optimal decisions. Among such tools are the *calculus of fuzzy decisions, cost-effectiveness analysis, benefit-effective analysis, cost-benefit analysis*, and some elements of statistics and probability. Each of these tools may only be appropriate for certain class of non-classical problems. For a class of some problems, a hybrid of these tools may be required if we need to consider all possible and feasible alternatives of the choice.

From the viewpoint of social decisions that have important implication in terms of their effects on general society or a section of it we need a tool or a set of tools that allows us the flexibility of producing a *matrix* of *net effects* for each feasible alternative. These net effects may be specified abstractly to be composed of *benefits* and *costs* that are likely to be generated as a result of any alternative *considered.* When a social choice problem is posed in this way, one may use the methods of cost-effectiveness, benefit-effectiveness or cost-benefit analysis. All these tools have their foundations in economic theory of social choice, where the objective is to optimize the 'effect' or the satisfactions. Let us consider each one of these three methods.

1.6.3.1 Cost-Effectiveness Analysis (CEA)

Cost effectiveness is a tool of decision making that is used when the objective is specific with benefits known and given and alternatives of reaching the objectives are many where the benefits of these objectives are the same. The analysis is usually an attempt to select a minimum cost alternative from a set of feasible alternatives. It is usually directed towards choice problems in which the *output cannot be evaluated* at the market price. In other words, *prices do not exist* in the market to reflect the utility of the choice or the benefits are given and constant over all the feasible alternatives. However, prices do exist for the inputs and hence the *costs* of alternatives can be evaluated or estimated and ranked. The *criterion* for choice in this case is: select the *alternative that allows the objective with given benefit to be achieved at the least cost.* As posed, the problem is static in that the costs are estimated at a point in time. It may easily be extended into a time domain. An example of areas application of cost-effectiveness analysis is, military operation decisions or space decision projects.

The Cost-effectiveness analysis may, then be said to be appropriate if:

a) Outputs as benefits cannot be evaluated by market prices and if they can be evaluated, they are fixed and constant over alternatives
b) Inputs on the other hand can be valued by market prices or at least quantifiable.

The problem in the logical form is simply: minimize cost subject to a given objective with known benefit $\overline{\mathrm{B}}$. The problem may then be written as:

$$\min_{x \in H} \mathrm{C}(x)$$
$$\text{S.T. } H = \left\{ \mathrm{C}(x) - \overline{\mathrm{B}} \leq 0, x \in \mathbb{A} \right\}$$

(1.6.3.1)

1.6.3.2 Benefit-Effectiveness Analysis (BEA)

There is a class of decision problems where the objectives, as well as the benefits are not known. The benefits vary with the chosen decision alternative relative to the objective. However, the resource in terms of cost is known for whatever the decision alternative is. The problem is simply given the *resource costs* we must select an alternative decision element with the corresponding benefits among a *set of destination* (objectives). The criterion for choice is that the selected alternative should yield the *highest benefit* relative to the given cost.

Examples of such decision problems abound in social decisions (either group or individual). A case is where we do have manpower and the problem is to choose an alternative activity such that the greatest benefit is realized. The benefits need not be directly measurable. The analysis may be directed towards choice problems in which the cost cannot be evaluated at the market prices. In other words, prices do no exist in the market to reflect the cost of the choice. The benefits, on the other hand, are such that they may be evaluated directly or indirectly on the basis of market prices. Under such circumstances the *criterion* for choice is on the basis of the *highest benefit* with a given resource cost.

The benefit-effective analysis, therefore, may be supposed to be appropriate if:

a) The resource is given and the benefits are unknown and vary with alternatives;
b) The output (or benefits) can be assessed by market prices or at least measurable in some meaningful sense; and
c) Cost cannot be evaluated by market prices and if it could it is fixed and constant over alternatives.

In a logical form the problem is simply choose the highest benefit alternative subject to a given cost, $\bar{\mathrm{C}}$

$$\max_{x \in D} \mathrm{B}(x)$$

$$\text{S.T.} \quad D=\left\{x \mid \mathrm{B}(x)-\bar{\mathrm{C}} \geq 0 \,, x \in \mathbb{A}\right\}$$

1.6.3.3 Cost-Benefit Analysis (CBA)

Cost-benefit analysis is an analytically logical system designed to aid decision makers in identifying preferred choice among possible alternatives involving a wide range of considerations, many of which are political, natural, environmental, or social in character. The method is to help the decision maker to identify the stream of benefits and costs that may be associated with a particular choice and how such benefits and costs are likely to be distributed among the populace. Thus cost-benefit analysis incorporates cost-effective and benefit-effective analysis. The differences may be simply stated as:

1. When the problem is stated as a minimization of cost subject to a given specified benefit then the appropriate method is *cost-effectiveness.*
2. When the problem is stated as choose an alternative from the feasible set so as to maximize the benefits subject to a given specified cost, then the appropriate method is *benefit-effectiveness.*
3. In the case of cost-benefit analysis both the benefits and costs may vary not only over all possible alternatives but even over a given alternative through, for example, its linkages. The power of cost-benefit analysis lies in its analytic advantage in placing the choice problem in more or less global context or in systems consideration if the choice is a group one but not individual. As such it can be used to analyze the "net effects" of, for example,

changes in existing laws, or governmental social and economic regulations, etc.

The general principles of cost-benefit analysis may be said to constitute a search for answers to the following broad series of questions such as:

1. What kind of real costs and benefits must be included in the analysis? *This is the identification problem of real cost and benefit characteristics.* The solution to this problem leads to a subset of characteristic (attribute) set, $\mathbb{X}$ specific to the project in question.
2. Whose costs and benefits must be included in the cost-benefit assessment and evaluation process? This is the *identification problem* of affected communities and individuals. The solution to this problem leads to the construction by the method of cluster analysis of a set of social classes defined in a particular sense that is specific to the project in question.
3. Should the society attach equal value (weight) to a unit of real cost (benefit) across social classes and across real cost (benefit) characteristics? This is the *problem of distributive weights*, the solution of which provides us with social assessment of the burden of real cost and benefits to society.
4. What are the relative weights that must be assigned to the cost attributes and benefit attributes to arrive at the total cost and benefit to be considered? In other words, how are these costs and benefits to be valued? This is the *identification problem of appropriate prices* that is linked to the theory of efficient prices. It is also a *static aggregation* problem where the solution allows heterogeneous real costs and benefits to be reduced to common measurable unit for aggregation.
5. Should we consider streams of benefits and costs? This is the *intertemporal benefit and cost identification problem.* It is also the forecasting problem of costs and benefits of the project if the lifetime is more than a year. The solution to this problem provides us with intertemporal *dataset* on costs and benefits in common units.
6. Should equal weights be assigned to benefits and costs at each time point if the benefit and cost are in streams? This is the problem of *discounting.* It is also the problem of *intertemporal aggregation* of dated values. The solution yields present value equivalences of dated values.
7. What value should be used to construct the intertemporal weights attached to dated values for discounting and aggregation? This is

identification problem of an appropriate discount rate. The solution provides us with social discount rate.

8. What are the relevant constraints if any on the decision alternatives? This is the *limitation identification problem.* The solution generates the feasible set of the decision alternatives
9. What is the criterion for choice among the decision alternatives? This is *criterion identification problem* based on benefit-cost relationships. The solution to this problem yields a set of rules for ranking decision alternatives.
10. What alternative decision element must be selected? This is the optimal choice problem on the basis of relative cost-benefit values.

The answers to these questions require a good and sound judgment on the past of the analysis and the decision maker. We hope that such judgment through a logical system of analysis to answer the above question would lead to a preferred and wise choice. Despite the scientific element of the logic of the analysis, the cost-benefit analysis is made up of an art and science of decision-making. In order to lay down the calculus of cost-benefit analysis, we shall start with a *crude criterion* for choice. The criterion simply states that choose all those alternatives whose "net benefit" is positive (in some specific sense) subject to the available resource. In order to develop this idea further we need to consider the *criteria* for judging *"net benefits"* from the point of view of the individual and "net welfare" from the point of view of society.

1.7 Introductory Framework of the Theory of Fuzzy Decisions for Cost-Benefit Analysis

When one examines the choice and decision activities of humans, one finds that these activities are characterized by imprecision, vagueness, ill-definedness, ill-posed problems and many other similar characteristics, all of which are subsumed under the general concept of fuzziness. This fuzziness is an inherent characteristic of human linguistic systems, perceptions and thought process. As such an exact description of any reality is not only difficult but unattainable. This fact must be accepted as an inherent character of nontrivial scientific analysis in general. It is not, therefore, surprising that the more precise a model becomes in representing reality, the less useful it becomes in solving problems that relate to such reality, particularly when the reality increases in complexity. Ironically, the more complex is the reality and the harder we try to achieve model precision the more we succeed in achieving irrelevance in models of problem solving. This is precisely the *principle of fuzzy incompatibility*. In general relevance and fuzziness are mutually nonexclusive in the world of human cognitive operations. Thus non-fuzziness is a separability condition between *relevance* and *precision* which partition relevance from irrelevance in model application. The point of such a cognitive separation is the *fuzzy threshold.* The general situation is that fuzziness is a characteristic of human perception which is a foundation of cognitive models that are relevant to human problem-solving activities. Thus, relevance slowly devoices itself from precision as more exactness is sought in a complex domain. At some critical point of the fuzzy threshold, relevance and precision become almost mutually exclusive characteristics of complex systems. Thus, at this fuzzy threshold, it becomes necessary to choose between precision and relevance since the two are no longer compatible in the models' representation of reality. The *fuzzy incompatibility principle* is even of more serious nature when a model representation of humanistic system is sought. In humanistic system, perception and reality interact naturally to create cognitive complexities. Models of reality are human perceptions and approximations. Such perceptions may diverge from reality. This divergence is due to *fuzziness* and *incomplete information* that must be treated within the cognitive model.

Social choice and decision cannot take exception to the principle of fuzzy incompatibility, as it relates to precision and relevance. All decisions and choices are made in different degrees of fuzzy environment and information incompleteness. Thus to model human decisions and choice solely on the basis of classical logic (or classical mathematical logic) may lead to production of models that lack flexibility and greater relevance in applications of human-problem solving. The utilitarian methodological approach to the analysis of decision and choice therefore suffers from lack of flexibility in order to account for the general problem of fuzzy characteristics in complex systems of human decisions and choice since such approach is intimately based on the classical logic and its corresponding mathematical structures that seek ultimate precision. Subjectivity is an inherent characteristic of human judgment, decision, choice, policy, evaluation and others. Such subjectivity is defined by the concept of preferences based on subjective taste and value in terms of utility as we have discussed in section 1.6. Preferences, subjective taste and utility are all fuzzy concepts that cannot be precisely analyzed within the classical mathematical logic without extreme assumptions of human nature.

From the perspective of economics of decision, one finds that much of the optimal decision rules have been developed through the method of welfare economics or generally normative economics all of which are based on people's evaluative activities in the utility space [399b]. While this space is fuzzy, the techniques and methods of either welfare or normative economics or explanatory science impose a certain required precision that is not only unattainable but irrelevant representation of certain evaluative activities of humans in this space. The problem is not the concepts of utility, welfare, subjectivity and normative. These are natural linguistic representation of human thoughts which are translated into evaluative activities in the utility space. It is a poor scientific practice if we force realities into mathematical straight jackets. This is precisely what the marginalists have done with either the welfare or normative economics through the mathematics of classical calculus. The classical logic and the corresponding mathematics are incapable of dealing with the problems of fuzziness that are characteristics of human systems. These statements may be extended into problems of new generations and areas of artificial intelligence and robotics and prescriptive science as more complex systems are designed and built to mimic human cognitive process.

The kind of precision required by classical mathematics used in the development of either welfare or normative economics is such that the construction of *cost-benefit analysis* from their foundations carries with it all the problems inherent with the classical logic. A new logical paradigm is required if we are to represent reality in such a way as to naturally account for vagueness, imprecision, subjective evaluation, ill-posed problems and other. This logical paradigm is the *fuzzy logic* and the corresponding mathematics is what is currently called *fuzzy mathematics* and algorithms to arrive at optimal decision element with exact equivalence is *fuzzy mathematical programming* and the computations in fuzzy data is called *soft computing*. Decision with fuzzy characteristics will be known as *fuzzy decision problems*. The logical structure and techniques of analyzing fuzzy decision problems will be called *the theory of fuzzy decision* [661][694][735][806].

We must point out that from the viewpoint of fuzzy logic, fuzziness cannot be subsumed under randomness. As such problems with fuzzy characteristics cannot be treated with probabilistic logic and calculus of randomness. To convince oneself, one must try to answer the following question on probability. What is the chance that it will rain heavily tomorrow? No probabilistic statement can answer this question without a restriction on the linguistic variable "heavy". Such a restriction cannot be derived from our current knowledge of probabilistic logic.

It is precisely the inability of our current knowledge of the classical logic and its corresponding mathematical structures to provide answers and solutions to certain new questions of our current scientific age and problems at the frontier of scientific and technological investigations that fuzzy logic and the corresponding mathematical structures were born. It is precisely on this basis also that this study is conceived. The aim of this study is to present cost-benefit analysis in a framework of the theory of fuzzy decision based on the logic and calculus of fuzziness while simultaneously presenting the essential element of classical cost-benefit analysis. The critical objective in this study is to derive crisp statements about fuzzy systems of human thinking and decision-choice practices based on costs and benefits. The fuzzy logic and the corresponding mathematics will not only provide us with the numerical tool to analyze and solve vague and ill-defined problems of decisions in general but their basic analytic framework would allow us to formulate social decisions and choices where costs and benefits are crucial, in an analytical way that is more or less natural to humanistic systems. The basic scientific premise underlying this approach is that fuzziness can be modeled mathematically in such a way as to derive crisp statements about vague notions of human though of costs and benefits in the field of decision. The fuzzy-decision-theoretic approach introduced here will enhance and complement the classical cost-benefit analysis. The approach will simply make explicit the fuzzy characteristics associated with the essential elements of benefit and cost estimates and how such estimates result in decision making over a wide range of social and private decisions including humanistic and non-humanistic ones.

Since this is an attempt to develop cost-benefit analysis with combined logical systems, not all aspects of benefit-cost analysis will have complete treatments. Nonetheless, we hope to provide a balance treatment that will be useful for general application. The advantages of this approach are enormous. It allows us to present benefit-cost analysis in the logic of normative science [441] and deal effectively with the critical problems of interpersonal and intertemporal aggregations. Furthermore, the framework allows us to extend the application of cost-benefit analysis across the domain of purely economic decisions and into other areas of human decisions and policy. All kinds of social decisions and choices can be analyzed with this framework of benefit-cost analysis. The quality of the decision will depend on the personality characteristics of the decision agent and his or her understanding of treatments of fuzzy characteristics as inherent properties of human thought, perception and subjectivity. The basic requirements of fuzzy mathematics and mathematics of fuzzy decision are given in Chapters 5 and 6 respectively. The whole analytical works is about theories of algorithms for cost-benefit rationality under fuzzy environment. The process of development is based on the epistemic notion that cost-benefit analysis proceeds on the acceptance of the principle of duality. In line with this principle, each decision alternative is simultaneously potential and actual. The same decision alternative has cost and benefit in itself. The cost and benefit are inseparable. The characteristics that define the cost and benefit in real terms can be identified, measured, valued and computed into cost and benefit information sets in support of decision making. The epistemological foundation of *cost-benefit rationality* is that every decision element is *potential-actual duality*. The potential is a *cost-benefit duality* and the actual is also a *cost-benefit duality*. In support of the cost-benefit duality is the concept of *category formation* over the decision space into definable partition where the conflict in the cost-benefit duality induces *categoral transformations* from the potential to actual and from the actual to the potential. It is this foundation that gives meaning to the *with-and-without* decision problem which then acquires important scientific relevance in human thought and action.

2 The Theory of Computable Cost-Benefit Identification Matrices

The discussions in Chapter 1 point to the fact that the application of cost-benefit analysis to decision and choice demands reliable data on costs and benefits given the sets of social objectives, decision alternatives and social structure. Developing cost and benefit information on decision alternatives is perhaps the most expensive and most difficult part in the use of the method and technique of cost-benefit analysis. The difficulty and high expenses that come with it are due to a number of factors that make the problem complex. These factors include the idea that:

1. The society is composed of different groups with different interests, social objectives and preferences.
2. The project's real costs and benefits usually have differential effects on different groups, and even within a group, perceptions about costs and benefits and the *burden* of costs and intensity of benefits vary over the members.
3. Intertemporal costs and benefits shift over groups and over individuals in a group in time and over time.
4. Variety of real cost and benefit characteristics are associated with each decision alternative.
5. Interdependence of costs and benefits as well as interdependence of group perceptions of cost-benefit impacts on different social groups are characteristics of collective choice and decisions.
6. Bias and investigative blindness that arise on the part of the analyst due to his or her interactions with the problem of accountability of costs and benefits of the project come into play in the cognitive process.
7. Interference by bureaucrats and politicians with different competing interests as well as interest groups whose preferences are driven by non-economic forces and rent-seeking activities are factors of decision making.

All these and other factors combine to make the job of cost-benefit analyst a difficult one.

The main objective in this chapter is to develop a logical map that will assist in the construction of quality cost data and benefit data that may be processed for decision making. A rational process must be designed to identify the relevant real cost- and benefit characteristics including primary and secondary characteristics and the segments of society that they may impact if we are to generate the required cost and benefit aggregates. In this connection a question arises as to whether the identification problem of costs and benefits over different segments of society must be left solely to the experts and analysts or should it involve members of society including politicians and policy makers. For the purpose of democratic ideals in decision making and good governance other segments of the society should be involved in the process of developing the cost and benefit data that reflect the wills of the individuals, interest groups and stakeholders. From the viewpoint of efficient governance and socio-political responsibility the identification of real costs and benefits should include information of specific and general assessments by the members of the elected body, the governing masses and experts. In other words the computable cost and benefit databases must be comprehensive.

2.1 Multi-Objective, Multi-Alternative, Multi-Attribute and Multi-Group Problem in Cost-Benefit Analysis

The proceeding discussions bring us to important elements in systemically, information-generating process and decision making. In some important decision making problems we are faced with multiple criteria in selecting alternatives for program implementation. In cost-benefit analysis, however, we have one criterion of choice where such a criterion is abstracted from the relationships between cost data and benefit data. Given the cost-benefit criterion for social decision making the society must then be seen as a system of *interest groups* with different objectives. These objectives induce a partition of the members of the society into *clusters* of interest groups. In this respect the social decision process is governed by *multi-objectives* that may be viewed as an organized set or vector. The objectives have relationships among themselves as viewed by the members of the social system. Such relationships may be conflicting, complementing, supplementing, interdependence and many others.

The nature of the relationships and the possible social partition that such relationship may induce depend on the number of social groups, their preferences and the group interests. Each element in the set of objectives may also be reached by many decision alternatives defined in terms of projects and actions. Thus corresponding to each objective we can develop a set of decision alternatives. The collection of these sets of decision alternatives may be required for choice in order to reach a social objective. In other words, any social objective must be accomplished by complementary or supplementary decision alternatives. The analytical framework projects a complexity of *multi-alternative problem.*

Given any set of decision alternatives defined in terms of projects or programs, each element in the set generates a set of real benefit characteristics and a set of real cost characteristics with complex interdependencies, linkages, trade offs and transformations within each set. Together these real benefit and cost characteristics constitute the *characteristic set* that provides identity to each alternative. The problem of identifying and accounting for these characteristics constitutes the information collection problem in *multi-attribute decision* making in cost-benefit analysis. Here, detailed analysis of the characteristic set is useful and required to construct cost-benefit database for the decision alternatives. Corresponding to each alternative, we have two sets of real cost and benefit characteristics (attributes) that must be accounted for, computed and combined into cost and benefit aggregates. A problem arises in accounting and aggregation processes in that these cost and benefit characteristics spread over social groups and individuals whose group evaluations of the impacts may differ from one another and from experts, policy makers and the members of the governing class. The identification process of these characteristics is further complicated by intertemporal cost and benefit shifting that may have differential impact on different generations.

The existence of social groups with different cost and benefit assessments based on their individual and group diverse interests involving interlocking of dissimilar social objectives (goals) and differential preferences for decision alternatives calls for a development of a reasonable degree of cost-benefit data completeness. The development of such data completeness requires a comprehensive accounting of real costs and benefits which a creative and judicious combination of cluster analysis, characteristic sets, attribute vectors and partition of the social space will allow. The framework for using the creative and judicious combination of sets and vectors of alternatives, cost and benefit characteristics and distributive weights established over differential benefits, costs and impacts on social groups to construct cost-benefit data is what is being referred to as the *comprehensive accounting* in cost-benefit analysis and cognitive decision process that requires a matrix display to identify the key factors of decision alternatives. The cognitive process for the identification and display of the distribution of the characteristics or important factors associated with decision alternatives over social groups is what is being referred to as *the theory of computable cost-benefit identification matrices*. The theory must lead to a comprehensive accounting of costs and benefits that we sought.

2.2 The Structure of Identification Matrices

The identification matrices may be viewed as a systematic approach to identify clusters of costs, benefits or factors in general analysis of decision alternatives. It may be pointed out that the theory of computable identification matrices has broad applications beyond cost-benefit analysis. Among areas of useful application are analysis of quality of life, environmental quality, benefit-risk studies and auditing of physical and social systems. The computable identification matrices may be used to reveal clusters of problems in cross-functional analysis of management and administration of organizations and social institutions. They extend to the logic of matrix management where costs and benefits are the basis of actions and institutional restructuring. The basic idea is based on relations and interactions among vectors of dissimilar characteristics whose cohort values may be cardinal or ordinal. The identification matrices are thus developed from the interaction of two sets that are organized in vectors to generate *relational values* that are assigned to specific cohorts. The traditional matrix arrangements hold. The nature of the matrices, however, is such that the traditional manipulations and computation are restricted or may not apply. The process allows one to identify quantitative and qualitative characteristics that would lead to construction of hard and soft data.

Definition 2.2.1

A matrix identification relation, $\mathfrak{R}$ is such that given two characteristic sets $\mathbb{A} = \{a_1 , a_2 , \cdots , a_n\}$ and $\mathbb{B} = \{b_1 , b_2 , \cdots , b_n\}$, $\mathfrak{R} \subset \mathbb{A} \otimes \mathbb{B}$ we may define

$$\mathfrak{R} = \{\mathbb{A} \otimes \mathbb{B} \mid M_{\mathfrak{R}}(a , b) = c , a \in \mathbb{A} \text{ and } b \in \mathbb{B}\}$$

where $c \geq 0$ is a relational value for the a-b cohort and $M_{\mathfrak{R}}$ is a matrix of the relational values.

Definition 2.2.2

Given a matrix identification relation, $\Re$, the identification matrix, $\mathbb{M}_{\Re}$ is defined by an (m x n) cohort relational values where there are m-roles of defined elements and n-columns of other defined elements.

Definitions 2.2.1 and 2.2.2 may be represented as in Figure 2.2.1.

$\Re(\bullet)$

$\mathbb{M}_{\Re}(\bullet) =$

b/a	a_1	a_2	$\cdots$	a_j	$\cdots$	a_{n-1}	a_n
b_1	c_{11}	c_{21}	$\ddots$	c_{1j}		$c_{1,n-1}$	c_{1n}
b_2	c_{21}	c_{22}		c_{2j}		$c_{2,n-1}$	c_{2n}
$\vdots$	$\vdots$	$\vdots$	$\ddots$	$\vdots$	$\vdots$	$\vdots$	$\vdots$
b_i	c_{i1}	c_{i2}		c_{ij}		$c_{i,n-1}$	c_{nj}
$\vdots$	$\vdots$	$\vdots$	$\vdots$	$\vdots$	$\vdots$	$\vdots$	$\vdots$
b_{m-1}	$c_{m-1,1}$	$c_{m-1,2}$	$\cdots$	c_{ij}	$\cdots$	$c_{m-1,n-1}$	$c_{m-1,n}$
b_m	c_{m1}	c_{m2}	$\cdots$	c_{mj}	$\cdots$	$c_{m,n-1}$	c_{mn}

Figure 2.2.1 A Conceptual Structure of Identification Relation Matrix

where a_j = j-social group or department in an organization and b_i = name of the ith characteristic that is applicable to the case at issue.
The table may be written as

$$\widehat{\mathbb{M}}_{\Re}(\bullet) = \begin{pmatrix} c_{11} & \cdots & c_{1n} \\ \vdots & c_{ij} & \vdots \\ c_{m1} & \cdots & c_{mn} \end{pmatrix} = \left[c_{ij}\right]$$

(2.2.1)

The vector $\mathbb{A} = \{a_1\ a_2, \cdots a_j \cdots, a_n\}$ is called a *social partition vector*, the vector $\mathbb{B} = \{b_1, b_2, \cdots b_i \cdots, b_m\}\}$ is called a *characteristic vector* where $\mathbb{M}_{\Re}(\bullet) = \left[c_{ij}\right]$ is the *identification matrix*. The identification matrix may be used in many decision problems such as finding the sources of profit loss, operational inefficiencies, and resource requirements in an organization with multiple departments. In cost-benefit analysis the logic of the identification matrices is a powerful and convenient tool for pulling together the costs and benefits needed for the analysis of decision alternatives. For dispensation of social help on the basis of fairness it allows clusters of impacts and problems to be identified.

The relational values c_{ij}'s must be found or computed. The manner in which the cohort relational values are obtained will depend on the nature of the decision problem and the process of decision making. It is possible that some b-characteristics may not be quantitatively measurable and hence some c's may not be numerical values and horizontal summation may not be operationally admissible. The relational values are not vertically summable even if all the elements of the characteristic vector are quantitatively measurable. In cost-benefit analysis the logic of the computable identification matrix provides the means of comprehensive accounting for quantitatively and qualitatively affected cohorts in terms of costs and benefits as they may relate to a decision alternative. It further broadens the decision framework that allows an incorporation of some essential elements of democracy in the decision-making process in terms of preferences.

The logical development of the computable identification matrices and the corresponding matrix identification relation require a) the partitioning of the social space into social groups that constitute the *social partition vector*; b) the identification of the vector of costs or benefits or risk or factor characteristics that constitutes the *characteristic vector* associated with any of the decision alternative considered and c) the identification and partition of the characteristic vector into those that are tangible and intangible as well as those that are measurable and non-measurable. The nature of partition of society into social groups will be project-specific as well as helping to bring into the decision process distributional impacts of decision alternatives. The set of computable identification matrices may be divided into two. They are 1) a set of *primary* identification matrices and 2) a set of *supporting* or *derived* identification matrices that may be used for the analysis of any given decision alternative.

The essential structure of the logic of the theory of computable identification matrices is that a) society is composed of interest groups that may be conceived of in terms of *fuzzy partition* whose fixed-level cut will provide a crisp partition of well defined social groups (this is the *social partition vector*); b) the interest groups have different social objective which may be relationally established through their preferences (this is the *social objective vector*); c) Each objective may be accomplished by means of different decision alternatives that may have differential impacts on the social groups. This is the *program vector*; d) The decision alternatives are identified by characteristics that constitute the *characteristic vector*; e) the characteristic vector is partitioned into real cost characteristics and real benefit characteristics that are referred to as *real cost characteristic vector* and *real benefit characteristic vector* that are related to the partition vector with any given social objective.

When the society is considered in relation to decision that may alter the general social welfare two simple but useful relational matrices may be formed. Let $\mathbb{G} = \{G_1 , G_2 , \cdots ,G_k\}$ and $\Omega = \{\omega_1 , \omega_2 , \cdots , \omega_n\}$ be the social partition vector and social objective vector respectively. In this respect two simple matrices may be represented as in Figure 2.2.2. The first observation is that $k \leq n$ or $k \geq n$ and that the social objectives are not in one to one relationship with the social groups. The objectives are abstracted from the social groups and hence for a democratic decision making any social group agrees with or supports at least one social objective. The construction of the social objective vector and the democratic ranking of the social objectives are themselves collective decisions that require the help of an important logical framework. We shall now turn our attention to the essential logic for the construction of the computable identification matrices that will provide the required cohort values of Figure 2.2.1.

Given the set of social objectives, each objective corresponds to a program vector (that is vector of alternatives) to reach the objective. The relationship between a social objective and the alternatives to reach it produces two simple relational matrices that involves program vector and partition vector that the given

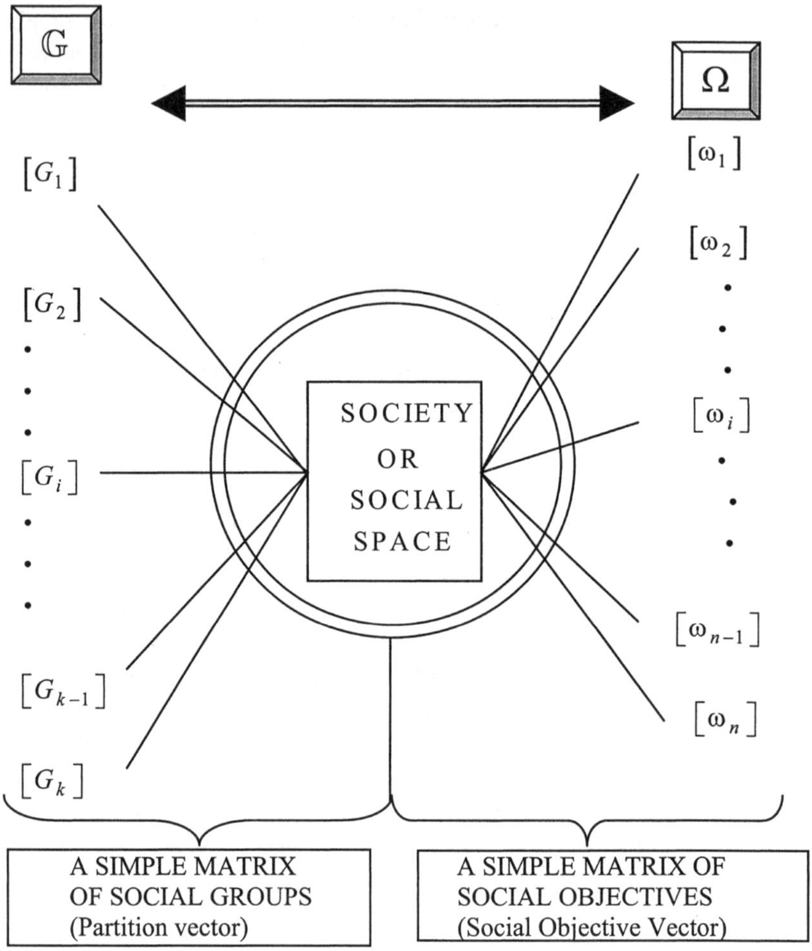

Figure 2.2.2
Social Group and Objective Structure

objective will impact. The logical structure of such relational matrices is presented in Figure 2.2.3.

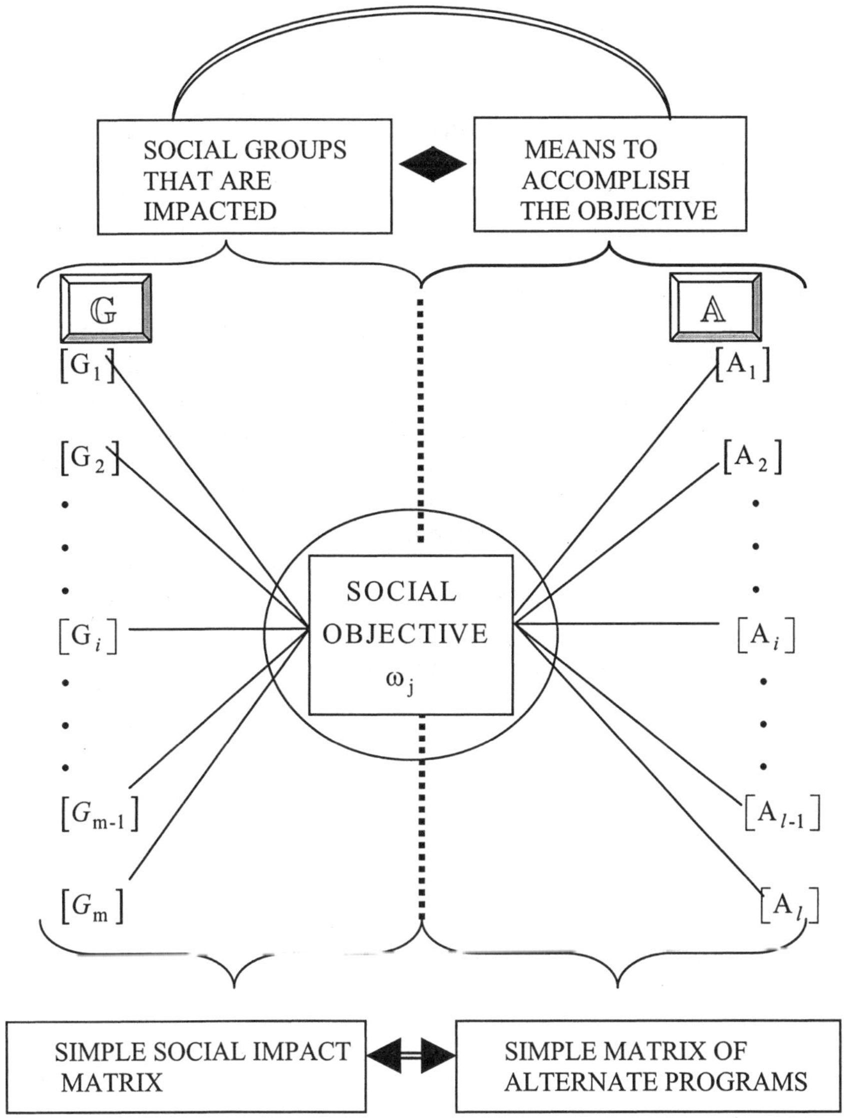

Figure 2.2.3 Group and Action Structure

From Figures 2.2.2 and 2.2.3 we observe that we have two social partition vectors. The first social partition vector is made up of groups with some interest in relation to social objective formation. The second social partition vector is composed of groups that experience similar and equal impact as a program alternative is used to arrive at the objective.

The size and nature of the two partition vectors are not necessarily the same. The group members may also not be the same. The simple matrix of program alternatives may also be viewed as an *action vector* since an implementation of any of its elements is an action to realize the objective.

Now let us consider the vector of program alternatives (that is, the action vector). Each of the decision alternatives is identified by a set of *real characteristics*. Such real characteristics may be technical, or social or both. The collection of these characteristics constitutes the characteristic set. The preferences established over the vector of program alternatives are actually preferences established over the corresponding sets of characteristics. These real characteristics may be identified to be either real cost or real benefit but not both. Thus associated with each alternative are two sets of real cost and benefit characteristics whose identification and accountability lead to the development of a set of simple matrices. The dimension of the real cost characteristics may not be equal to that of real benefit characteristics. The associated real cost and benefit characteristics affect social groups whose identification generates *social impact matrix* (that is a vector of groups with similar cost and benefit effects). The identification of the real cost and benefit characteristics also generate two simple relational matrices associated with the decision alternative. The input structure of these identification matrices is given in Figure 2.2.4.

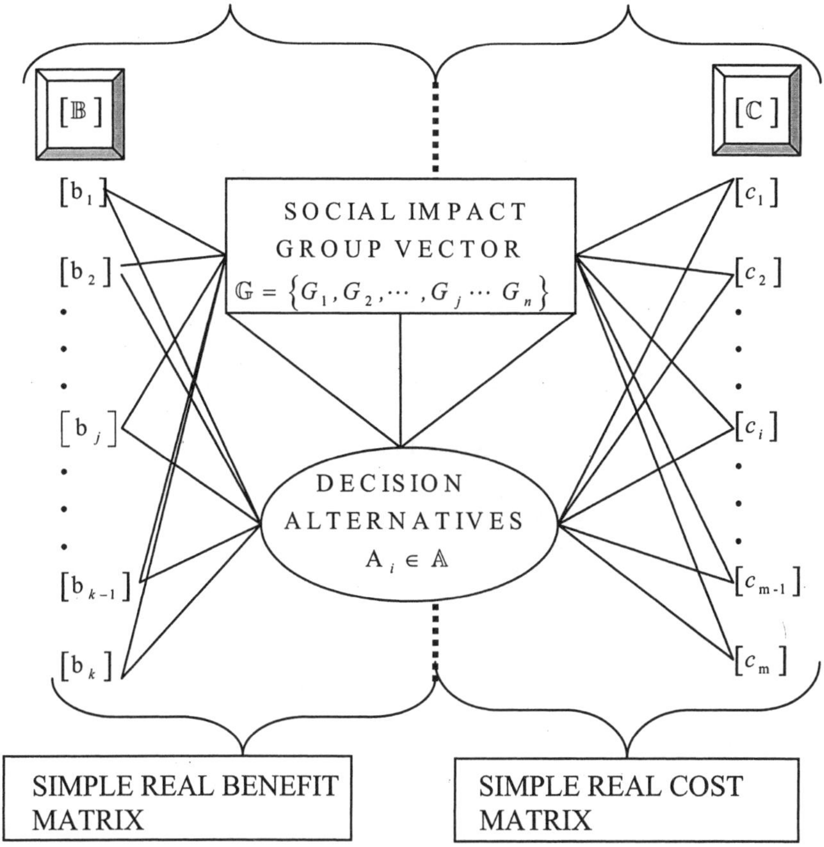

Figure 2.2.4 Organic Structure of Benefit-Cost Identification Matrices

The figure 2.2.4 presents real cost-characteristic vector of the form $\mathbb{C} = \{c_1 , c_2, \cdots, c_i , \cdots, c_m\}$ and a real benefit-characteristic vector of the form $\mathbb{B} = \{b_1 , b_2, \cdots b_j \cdots , b_k\}$ and a general social impact group vector $\mathbb{G} = \{G_1 , G_2, \cdots , G_j, \cdots , G_n\}$. Since the groups that may be affected by costs may be different from the groups that may be affected by benefits two different simple matrices may be abstracted from Fig. 2.2.4. They are simple real cost and benefit matrices relative to the vectors of social impact groups given the decision alternative. The simple cost matrix is shown in Figurc 2.2.5

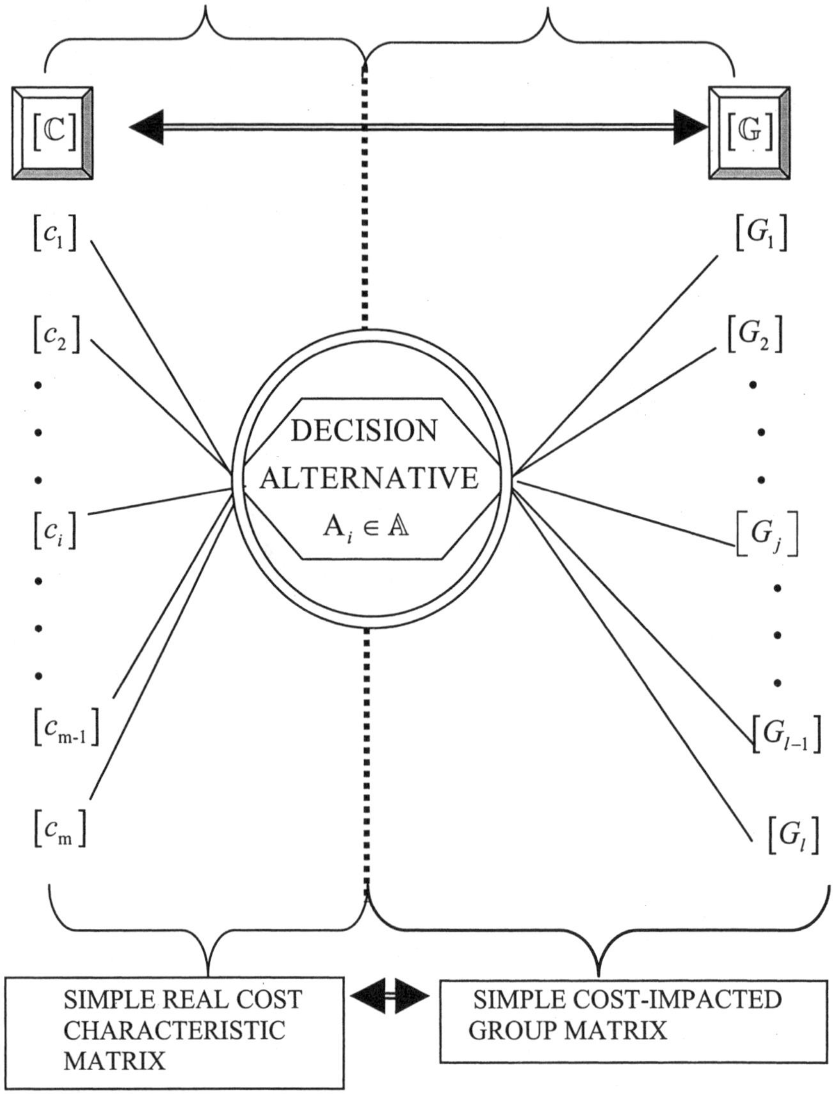

Figure 2.2.5 A Conceptual Structure of Cost Identification Matrix

The simple real benefit matrix is shown in figure 2.2.6.

Let us turn our attention to another set of computable identification matrices besides those for real cost and benefit characteristics given the impacted groups. The identification matrices of cost and benefit characteristics are logical devices to identify the benefit and cost clusters and the groups that are affected by different characteristics. They do not tell us about the degree of impacts.

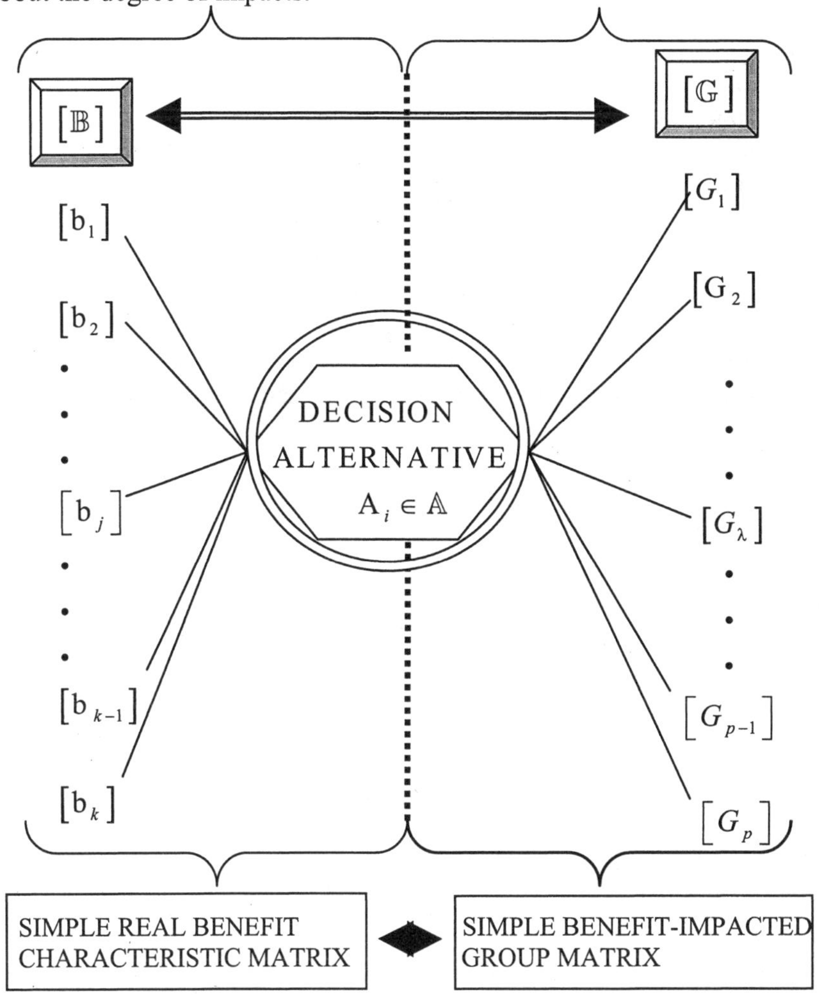

Figure 2.2.6 A Conceptual Structure of Benefit Identification Matrix

In other words, they reveal the distribution of cost and benefit characteristics over the social groups but not the intensities of their effects. To incorporate the intensities of cost and benefit characteristics we need a computable identification matrix for *distributional weights* where such weights identify the degrees of *benefit intensity* and *cost burden* associated with each social group relative to either a benefit or cost characteristic.

The construction of the matrices of cost burden and benefit intensities over different social groups implies some level of inter-group comparison of utilities. For democratic decision making these distributional weights must reflect group cross preferences in addition to those of experts, policy makers and politicians. The nature of computable distributional weights must combine group cross assessments in addition to those of experts, policy makers and politicians in accordance with their preferences. The nature of computable distributional weight identification matrices is of the form shown in figure 2.2.7. A similar matrix structure may be constructed for weights that capture the intensities of benefit characteristics. It may be noted that there are as many matrices of distributional weights as there are real cost and benefit characteristics. The size of each matrix will depend on the size of the vector of social groups given the cost or benefit characteristics. These sizes may be made smaller by the use of samples.

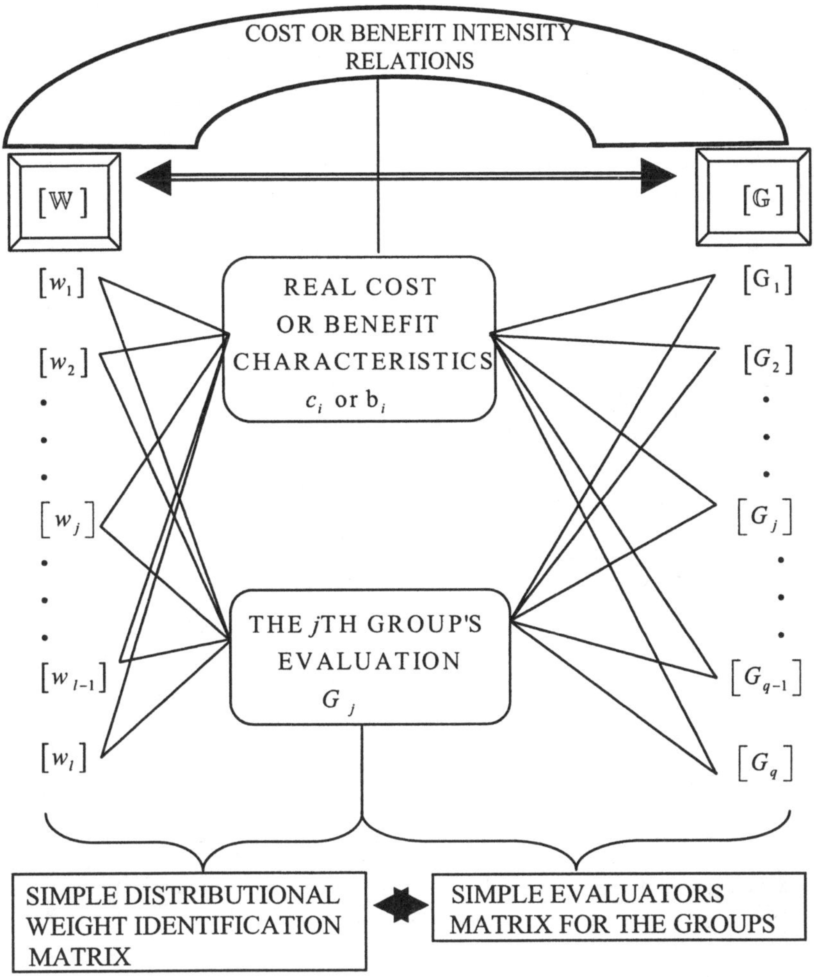

Figure 2.2.7 Identification matrix For Cost Burden Or Benefit Intensity Distributional Weights and distributional impacts on social groups

The structure of the identification matrices reveals two sets of matrices of final matrices and supporting matrices. In the case of cost-benefit analysis the final matrices must reveal cost and benefit data and the weights of their distributional impacts on social groups: The process of constructing the final computable matrices and the supporting matrices leading to required information for decision making may be represented as in Figure 2.2.8. The process begins with the general society which is divided into public and private segments in addition to social groups that are formed on the basis of interest clustering. Social objectives are then abstracted from the preferences of interest groups. From the social objectives, program alternatives are constructed relative to their abilities to realize the objective of interest. From each program alternative, characteristics that are divided into cost and benefit are identified through relational matrices.

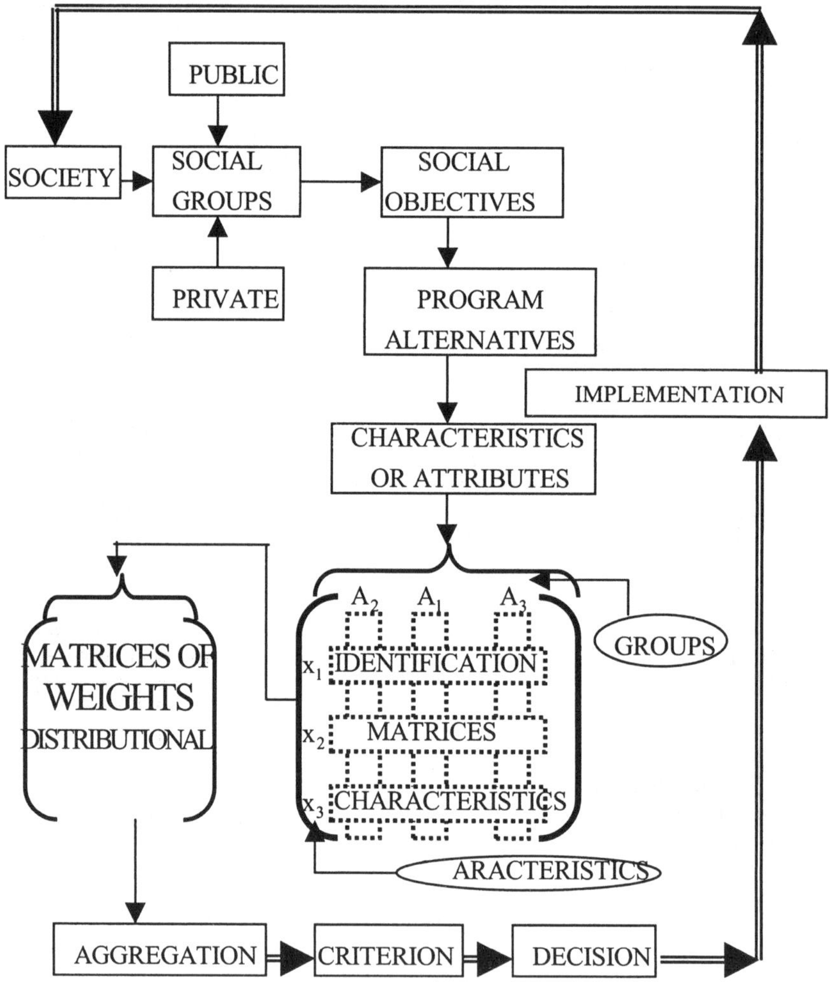

Figure 2.2.8 The Logical Process of Identification Matrices and Decision

Given the characteristic matrices of costs and benefits, distributional weights are constructed through cross-group relational matrices.

Both matrices provide us with cost and benefit data that must be valued and aggregated for each program alternative into cost aggregate and benefit aggregate. From the characteristics, values must be computed into the same unit of measure and then processed into aggregate data. Three sub-data sets are possible for each decision through data partitioning. The cost and benefit data may be partitioned into quantitatively measurable and non-measurable sets. The quantitatively measurable set of characteristics may be partitioned into pecuniary measurable and non-measurable sets. These sets together constitute the information support of decision making through cost-benefit process. The cost and benefit imputations must be repeated at each point in time if more than one period is the lifetime of the project alternatives. The dated benefit and cost aggregates may also be forecasted on the basis of initial imputations and other available information. From the cost and benefit aggregate data a criterion is constructed for decision making on the program alternatives. The monetary aggregates require value imputations of quantitatively measurable heterogeneous real characteristics. Such value imputations and the methods of their realization have been the center of rational debate in economic theory of decision. We shall turn our attention to reflect on the essential characters of the existing methods and how they relate to the method of fuzzy logic and decision.

2.3 Reflections on Contingent Valuation Method (CVM), Revealed Preference Method (RPM), Computable Identification Matrices and the Role of the Theory of Fuzzy Decisions

There are two established major methodological approaches to deal with value computations of characteristics of socioeconomic alternatives that require comparative analysis for ranking and choice. The imputed values are costs and benefits and the two methods are *revealed preference method* **(RPM)** and *contingent valuation method* **(CVM)**. The unit value is what we refer to as the price. The RPM is an indirect approach based on actual and potential outcomes in a system of markets. The CVM on the other hand is a direct approach in value computation based on information elicitation through an instrument of sampling and questioneering. The total value if computable and whether it is either cost or benefit is measured in the price-quantity space that reflects the decisions of individual and the social collectivity. The computations of total cost and benefit values require the creation and/or availability of *three* data sets composed of real cost characteristics, real benefit characteristics and corresponding prices.

The real cost characteristics and benefit characteristics are technically the inputs and outputs respectively of either individual or collective decision. The real cost and benefit characteristics are heterogeneous in nature. They must be identified for every given decision alternative before any form of aggregation can be taken. Given the sets of real cost and benefit characteristics associated with decision element, the computations of the total values of costs and benefits require us to know the price associated with each characteristic. The set of prices, no matter by what method they are obtained, must satisfy the *equivalence principle* of unit values. The equivalence principle implies that given a set of heterogeneous entities equal unit values are computable through proportionality processes such that the conditions of heterogeneity vanishes into homogeneity with the same unit values abstracted from equivalent proportions. The satisfaction of the unit-value equivalence principle allows comparability and summability of heterogeneous real cost and benefit characteristics. The real characteristics may be conceptually viewed in terms of Lancaster's approach to demand theory [468] [469]. The value equivalence principle of characteristics is a relative concept that allows each unit value, price, of characteristics to be compared and reduced to one common denominator that would qualify for aggregation [929].

Each element of the set of prices, besides satisfying the unit-value-equivalence principle must also satisfy the *principle of willingness*, at least, at the level of theoretical construct. A clarification on the willingness concept as applied to unit value computation is required. The willingness concept appears as dual in the form of willingness to pay (WTP) and willingness to accept (WTA). The principle of willingness is linked to the general welfare comparative position of the individual and the collective. As such it is defined in both consumption and production spaces when a socioeconomic system or subsystem is under consideration for a given individual and collective preferences that reflect the levels of individual utilities and collective welfare.

The individual willingness to pay and willingness to accept as defined in the consumption space generate the set of individual prices (unit values) for each input and output characteristic at a given information support of the decision. Under the condition that for any ith individual and relevant information support the WTP is equal to WTA, the individual prices become equilibrium prices for the cost and benefit characteristics. Corresponding to the set of individual equilibrium prices is a set of consumer surpluses if the decision data set can generate individual demand functions. Technically, we must expect that the functional path of the WTP will exhibit downward slopedness while the WTA will follow upward slopedness when one examines the quantity-willingness plane. As presented to the individual, the willingness to pay as a measure of unit value reflects unit cost appropriately conceived. The willingness to accept as a measure of unit value on the other hand reflects unit benefit for the same characteristic as appropriately conceived. The equilibrium price as a unit value for any characteristic is that for which unit-benefit value is equal to unit-cost value (WTA = WTP). The WTA and WTP result in demand and supply behaviors as exhibited by the individual acting as producer and consumer simultaneously.

The resulting demand *curve*, is the functional path for which unit benefit equals unit cost for the same characteristic and for any given individual and corresponding information support for his or her decision. When aggregates are constructed from the set of individual equilibrium prices we obtain aggregate equilibrium prices for the characteristics involved. On the side of consumption the path that connects these individual prices on one hand and aggregate prices on the other over different allocations may be viewed as *consumer price offer curve* (C-POC) (WTP) on the aggregate and at a given level of social welfare improvement. A similar analytical procedure carried on in production space leads to a set of individual producer supply prices where the producers' WTA equals WTP at a given information support. This leads to the individual and aggregate *producer price offer curve* (P-POC) for a given characteristic.

Both the C-POC and P-POC are defined over different allocations at a given information support for either producer or consumer decision. The WTP and WTA as underlying forces of C-POC and P-POC must always be related to cost and benefit in the characteristic space. Interpretation of WTA and WTP should not be confined to inter allocative change that may be linked to consumer surplus. This is inter-allocative interpretation that may be linked to compensating and equivalence variations. There is another interpretation that can be offered within any allocation. This is intra-allocative interpretation of WTP and WTA. Here the question is not, "to change or not to change". The question is how much is one willing to accept (WTA) in order to produce a benefit or how much is the same individual willing to pay (WTP) someone to produce the benefit characteristic within the same allocation.

The task faced by a cost-benefit analyst is first to develop the dataset of real cost and benefit characteristics for any allocation or social project. Given the data set of real cost and benefit characteristics how does one construct information on both the WTP and WTA and to develop from it the needed unit values for any given benefit or cost characteristic or a decision element. The substance of this is that of obtaining aggregate unit values. For some given characteristics the measure of willingness can be abstracted, no matter how imperfectly, from the data as revealed by the market imputations to the extent to which such characters are market tradable and under the restrictions of perfect markets. This approach of obtaining unit values and total values of cost and benefit characteristics from market outcomes is the RPM. It is indirect and the accuracy of the values depends on the efficiency of market institutions. The market imputations of the values of these characteristics are taken as equilibrium prices that are the measures of willingness to pay and willingness to accept simultaneously.

The individual WTP and WTA values are linked to the social welfare of community composed of two or more individuals for any given allocation. Let us illustrate the concepts of WTP and WTA and the difficulties they present in value computations and cost-benefit analysis. Consider a whole society, $\mathbb{S}$, that is partitioned into mutually exclusive and collectively exhaustive groups of A and B. Two products x and y are produced using factors, labor, L, and capital, K. Let the utility levels of A and B be measured by $u^A(x^A, y^A)$ and $u^B(x^B, y^B)$ depending on the levels of x and y consumed. The total welfare of the community for any given allocation is measured by

$$W = W\left(u^A(\bullet), u^B(\bullet)\right)$$

(2.3.1)

The change in welfare for a movement from one allocation to another is a weighted aggregate of individual welfare improvement in the form

$$dW = \frac{\partial W}{\partial u^A} du^A + \frac{\partial W}{\partial u^B} du^B$$

(2.3.2)

The examination of the welfare improvement of a social system and the distribution of such improvement is equivalent to social system's auditing before

Table 2.3.1 Simple Cost-Group identification Matrix, (TC=Total Real Cost)

C \ G	A	B	TC
L	L^A	L^B	L
K	K^A	K^B	K
TC	$L^A + K^A$	$L^B + K^B$	L+K

$$L = L^A + L^B, \quad K = K^A + K^B$$

or after social decision. The comprehensive accountability of this simple logic of computable WTP and WTA leading to equations (2.3.1) and (2.3.2) may easily be cast in terms of cost-benefit identification matrices in simple forms. The cost identification matrices are shown matrices are shown below.

Table 2.3.2 Simple Cost-Product Identification Matrix

C \ B	x	y	TOT
L	L^x	L^y	L
K	K^x	K^y	K
TOT	$F(K^x, L^y)$	$G(K^y, L^y)$	x+y

Table 2.3.3 Simple Group-Benefit Identification Matrix

B \ G	A	B	TB
x	x^A	x^B	$\hat{W}_1\left(x^A, x^B\right)$
y	y^A	x^B	$\hat{W}_2\left(y^A, y^B\right)$
TB	$u^A\left(x^A, y^A\right)$	$u^B\left(x^B, y^B\right)$	$W\left(u^A(\cdot), u^B(\cdot)\right)$

(Where $\hat{W}_1(\bullet) = \hat{W}_1\left(u\left(x^A \mid y^A\right), u\left(x^B \mid y^B\right)\right)_1$ and similarly $\hat{W}_2(\bullet)$)

These three simple matrices can then be supported by identification matrices of distributional weights of cost and benefit intensities as they are revealed by equations (2.3.3) and (2.3.4). The family of social welfare levels is represented in Figure 2.3.1 as social welfare indifference curve

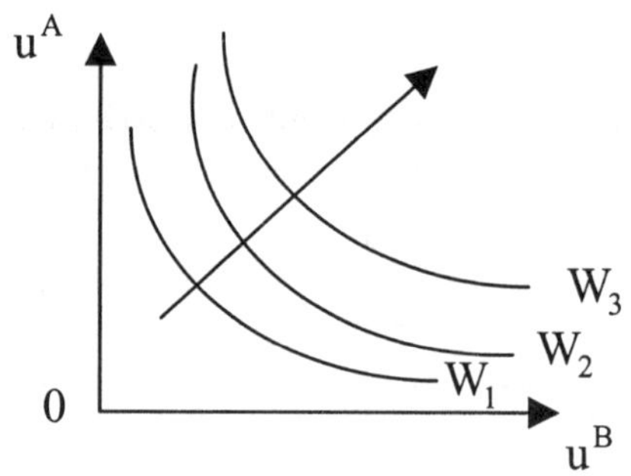

Figure 2.3.1, Iso-social Improvement Curves, where $W_3 > W_2 > W_1$

Equation (2.3.2) may be reorganized in terms of unit measure of a numerier. Let such a numerier be in terms of y. Thus if $u_y = (\partial u / \partial y)$ marginal utility of y then

$$dW = \frac{\partial W}{\partial u^A} u_y^A \frac{du^A}{u_y^A} = \frac{\partial W}{\partial u^B} u_y^B \frac{du^B}{u_y^B}$$

(2.3.3)

Each term in equation (2.3.3) is composed of two multiplicative values. The value du/u_y is the value of a change in individual total welfare as measured in terms of the numerier good, y. It indicates approximately the number of units of y that each decision agent is either willing to pay or accept in order to change or remain in a given allocation. The value $(\partial W/\partial u)u_y$ is the weight that society attaches to each unit of WTP or WTA as assessed by individual decision agent in the community. Equation (2.3.3) may simply be written as

$$\Delta W = \alpha \Delta u_A + \beta \Delta u_B$$

(2.3.4)

where α and β are distributional weights that express social importance.

The welfare change, ΔW, is measured in terms of the units of numeraire good for a change in allocation. Both ΔW and Δu are net benefit imputations. The parameters α and β are measures of social importance (weights) attached to each WTP or WTA. The concept of social importance is vague and imprecise so also its measurement. Similarly, the concepts of WTP and WTA are vague and Δu's are approximations of real WTP and WTA.

In addition to the problem of vagueness, there are certain real characteristics of choice-decision elements that have no markets to reveal unit values even if market imputations are our preferences. In this case the option of market imputations of unit values and their utilization are not open to us in the decision making process. For example, individuals cannot buy clear air or safety of zone depletion in the market for himself or herself alone except through collective social action outside the market place. The question that arises in cost-benefit analysis is what do we do in situations where there are no markets or the markets are substantially distorted by externalities that hamper efficient imputations. These conditions are prevalent in situations of collective good, general health, statutes, national security investments, social risk involving environmental disasters, and collective disasters.

In these situations, that is, the absence of markets, two approaches have been traditionally advanced. They are *hedonic pricing approach* (HPA) and *contingent valuation approach* (CVM) to estimate unit and total value that may be associated with decision element. The hedonic pricing approach can be used to approximate some values to WTP or WTA. It ing approach can be used to approximate some values to WTP or WTA. It cannot, however, be used (as we know it now) to approximate the value of social importance that may be attached to either WTP or WTA.

In the case of HPA when prices do not exist for a characteristic of the decision element the value of either WTP or WTA is calculated by using the values of characteristic of decision element made or defined in different context in order to infer the values of characteristics associated with decision elements in different environment. For example, consider a problem of finding how much is one's WTP or WTA for a clean air or air pollution as measured in some units. To find the value of these WTP and WTA we consider two identical houses X and Y in different location A and B in the same city relatively. The only difference in the two locations, A and B is the air quality. Let the prices of the houses be p_x and p_y where the difference is given as $p_x - p_y = 20{,}000$. Location A has an excellent air quality while location B has relatively poor air quality measured in terms of units of US PSI (Pollution Standard Index). Let the air quality difference between locations A and B be 200 PSI units. If a person confronted with this information selects to buy house X at location A then we infer that the individual is willing to pay a price of $100.00 to reduce air pollution by one PSI unit. The value $100 is the hedonic price for one PSI unit. If a reduction in air pollution by one PSI unit leads to a reduction of 0.001 probability of death then we may infer that the individual is willing to pay $100.00 for a reduction of 1% risk of death.

These hedonic prices are individual unit cost, benefit and willingness values. The cost and benefit to society must account for the weight of social importance that must be attached to the individual willingness values. The HPA does not provide us with a way for compiling these social welfare distributive weights. We therefore, must look somewhere for the methods that would allow us to assess the values of the social weights. Additionally, the HPA like RPM is indirect depending on inference from market imputation and outcomes of subjective assessments.

In the case of CVM the values of willingness are obtained directly by a process of information elicitation through instruments of survey and questioneering. The data as obtained is processed to obtain the relevant values. The same method and type of instrument may be used to elicit information relevant to compute the values of weights of social importance. A problem arises, however, with CVM as to the degree of accuracy that must be attached to the elicited information. Similarly, some questions arise as to what method should be used to abstract aggregate values and what adjustment if any must be made to the aggregate values before used.

[33] [38] [98]; the second question belongs to the class of aggregation problems while the third question centers on the problem of accuracy and the stability of the aggregates.

In a general sense, the concept of willingness comes with vagueness, the utility measures are characterized by ambiguities while the computations of values of willingness and weights of social importance are either imprecise or approximations. The optimal decision calculus obtained with the use of such vague information is thus driven by the classical decision choice rationality that is more or less bounded in the sense described in [445]. However, the decision rationality and the method of analysis are driven from classical logic of binary outcome that does not allow shades of accuracy, truth and others. Human decisions follow paths of approximations, sequential adjustments and imprecision that are assumed away in classical decision modeling. Given these characteristics of decision environment, a question arises as to whether we can improve upon the classical decision-choice process as see through the prism of cost-benefit analysis. If improvement is possible how and what are the methods and techniques of the improvement?

The concept of bounded rationality offers us some insights into the limitations of classical rationality. It does not, however, offer us an approach to model the decision-process in an environment where vagueness, ambiguities and imprecision are important characteristics that further increase complexities and constrain cognitive activities in the choice process. Discussions in [446] [519a] offer some possible direction for modeling bounded rationality. There is, however, a need for an alternative approach. This alternative approach that will allow one to deal with decision environment that has the presence of vagueness, ambiguity, inexactness, subjectivity, approximations and ill-posed problems in the framework of cost-benefit analysis is *fuzzy logic* [974] [985] [1020] [1033] with corresponding *fuzzy mathematics* [1058] [1118] [113b] [1151] and *fuzzy decision theory* [661] [694] [707]. The method of the theory of fuzzy decisions allows one to deal with the problems of vagueness, ambiguities, subjectivity and inexactness as part of logical and mathematical problem formulation and solution realization. The general fuzzy analytical framework can be used to deal with direct or indirect value computations in the absence of markets or perfect markets. It can also incorporate random phenomena into the analytical structure of the cost-benefit decision making process. The *decision-choice intelligence* that emerges out of the fuzzy analytical framework is *fuzzy rationality* [446], [661] [694] as compared to both *classical* and *bounded rationalities* [446] [519a]. The general analytic approach will be called fuzzy decision method (FDM).

Fuzzy rationality is an abstraction based on the notion that human reasoning is approximation, truth comes in shades, meanings are contextual, subjective actions are perception-dependent, collective social actions are more ideology-based than on science and others [446] [661] [735] [752]. The problem of general fuzziness is done away within the classical logic through assumptions. Fuzzy rationality allows us to compute *reasonable decision* solution from contaminated environment where perfect and exact information is unattainable. The basic logical structure of method of fuzzy decision theory *is fuzzification* and *defuzzification* of the choice-decision process to obtain a crisp solution that is more or less approximation called *exact equivalence* in the same notions as certainty and present value equivalences are the results of stochastic and discounting processes [1250] [627] (See also Chapter 6 of this volume for detailed discussions).

As applied to cost-benefit analysis, the methods of fuzzy decisions can be used to compute the weights of social importance that may be attached to the values of individual WTP and WTA. The information required for such computing is obtained from fuzzy clustering and/or partitioning, sampling and instrument of questionnaire. The fuzzy logical framework can stand by itself as we shall find out. It may also be used to complement RPM in value computation where there are presence of externalities in the market and where subjective adjustments are required for corrections. It may also be used to supplement and complement CVM in the areas of information processing, aggregation and combination of expert judgments. In all these, FDM stands by itself in developing fuzzy computing process for cost-benefit values of subjective phenomena through the channels of *approximate reasoning* whose scientific importance has lead to the establishment of the International Journal of Approximate Reasoning [9298] [710] [970]. The values that may be computed include prices that allow heterogeneous cost-benefit characteristics to be aggregated; and subjective individual or social discount rates that allow the future values to be subjectively connected to the present in decision [364] 379].

The information requirement to implement *optimal fuzzy decision rules* is conceived in terms of elicitation through approximate instruments. In other words, the cost-benefit data required for (fuzzy decision approach) in cost-benefit analysis may be derived from some experimentation or from sampling and questionnaires. The resulting information constitutes the normative database. The quality of the acquired database will depend essentially on the nature of decision experiment and the structure of sampling designed. From the database, relevant fuzzy sets with corresponding membership functions are constructed. This is the *fuzzification process.* The collection of these fuzzy sets constitutes a family of fuzzy sets.

Given the family of fuzzy sets abstracted from different groups that are made up of policy makers, experts and members of social groups, a fuzzy decision problem is formulated and solved to obtain crisp solution that yields values that are called *fuzzy_equilibrium values*. The *process is defuzzification* and involves fuzzy aggregation, weighting and optimization among other things [661] [665] [682] [694] [809]. The analytical structure and the cognitive path of fuzzy decision process leading to optimal prescript rules are represented in Figure 6.6.1 of Chapter 6 of this volume.

The fuzzy decision approach (FDA) brings with it some important analytical tools into cost-benefit analysis. In this approach, judgments of different groups can be brought together, in terms of fuzzy domain structure, combined and or aggregated through any appropriate technique of fuzzy aggregation to form a single or interval decision value. Furthermore, it provides us with a channel to compute social decision weights that may be attached to quantitatively measurable and non-measurable as well as pecuniary measurable and non-measurable real characteristics obtained through either RPM or CVM. Completing all these approaches and information requirement is the role of *cost and benefit identification matrices* that allow a comprehensive accounting to be imposed on the decision space and for each decision element. An elementary example of cost-benefit identification matrix may be found in [94] [122] and more complex ones in [928] [930].

2.4 Computable Cost-Benefit Identification Matrices and other Areas of Decision Applications

As we have suggested, the logic of the computable cost-benefit identification matrices may be used in different areas of decision in both social and physical systems. The effectiveness of the use and the efficiency of outcomes of the matrix logic will depend on the type and structure of the decision problem. To extend the theory and explore the structure of areas of application, at least, we need to establish some conceptual understanding of the general form of the theory of computable cost-benefit identification matrices and point to the logic of the areas that we might find uses.

This general form of the conceptual understanding demands that we know 1) the capabilities and limitations of the theory, 2) the goals and objectives of the cost-benefit identification matrices; 3) the criteria for applying the theory and constructing the identification matrices and 4) coordination and combination of the main cost-benefit identification matrix and supporting matrices to obtain databases for costs and benefits. We shall discuss the main ideas of these demands of the theory of computable identification matrices. It must be kept in mind that the outline of the logic of the extension of the applications of the theory will also induce sub-theories of cost-benefit identification matrices that are areas specific.

2.4.1 The Logic of Computable Cost-Benefit Identification Matrices

In order to fully appreciate the capabilities and limitations of the theory of computable cost-benefit identification matrices and the range of its possible applications we need a basic understanding of the premises on the basis of which the theory is constructed. The logical structure of the theory is based on and derived from the scientific gains in the development of systemicity and organicity to arrive at a cost-benefit comprehensive accounting system. Here the matrix structure and functional description of benefit and cost characteristics are identified. Decisions are system oriented in the sense that cost and benefit outcomes due to decisions are essentially systems-induced. The breadth and depth of effects of the cost and benefit outcomes depend on the decision type and the nature of the system under consideration.

The logic of computable cost-benefit identification matrices is to integrate collective decision, multicriteria decision attributes and relevant mathematical apparatus in addition to available and useful fields of expert knowledge to construct its own formalized calculus for building information and knowledge bases in support of decision making. The thinking here is that the laws of organization of cost and benefit systems are the same for all possible decision alternatives for both humanistic and non-humanistic selections. As such there must be a general method for cost and benefit identification, assessment and computation to obtain the required cost and benefit information for dealing with the ranking of decision alternatives on the basis of costs and benefits broadly defined.

The entry point of the theory of computable cost and benefit identification matrices is the notion of interdependence of costs and benefits and organicity of projects and actions under decision. In general a project or an action may be viewed broadly as control from the organic structure of the system. Thus we may speak of decision alternatives or action alternatives or control alternatives depending on whether we are dealing with humanistic or non-humanistic system. All these appear in terms of dualities polarities and categories in grades of belonging.

Each alternative has cost sources and benefit sources. These sources may be viewed as characteristics that must be identified. Their effects are distributed over elements or parts of the system. In social system these elements or parts appear to us as social groups with common characteristics as we have discussed. In mechanical system these element appear to us as different parts whose performance may be enhanced or hampered through control or decision action. In a situation where the human and non human systems are integrated into an organic whole such as man-machine systems, the elements may appear as mechanical parts and social groups. In the basic logical form, the wholeness of the decision system is made up of cost and benefit sources, control alternatives and affected elements, all of which must be identified. The wholeness is conceptually seen as relations among cost-benefit sources, control alternatives and elemental parts (social groups). The structure is such that there is cost alternative and group relation on one hand; and benefit alternative and group relation on the other hand. These cost and benefit relations present themselves as information that reveals the state of welfare of the system as well as for controlling adjustments to the systems welfare.

The identification, assessment and computation of the cost and benefit information for designing the controllability of the system's welfare are based on the distribution of the cost-benefit relations over the social groups or elemental parts or both. The construction and development of the cost-benefit information bases require a comprehensive accountability of benefit and cost characteristics over all relevant elemental parts. Such a comprehensive accountability is made explicit and easier by matrix displays of cost components and groups (elemental parts) as well as benefit components and corresponding groups.

The methodological approach is a conceptual separation of the whole into relevant parts, definition of cost and benefit characteristics, specification of the relations among sources and parts for each alternative, and the conceptual reunification and synthesis of the analysis into a whole. The analytical core is category formation on the structure, functions and relationship among the organic parts of the system. The structure encompasses categories of attributes that define quality and quantity of the system's welfare. Function relates to groups or categories of parts of the decision system. Relations are established by cost and benefit connections that link functional controls to systems structure to establish organic unity and system's welfare as measured in some units.

The theory of computable cost-benefit identification matrices, thus, derives its logical justification from the wholeness of systems. The system's wholeness is viewed to be characterized by cost-benefit qualities and attributes that may not be inherent in the elemental parts but are revealed from interactions as decision (control) alternatives are selected and applied [441]. In this logical structure, analysis is pushed to identify the cost and benefit attributes of the system's relations, the sources and the affected parts. Synthesis is carried on to unite the cost and benefit characteristics in order to access the positive or negative net contribution that is made by each project or control to the welfare of the system.

As conceived and perceived, every synthesis of the computable cost-benefit matrix is a projection through the eyes of analysis of costs, benefits, decision alternatives and the elemental parts of the system while every analysis is projected through cost-benefit synthesis to complete the process of the wholeness and unity of the system [441] [465a] [928] [929] [930].

2.4.2 The Goals and Objectives of the Theory of Computable Cost and Benefit Identification Matrices

An epistemic reflection of the logic of computable cost and benefit identification matrices was presented in section 2.4.1. The cognitive process and its construct are intended to accomplish some goals and objectives. It is important that these goals and objectives be made explicit and understood by those that may employ the matrix logic as it is presented. It must also be understood that the theory of computable cost-benefit identification matrices just like any theory about economic phenomenon is an artifice of cognition and simply a thinking technique that will lead the user to a comprehensive accountability of costs and benefit associated with different decision alternatives.

The intention for the construct of the theory is simply to present a method or an algorithm for developing a comprehensive cost-benefit information base in support of decision making. In this respect a synthetic and analytic apparatus of the mind is constructed to provide a framework for generalized cost and benefit analysis. The goals and objectives of the framework are many folds that when attained will provide the utility of rational decision-making on the basis of relative positions of costs and benefits. There are grand goal and objective in addition to sub-objectives that include the development of instruments for identifying costs and benefits and the manner in which these costs and benefits are distributed over different segments of the society.

The grand goal of the theory of computable cost-benefit identification matrices is to construct a cognitive structure that is made up of logical and methodological instruments for developing analytic and synthetic models that present distribution of cross relational effects of costs and benefits in a manner that allows one to account for specific cost and benefit features of the decision system for policy and control.

The logical aspect of the theory is to create a synthetic and analytic unity of cost and benefit computations for systems in general. The methodological aspect covers objectives of:

a) Explication of basic concepts of cost and benefit characteristics;
b) The classification of cost and benefit characteristics;
c) The partitioning of the system into affected (or will-be-affected) parts or groups;
d) The identification of sources of cost and benefit characteristics from potential alternatives;
e) The matrix displays of relations between cost characteristics and cost sources on one hand and between benefit characteristics and benefit sources on the other hand for any given decision alternative;
f) The matrix display of value relations between cost characteristics and affected groups on one hand if social (mechanical) system is under analysis, and value relations between benefit characteristics and recipient groups (parts) if social (mechanical) system is under decision assessment; and
g) The development of instruments of cost-benefit information base elicitation or assessment; and the development of mathematical and computation analysis for dealing with the problems of distortions and aggregations of heterogeneous cost and benefit characteristics associated with decision alternatives.

The basic structure to fulfill these goals and objectives may be presented as in Figure 2.4.2.1. The cost-benefit information system for decision making is geared towards facilitating the control process in the decision system and to enhance the efficiency of managing the organic improvement of the welfare of the system. The four quadrants present a way for viewing the assessment and computations of the cost-benefit information through an appropriate logical process.

The center of Figure 2.4.2.1 is the decision or control system. The *project space* presents decision or control alternatives that must be ranked by a cost-benefit index for efficient selection. The *distribution space* presents the structure as to how costs and benefit effects are distributed over groups and parts of the organic system. The *cost space* reveals the cost characteristics and the sources of these cost characteristics. The *benefit space*, on the other hand, is where the benefit characteristics and the sources of such characteristics may be identified and computed.

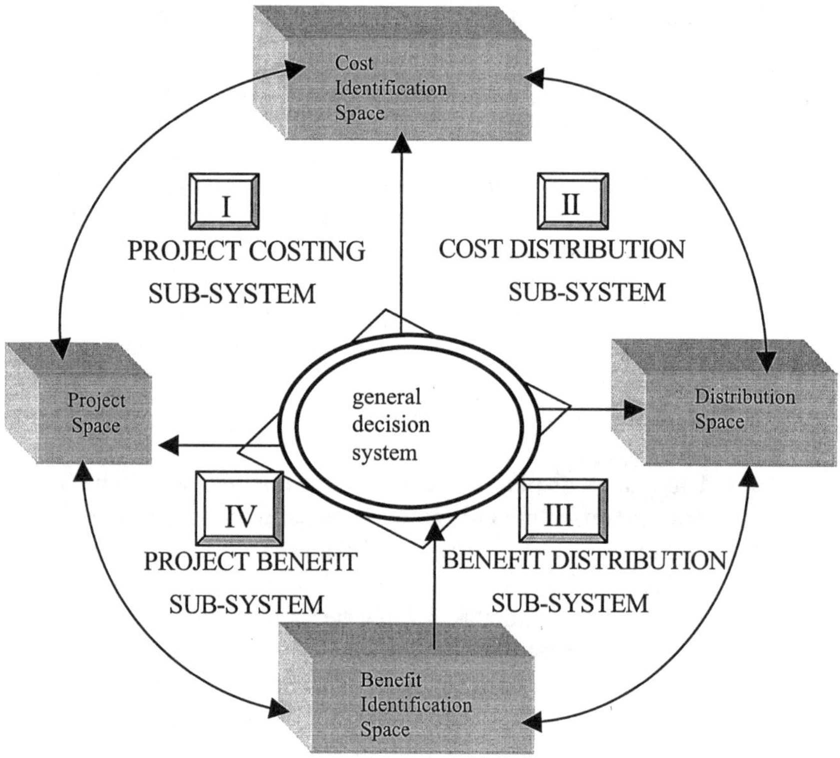

Fig. 2.4.2.1.The Structure of Computable Cost-Benefit Information System.

The *cost distribution subsystem* (II) presents a structure where real cost characteristics and affected groups are identified, accounted for and cohort values computed. This subsystem leads to a relational matrix like the form in Figure 2.2.5 where the elements of vector will be parts, if mechanical system is being considered. The *project costing subsystem* (I) defines a structure where cost characteristics and their sources are identified for any given decision or control alternative. The project costing subsystem (I) and the cost distribution subsystem II may be brought together to present a cost relation matrix that can display a comprehensive cost accounting system in terms of real cost characteristics, sources and affected groups or parts of the organic system. The resulting cost relational matrix and the supporting sub-matrices will constitute the *computable cost identification matrix system.*

Similarly, the benefit distribution subsystem (III) is a structure where the benefit recipients are identified and accounted for given the set of benefit characteristics. The project benefit sub-system (IV) on the other hand defines a structure where the real benefit characteristics and their sources are identified and assessed for any given decision or control alternative. The project benefit subsystem (IV) and the distribution subsystem (III) can be combined together to construct a benefit value relational matrix that presents a comprehensive benefit accounting in terms of real benefit characteristics, resources, beneficiaries and cohort values. The resulting benefit relational matrix and the supporting sub-matrices will constitute the *computable benefit identification matrix.*

The subsystems I, II, III and IV constitute the organic information logical system where benefit and cost information bases are developed in support of the decision process. They constitute the foundation of the theory of computable cost-benefit identification matrices. The theory is constructed to take advantage of democratic ideals and to enhance participatory democracy in decision and control systems irrespective of whether the system is humanistic or non-humanistic.

The identification, assessment, computation and accounting process of cost and benefit characteristics are to solicit cost and benefit information from members of the society, experts and policy makers. The information obtained is then processed and aggregated by acceptable mathematical methods to arrive at the cohort values in real terms. The other step in the computational process is to aggregate the cohort cost and benefit values into grand totals for any decision alternative of concern through an acceptable mathematical method of aggregation.

The first aggregation process to obtain the cohort values falls under the mathematical technique and logic of combination of differential judgments and assessment of cost-benefit information for any given characteristic. The second aggregation process, to obtain the overall cost and benefit totals, involves the mathematical technique of summability of heterogeneous entities (characteristics or attributes) for any given control or decision alternative.

The logic of the theory of computable cost-benefit identification matrices is such that there are many systems of cost-benefit identification matrices as there are decision alternatives. For each decision alternative and at any given time there are two matrices in support of decision making. There is one system of matrices for cost and another for benefits for any given time. When the project's life extends through time then we have a matrix of matrices (a power matrix) where the size is defined by the number of decision or control alternatives and the number of relevant time points.

If $\mathbb{A}$ and $\mathbb{B}$ are time point cost and benefit computable identification matrices and the number of decision alternatives, $\#\Pi = \mathrm{m}$, and the number of discrete units of the time set $\#\mathbb{T} = \mathrm{n}$ respectively then there are two grand systems of matrices $\mathbb{A}$ and $\mathbb{B}$ that present a comprehensive information on costs and benefit in support of the decision process. The entry values of the system of cost and benefit matrices are $\mathbb{A}_{it}$ and $\mathbb{B}_{it}$, $i = 1,2,\cdots,\mathrm{m}$; $t = 1,2,\cdots,\mathrm{n}$. The grand matrices $\mathbb{A}$ and $\mathbb{B}$ may be written in identification matrixes as

$$\underset{\mathrm{mxn}}{\mathbb{A}} = \begin{array}{|c|cccccc|} \hline \pi \backslash T & 1 & 2 & \cdots & t & \cdots & \mathrm{n} \\ \hline \pi_1 & \mathbb{A}_{11} & \mathbb{A}_{12} & \cdots & \mathbb{A}_{1t} & \cdots & \mathbb{A}_{1\mathrm{n}} \\ \pi_2 & \mathbb{A}_{21} & & & & & \\ \vdots & & \ddots & & & & \\ \pi_i & \mathbb{A}_{i1} & & & \mathbb{A}_{it} & & \vdots \\ \vdots & & & & & \ddots & \\ \pi_{\mathrm{m}} & \mathbb{A}_{\mathrm{m}1} & & \cdots & & & \mathbb{A}_{\mathrm{mn}} \\ \hline \end{array}$$

(2.4.2.1a)

Equation (2.4.2.1a) may simply be written as

$$\underset{\mathrm{mxn}}{\mathbb{A}} = \begin{pmatrix} \mathbb{A}_{11} & \cdots & \mathbb{A}_{1\mathrm{n}} \\ \vdots & \mathbb{A}_{it} & \vdots \\ \mathbb{A}_{\mathrm{m}1} & \cdots & \mathbb{A}_{\mathrm{mn}} \end{pmatrix}$$

(2.4.2.1b)

Similarly the computable identification matrix for benefit may be written as eqn. (2.4.2.2) Given the aggregation of differential evidence that leads to the cohort values and aggregation of the heterogeneous benefit and cost characteristic we then perform the last aggregation of intertemporal values through an acceptable mathematical method. This aggregation leads to cost and benefit vectors that have many entries for both cost and benefit aggregates as there are decision or control alternatives.

$$\underset{\text{mxn}}{\mathbb{B}} = \begin{array}{|c|cccccc|} \hline \pi \diagdown T & 1 & 2 & \cdots & t & \cdots & \text{n} \\ \hline \pi_1 & \mathbb{B}_{11} & \mathbb{B}_{12} & \cdots & \mathbb{B}_{1t} & \cdots & \mathbb{B}_{1\text{n}} \\ \pi_2 & \mathbb{B}_{21} & & & & & \\ \vdots & & \ddots & & & & \vdots \\ \pi_i & \mathbb{B}_{i1} & & & \mathbb{B}_{it} & & \\ \vdots & & & & & \ddots & \vdots \\ \pi_{\text{m}} & \mathbb{B}_{\text{m}1} & \cdots & & & \cdots & \mathbb{B}_{\text{mn}} \\ \hline \end{array}$$

(2.4.2.2a)

$$\underset{\text{mxn}}{\mathbb{B}} = \begin{pmatrix} \mathbb{B}_{11} & \cdots & \mathbb{B}_{1\text{n}} \\ \vdots & \mathbb{B}_{it} & \vdots \\ \mathbb{B}_{\text{mn}} & \cdots & \mathbb{B}_{\text{mn}} \end{pmatrix}$$

(2.4.2.2b)

2.5 Areas of Potential Application

The theory of computable cost-benefit identification matrices has been developed to extend the application areas of cost-benefit analysis to encompass humanistic and non-humanistic systems that require internal decision and control to establish and maintain equilibrium and stability of the system. Most of the application areas of cost-benefit analysis has been restricted to social system and decision problems of natural system that have direct bearing on social system. Our current age of high technology for automation and automatic controls for excited and non-excited processes requires that the application areas of cost-benefit analysis be broadened and extended to include mechanistic systems. Here the techniques and methods of cost-benefit analysis is combined with artificial intelligent technology to design optimal automatic controllers in a mechanistic system where the system's welfare (composed of attributes such as stability, controllability and others) is viewed in terms of properly defined cost-benefit relations.

Conceptually, one can consider an autonomous complex mechanistic system that is to be designed. The mechanical system may be a complex space ship or complex ocean liner or a military combat ship such as shipboards, submarines that are self-contained. The operations of the self-contained mechanistic system require a number of autonomous controls where the optimal controllers are designed to deal with different contingencies of hazards that the mechanical system may confront during its operations. Such controllers may be conceptualized and architecturally designed in terms of fuzzy or non-fuzzy controllers. Whether the automatic controllers are fuzzy or non-fuzzy, optimal ones are always sought. In this respect the concept of optimality takes on many forms that need to be defined and specified.

The approach that is being suggested here is to use cost and benefit relations to specify conditions of optimality of the automatic controllers that may correspond to various contingencies. The procedure would be first to define and specify the set of attributes of the mechanical system (for example this will include, weight, mass, volume speed, number of persons on board and others). Given the attribute set, an optimal architectural design of the mechanical system with complex automatic controllers may be contemplated. Given the potential optimal construct of the mechanical system, a contingency set is constructed to account for the essential potential contingencies. The collection of such contingencies will constitute the contingency set, $\mathbb{K}$, with a corresponding set of probabilities $\mathbb{P}$. For each identified contingency, cost characteristics are assessed to obtain a relational cost identification matrix of the form that relates probabilities δ's of outcomes to cost characteristics for any given contingency. This is shown in eqn. (2.5.1a).

Table 2.5.1 Cost-Contingency Identification Matrix

$$\underset{\text{mxn}}{\Delta} = \begin{array}{c|cccccc} {}_{\mathbf{C}}\backslash^{\mathbf{K}} & \mathrm{K}_1 & \mathrm{K}_2 & \cdots & \mathrm{K}_j & \cdots & \mathrm{K}_n \\ \hline \mathrm{C}_1 & \delta_{11} & \delta_{12} & \cdots & \delta_{1j} & \cdots & \delta_{1n} \\ \mathrm{C}_2 & \delta_{21} & & & & & \\ \vdots & \vdots & & \ddots & & & \vdots \\ \mathrm{C}_i & \delta_{i1} & & & \delta_{ij} & & \\ \vdots & \vdots & & \cdots & & \ddots & \vdots \\ \mathrm{C}_m & \delta_{m1} & & \cdots & \delta_{mj} & \cdots & \delta_{mn} \end{array}$$

(2.5.1a)

Equation (2.5.1a) may also be written as:

$$\underset{\text{mxn}}{\Delta} = \begin{pmatrix} \delta_{11} & \cdots & \delta_{1n} \\ \vdots & \delta_{ij} & \vdots \\ \delta_{mi} & \cdots & \delta_{mn} \end{pmatrix}$$

(2.5.1b)

In this matrix, the entries, δ_{ij} 's are the probabilities that show that a particular cost characteristic C_i will occur at a given contingency, K_j.

From the set of contingencies, $\mathbb{K}$ a corresponding family of sets of controls $\mathbb{Z}$ are then developed where each of the control sets, $\mathbb{Z}_i$, for a given contingency has its own cost and benefit characteristics. Here the economic concept of opportunity cost enters where a prevention of a disaster or a reduction of a particular cost of a disaster is the benefit of the action of a controller when it is activated for a defined contingency. Thus given a contingency $\mathrm{K}_j \in \mathbb{K}$ we have a set $\mathbb{Z}_i \in \mathbb{Z}$ of fuzzy or non-fuzzy control alternatives that has cost-benefit relations. The occurrence of any contingency will affect the systems balance or equilibrium and hence the overall stability and welfare of the mechanical system. The activation of a controller, therefore, will depend on the type of contingency, degree of resulting instability, cost of contingency when it occurs and the benefits of applicable control and its cost of construction and application. Given the contingency set we can construct a *contingency index* either by probabilistic or *possibilistic* method (For possibility distribution and the corresponding mathematics see [675][697][806]). Such an index can be used to define and check the equilibrium and disequilibrium states of the mechanical system. Every disequilibrium state, produced by any contingency, reveals itself through the attributes of reliability, security and damage (real cost). Each of these attributes may be viewed in terms of a set of degrees as measured by appropriate indexes that may be related to the contingency set.

The overall welfare of the mechanical system may thus, be viewed in terms of security and captured in term of indices and translate into real cost for achieving and maintaining it. The set of required security indices may be constructed from the relationship between the sets of degrees of stability and damage viewed in terms of real cost. We may now define the security of the mechanical system.

Definition 2.5.1

Let $\mathbb{R}$ be a set of degrees of security, $\mathbb{Q}$ a set of degrees of stability and $\mathbb{D}$ a set of degrees of damage then $\mathbb{R} = \mathbb{Q} \otimes \mathbb{D} = \left\{ r = \varphi(q,d) \mid q \in \mathbb{Q} \text{ and } d \in \mathbb{D} \right\}$ defines the system's security.

The $\mathbb{R}$, $\mathbb{Q}$ and $\mathbb{D}$, will be referred to as security, stability and damage sets relative to the system's welfare respectively. The security may be related to the real cost and benefit of the overall system.

2.6 Security and Risk Analysis

To understand the security potentials of the overall system requires a security analysis. Such a security analysis itself requires analysis of contingency, stability, disequilibrium and damage in terms of their relationships. It may be obtained through the combination of expected degrees of stability and damage for any given contingency in order to obtain *security identification matrix* of the form:

Table 2.6.1 Computable Security Identification Matrix for a System

$$r_{ij} = \varphi(d_i, q_i)$$

$\underset{m \times n}{\mathbb{R}} =$

Q \ d	q_1	q_2	$\cdots$	q_j	$\cdots$	q_n
d_1	r_{11}	r_{12}		r_{1j}		r_{1n}
d_2	r_{21}	r_{22}				r_{2n}
$\vdots$			$\ddots$			$\vdots$
d_i	r_{i1}			r_{ij}		
$\vdots$					$\ddots$	$\vdots$
d_m	r_{m1}		$\cdots$		$\cdots$	r_{mn}

(2.6.1a)

Thus

$$\underset{m \times n}{\mathbb{R}} = \begin{bmatrix} r_{11} & \cdots & r_{1n} \\ \vdots & r_{ij} & \vdots \\ r_{m1} & \cdots & r_{mn} \end{bmatrix}$$

(2.6.1b)

Hence we may write

$$\mathbb{R} = [\mathbb{R}_1, \cdots, \mathbb{R}_i, \cdots, \mathbb{R}_n]$$

(2.6.1c)

where $\mathbb{R}_i$'S are the column vectors.

The security identification matrix is analyzed to develop the optimal security set, $\tilde{\mathbb{R}}$, and then related to the equilibrium or disequilibrium of the system. The optimal security set is then related back to the contingency set to abstract the optimal contingency set $\tilde{\mathbb{K}}$. Any element of the optimal security set constraints an important threat to the efficient functioning of the mechanical system. Any element of the optimal contingency set defines the intensity of cost outcome in addition to its probability or possibility.

To maintain the equilibrium of the system or to correct disequilibrium condition automatic or non-automatic controls must be applied. These controls must be conceived as part of the optimal architectural design, constructed and installed in the mechanical system. The controls must be the best in relation to the defined security. This relates to the problem of optimal controls. The optimal controls must be related to feasible technology and real benefit-cost outcomes. The benefits and costs for each technology and each element, *r*, of the set of degrees of security, $\mathbb{R}$, may be identified through a matrix display that shows the identified costs and benefits. The task, then, is to define the technological set from which the best controllers relative to contingencies may be constructed.

Let $\widehat{\mathbb{A}}$ be the set of potential technologies that may be developed for use in the construction of any given controller. Furthermore, let $\widehat{\mathbb{D}}$ be a set of currently available technologies that may be used to construct the controller for dealing with reduction of insecurity or enhancement of security or maintenance of equilibrium of the mechanical system. Let $\widehat{\Omega}(t,\alpha)$ be a general technological set with a generic element, $\omega(t,\alpha) \in \widehat{\Omega}(\bullet)$ where α is a parameter of general social knowledge conceived in terms of social capital and social technology and t is a time element that defines the path of history of the system.

Definition 2.6.1

The space of control technologies, Γ, for the system's stability and welfare is

$$\Gamma = \left\{\gamma \mid \gamma \in \widehat{\mathbb{W}} = \widehat{\mathbb{A}} \cup \widehat{\mathbb{D}}, \text{ where } \widehat{\mathbb{A}} \text{ and } \widehat{\mathbb{D}} \subset \widehat{\Omega}(\bullet)\right\}$$

called a set of γ-control technologies.

Each γ-technology, simply referred to as controller, is associated with technological benefit and cost characteristics. These characteristics are defined in terms of what they can positively and negatively affect the welfare of society and the mechanical system for which the controllers are constructed. For each element, *r*, in the security set there is a corresponding set of appropriate feasible control technologies. Let the feasible set of controllers be $\widehat{\Gamma} = \{\gamma_1, \gamma_2, \cdots, \gamma_l\} \subseteq \Gamma$ and corresponding to each $\gamma \in \widehat{\Gamma}$ is a set of technological benefit characteristics

$\widehat{\mathbb{B}} = \{b_1 \cdots b_i \cdots b_m\}$ and technological cost characteristics $\widehat{\mathbb{C}} = \{\sigma_1, \cdots, \sigma_i, \cdots, \sigma_n\}$. By combining $\widehat{\Gamma}$ with $\widehat{\mathbb{B}}$ a technological benefit identification matrix may be constructed. Similarly, by combining $\widehat{\Gamma}$ with $\widehat{\mathbb{C}}$ technological cost identification matrix may be constructed. The structures of the resulting identification matrices are shown as Tables 2.6.2 and 2.6.3.

Table 2.6.2 Technological Benefit Identification Matrix for Technological Control Alternatives

b \ γ	γ_1	γ_2	$\cdots$	γ_i	$\cdots$	γ_l
b_1	b_{11}	b_{12}		b_{1i}		b_{1l}
b_2	b_{21}					b_{2l}
$\vdots$			$\ddots$			
b_i				b_{ij}		
$\vdots$					$\ddots$	
b_m	b_{m1}	$\cdots$		b_{mj}	$\cdots$	b_{ml}

where

$$b_{ij} = \begin{cases} 1 \text{ , if } \gamma_j \text{ produces real benefit characteristic} \\ 0 \text{ , otherwise} \end{cases}$$

and hence we can write a matrix of the form

$$\widehat{\mathbb{B}}_{\Gamma} = \begin{pmatrix} b_{11} & \cdots & b_{1l} \\ \vdots & b_{ij} & \vdots \\ b & \cdots & b \end{pmatrix}$$

(2.6.2a)

In a column vector form we have

$$\widehat{\mathbb{B}}_{\Gamma} = \left[\mathbb{B}_1, \cdots, \mathbb{B}_j, \cdots, \mathbb{B}_l\right]$$

(2.6.2b)

Similarly we have a corresponding technological cost identification matrix of the form

Table 2.6.3 Technological Cost Identification Matrix for Technological Control Alternatives

C \ γ	γ_1	γ_2	$\cdots$	γ_j	$\cdots$	γ_l
c_1	c_{11}	c_{12}	$\cdots$	c_{1j}	$\cdots$	c_{1l}
c_2	c_{21}	c_{22}				
$\vdots$		$\ddots$				$\vdots$
c_i				c_{ij}		
$\vdots$					$\ddots$	$\vdots$
c_n	c_{n1}		$\cdots$	c_{nj}	$\cdots$	c_{nl}

where

$$c_{ij} = \begin{cases} 1\,, & \text{if } \gamma_j \text{ produces cost characteristic} \\ 0\,, & \text{otherwise} \end{cases}$$

and hence we can write the computable matrix as

$$\hat{\mathbb{C}}_\Gamma = \begin{pmatrix} c_{11} & \cdots & c_{1l} \\ \vdots & c_{ij} & \vdots \\ c_{n1} & \cdots & c_{nl} \end{pmatrix}$$

(2.6.3a)

Equation (2.6.3a) may be written in terms of the column vectors as

$$\hat{\mathbb{C}}_\Gamma = \left[\mathbb{C}_1 \cdots \mathbb{C}_j \cdots \mathbb{C}_l \right]$$

(2.6.3b)

The column vectors present the benefit and cost characteristics associated with the corresponding control technology.

These technological real benefit and cost characteristics may be assessed from their impact on the society and automatic system to be constructed. The assessment may be done with the help of experts, policy makers and relevant communities in order to obtain a complete accountability of costs and benefits.

From the structures of technological benefit and cost identification matrices we may abstract socioeconomic real benefit and cost characteristics by the method of identification matrices. The construction and identification process is to relate the technological cost and benefit characteristics to economic real cost and benefit characteristics and to account for the relevant socioeconomic costs and benefits of the society and the mechanical system. Let the set of benefit characteristics for the mechanical and social economic space associated with γ_j-control to deal with r_{ij} degree of security (or $\left[1-r_{ij}\right]$ degree of insecurity or threat) of the system be of size l. Thus given r_{ij} degree of security we have a set of controllers $\left\{\gamma_1,\cdots,\gamma_j,\cdots\gamma_l\right\}$ and corresponding real benefits that interact to maintain equilibrium of system's stability when the system is confronted with $\left(1-r_{ij}\right)$ degree of threat. The real socioeconomic benefit identification matrix associated with the set of technologies will be of the form of Table 2.6.4 where $\left[v_i\right]$ represents a benefit characteristic vector.

Table 2.6.4 Socioeconomic Benefit Identification Matrix For Technological Controllers

	γ_i	γ_2	$\cdots$	γ_j	$\cdots$	γ_l
v_1	v_{11}	v_{12}	$\cdots$	v_{1j}	$\cdots$	v_{1l}
v_2	v_{21}					
$\vdots$	$\vdots$	$\ddots$				$\vdots$
v_i	v_{i1}			v_{ij}		
$\vdots$	$\vdots$				$\ddots$	$\vdots$
v_k	v_{k1}	v_{k2}	$\cdots$	v_{kj}	$\cdots$	v_{kl}

where

$$v_{ij}\begin{cases}=0\text{, if the real benefit doe not apply to the cohort}\\=1\text{, if the real benefit characteristics apply but physically unmeasurable}\\\geq 1\text{, if the real benefit characteristic is physically measurable}\end{cases}$$

In short, we can represent Table 2.6.4 as a computable matrix in the form

$$\widehat{\mathbb{V}} = \begin{pmatrix} v_{11} & \cdots & v_{1l} \\ \vdots & v_{ij} & \\ v_{k1} & \cdots & v_{kl} \end{pmatrix}$$

(2.6.4a)

and in a column vector form as

$$\widehat{\mathbb{V}} = \left[\mathbb{V}_1, \cdots, \mathbb{V}_j, \cdots, \mathbb{V}_l \right]$$

(2.5.4b)

Similarly the corresponding socioeconomic cost identification matrix takes the form like Table 2.6.4. as shown in Table 2.6.5

Table 2.6.5 Socioeconomic Cost Identification Matrix For Technological Controllers

$\rho \backslash \gamma$	γ_1	γ_2	$\cdots$	γ_j	$\cdots$	γ_l
ρ_1	ρ_{11}	ρ_{12}	$\cdots$	ρ_{1j}	$\cdots$	ρ_{1l}
ρ_2	ρ_{21}					
$\vdots$		$\ddots$				$\vdots$
ρ_i	ρ_{i1}			ρ_{ij}		
$\vdots$					$\ddots$	$\vdots$
ρ_q	ρ_{q1}		$\cdots$	ρ_{qj}	$\cdots$	ρ_{ql}

where q is the number of socioeconomic cost characteristics, ρ, abstracted from the technological cost characteristics, c. The structure of the table is such that the elements must satisfy the following relation

$$\rho_{ij} \begin{cases} = 0 \text{ , if the real socioeconomic cost characteristicic not applicable} \\ =1 \text{ , if the real cost characteristic applies but not measurable physically} \\ \geq \text{, if the real cost characteristic applies and physically measurable} \end{cases}$$

From the Table 2.6.5, we can write the computable identification matrix as

$$\widehat{\mathbb{U}} = \begin{pmatrix} \rho_{11} & \cdots & \rho_{1l} \\ \vdots & \rho_{ij} & \vdots \\ \rho_{q1} & \cdots & \rho_{ql} \end{pmatrix}$$

(2.6.5a)

Equation 2.6.5a may be written in a vector notation as

$$\widehat{\mathbb{U}} = \left[\mathbb{U}_1, \cdots, \mathbb{U}_j, \cdots, \mathbb{U}_l\right]$$

(2.6.5b)

Equations (2.6.4) and (2.6.5) are the final computable real benefit and cost identification matrices. Each column vector such as $\mathbb{V} = \left[v_{1j}, \cdots, v_{ij}, \cdots, v_{kj}\right]'$ and $\mathbb{U}_j = \left[\rho_{1j}, \cdots, \rho_{ij}, \cdots, \rho_{qj}\right]'$ represents real socioeconomic benefit and cost characteristics that correspond to a particular control technology, *j*. These characteristics of each vector must be aggregated through some acceptable mathematical logic in order to arrive at benefit and cost aggregates for decision making in the cost-benefit space as we have discussed in a companion volume, devoted to *fuzzy value theory* [931b]. All entries in the identification matrices must be obtain through some acceptable methods of information elicitation and collection. The details are discussed in Chapters 3 and 4.

2.7 Computable Cost-Benefit Identification Matrices and the Study of Quality of Life

In policy analysis of social systems, a question always arises as to what is the level of quality of life associated with a social system. This question involves a number of critical and important issues and related sub-questions. To assess the quality of life requires social economic system's auditing. The socioeconomic system's auditing requires us to define levels of quality of life. Each level assessment must have cost-benefit information support. The required information for analysis and comparison of different levels of quality of life may be obtained through the method of computable cost-benefit identification matrices that we have developed.

The levels of quality of life in a collective sense are associated with many real socioeconomic cost-benefit characteristics of social production. The broad characteristics will include environmental quality, risks of all forms, occupational hazard, levels of unemployment, health, poverty, education, stress, and many others that may be captured by *quality index* or *misery index*. In developing computable cost-benefit identification matrices, each of these broad characteristics may be associated with vector of its own characteristics or even submatrices of values of the own characteristics. In all these situations the logic of computable cost-benefit identification matrices will be an aid to obtain a comprehensive accountability of costs and benefits that will be associated with each of the broad characteristics and different levels of quality of life.

The analysis of quality of life in the socioeconomic production space may be partitioned into problem subsets, each of which lends itself to the use of logic of computable cost-benefit identification matrices. Within the logical framework; comprehensive information needed for benefit-risk studies or physical and social systems' auditing may be developed. Keep in mind that if we take on the environmental quality as subproblem of quality of life we will be confronted with a matrix problem for assessing levels of quality associated with soil, air, water and natural ecology. The focus point is simple, in that cost-benefit information databases are needed to analyze and rank alternatives in order to make rational decision on the basis of cost and benefits. The needed cost-benefit information can be developed by using the method of computable cost-benefit identification matrices. The theory of computable cost-benefit identification matrices lends itself to many areas of application where costs and benefits reflect the foundation of decision making. It is an epistemic position in the development of in the theories being advanced that cost-benefit analysis constitutes a general theory of decision making. This epistemic position is central to the set of algorithms provided.

3 The Theory of Social Cost and Costing

There are two important sides to any decision and choice when alternatives are present. The two sides are simply the *costs* that may be incurred in order to receive the *benefits* that may be associated with a particular decision. As such in all types of non-trivial decisions we have to consider the costs and benefits that may be associated with decision alternatives. This applies to individual decisions as well as those of either a group or community. In this chapter, we shall concentrate on cost concept, measurement and estimates of cost of decisions while taken as known those of benefits. Thus our problems are to define costs, measure them and estimate them.

In many decision problems, the concept of cost has either an intrinsic ambiguity or imprecise measurement or both. In other words, the cost component of the decision problem is fuzzy. As such the traditional method of cost-benefit analysis as a decision tool will require some important modifications if it is to aid decision and choice under fuzzy conditions. These fuzzy conditions arise as a result of interpersonal comparison, interdependence of cost characteristics, burden of cost and others. An interesting and important question arises as to how one can incorporate the fuzzy elements in the process of estimating and assessing the cost component of decision. In order to account for the problem of fuzziness it is useful to know essentially the basic characteristics of decisions and cost.

From the viewpoint of individual decisions, the problem is to identify the costs and attempt to quantitatively estimate them since the corresponding benefits are assumed known. The problem of cost estimate, however, is compounded when a group decision is considered or when individual costs are interdependent. For not only do we have to identify and quantitatively estimate the costs of decision given the benefits but we also need to know how these costs affecting the individual members are brought together as an aggregate estimate. Furthermore, we have to know how these costs are distributed over or may be distributed among the members. Since substantial part of cost-benefit analysis deals with community decisions and choices (in terms of projects and programs) any serious theoretical and practical analysis must deal with the distributional aspects of both costs and benefits associated with social decisions or individual decisions in the presence of externalities, interdependence and degree of burden of cost.

The cost that may be associated with a particular decision may be either static or dynamic. The cost is said to be static when it occurs at a single time point and dynamic when it occurs at many time points. We shall consider the two cases and show how they differ and how they may affect the outcome of a decision. Before we turn our attention to the concept of cost, let us consider an important classification of decisions. This classification will affect the nature of cost and criterion chosen to rank decision alternatives for choice. Even though choice and decision may be different we shall assume that they are the same.

3.1 Categories of Choice or Decision in Cost-Benefit Analysis

There are two classes of choice problems that may be considered. One class has the characteristic of Bernoulli outcome. As such we shall elect to refer to any problem belonging to this class as Bernoulli choice problem. The other class may conveniently be referred to as non-Bernoulli choice (decision) problem.

Definition 3.1.1

A choice problem is said to be Bernoulli, $\mathbb{B}^*$, if the choice set has only two alternatives, (x, y) where one of the alternative is the relative complement of the other. That is $\mathbb{B}^* = \{(x,y) \mid y = x'\}$ where x' the complement of is x.

Examples of Bernoulli choice problem may be found in questions like: should a public park be either developed or not? Should a particular law be either enacted (repeal) or not? Each element in this class of choice problems usually but not always has the characteristics of either known benefits or costs. The choice problem, therefore, involves assessing either the costs or benefits and then deciding on yes or no answer. Such choice problems may be analyzed for solution by methods of cost-effectiveness or benefit-effectiveness or both. An alternative name in cost-benefit analysis is with-and-without decision problem.

Definition 3.1.2

A choice problem is said to be non-Bernoulli, $\mathbb{B}'$ if the choice set, $\mathbb{A}$ contains at least two alternatives and the set does not contain a relative complements. That is,

$$\mathbb{A}=\left\{\mathrm{x}_i \mid i\in\mathbb{I} \text{ and if } \#\mathbb{I}=2 \text{ then } \nexists \mathrm{x}_j=\mathrm{x}'_i\ ,\ j\in\mathbb{I}\right\}$$

where $\mathbb{I}$ is an index set.

Examples of non-Bernoulli choice problems are: the states of economy characterized by allocation, production and distribution indices; alternative routes of constructing a road to link two cities; and different cities that may be selected for an industrial site. The elements in this class of choice problems also have the characteristics of unknown benefits and costs.

It may be noted that a connected sequence of choice problems may be a mixture of both Bernoulli and non-Bernoulli. For example: 1) a decision to build a bridge is considered (Bernoulli choice problem). If the answer is yes (by politics or otherwise) we then consider all other possible alternatives relative to cost as to how to construct the bridge at a minimum cost. The class to which a choice problem belongs will determine the nature and structure of costs associated with the choice problem.

3.2 The Concept of Cost in Decision and Cost-Benefit Analysis

We have so far used cost as if we know what cost is, and that the concept of cost has no intrinsic element of *ambiguity*. This may not be the case and usually it is not the case as such we shall explore in some details the concept of cost in decision and choice. Generally, the concept of cost may be considered from two but interdependent viewpoints of *accounting* and *economics*. The accounting concept of cost involves the monetary value of payment to factors and materials used in the process of social production, broadly defined. The accounting costs may be divided into *historical, futuristic*, and *forward* (predicted) costs.

Definition 3.2.1

Given a time set $\mathbb{T}$ a time-dependent cost set, $\mathbb{H}_t$ is a collection of time-ordered costs, c_τ that may be associated with factor and material uses for the production of the same kind of product at decision time, t

$$\mathbb{H}_t = \{c_\tau \mid \tau \in \mathbb{T} \text{ and t is fixed in} \mathbb{T}\}$$

The cost set, $\mathbb{H}_t$ is said to be a *historic cost* relative to time decision time point t, if

$$\mathbb{H}_t = \{c_\tau \mid \tau < t,\ \tau \in \mathbb{T} \text{ and t fixed in} \mathbb{T}\}$$

It is said to be *future cost set* relative to decision time point, t, if

$$\mathbb{H}_t = \{c_\tau \ \tau > t, \tau \in \mathbb{T} \text{ and t fixed in} \mathbb{T}\}$$

It is said to be a forward (predicted) cost set

$$\mathbb{H}_t = \{(c_\tau, u_\tau) \mid \tau \geq t,\ \tau \in \mathbb{T}\text{ , t fixed in } \mathbb{T}\}$$

where u_τ is an uncertainty coefficient.

The historic cost set summarizes the historic information about outcomes of the process of socioeconomic activities involving decisions and choices. The future cost set provides information as to what cost will actually rule at the future date. The forward or predicted cost set, however, contains cost measures of expected future activities conditional on the historic cost data. All these costs are particular to a given enterprise. A systematic process of providing such cost data for either a given production or consuming unit may be referred to as *private costing*. Such costs are based on *prices* that become registered in the market. They thus fail to account for important costs of some economic activities whose values are not registered in the market. Historic cost is always ex-post concept while forward and future costs are ex-ante.

Besides the accounting concept, there is the economic concept of cost which is defined in terms of the best choice alternatives foregone. This is the well-known *opportunity-cost concept*. The economic concept of cost is more extensive and defined in terms of real resources and commodities. In a two-commodity world of x and y, the opportunity cost of producing $(x_2 - x_1)$ units is defined in terms of $(y_1 - y_2)$ units.

Definition 3.2.2

Given a vectors of n-tuple and m-tuple of commodities x and y arranged in ascending order of magnitude then the cost of producing

$x_j - x_i$ is $C(x_j - x_i) = y_i - y_j$, where **C** is a function.

Given the concepts of cost we must identify, measure and aggregate costs associated with the decision alternative in order to develop a cost data. There are a number of approaches to ascertain cost information for projects and social decisions in many areas of human choice activities [125, 135, 140, 147, 150, 151, 166, and 169]. These approaches are not comprehensive. They also fail to account for subjectivity and imprecision in the costing process. The works [122; 602, 603] adopt the general comprehensive approach. However, they neglect the element of subjectivity, imprecision and vagueness in the cost measurement. All these are different approaches to *social costing* whose objective is to develop *cost data* for social decision or project selection in cost-benefit framework. Most of the approaches are not only non-comprehensive but are decision-specific rather than a development of a general theoretical framework for social costing of human decision-choice activities. The objective of this chapter is to develop a general framework to obtain comprehensive cost information for social decisions on the basis of democratic assessments.

The main focus of this chapter is on the development of a theory of social costing with a larger area of application including decisions of engineering, environment, politics, risk, law and others. As such, a comprehensive rational framework for costing in cost-benefit analysis is advanced in a fuzzy decision space. The framework is developed to account for *qualitative* and *quantitative* cost characteristics as well as imprecision, vagueness and subjectivity in costing. It will be broadened to include the construct of cost distributional weights for the burden of cost which is usually neglected in traditional cost-benefit analysis. The objective of developing such a comprehensive fuzzy framework is to provide a logical structure that will allow for the accountability of all relevant cost variables in either social project or decision analysis even when they are either tangible or intangible.

3.3 A Note on the Concept of Cost in Decision and Choice

The economic concept of cost in any theory of decision and choice is defined in the utility space instead of in the commodity space. It is measured in terms of utility loss. As such to find a suitable measure for cost one must find a suitable measure for utility which is subjective in an essence, if the true cost component of choice is to be captured. From the viewpoint of individual decision, the assessment and valuation of private costs of decision require some behavioral consistency in decision-choice activities. Matters of cost and costing are different and more difficult when social costs of individual decisions and total costs of social decisions are considered. Here, two problems arise where not only must we find an index of individual valuations but also the process through which such individual cost values may be ascertained and aggregated into a total cost. The success in finding satisfactory solutions to these problems depends on how the computational concept of cost is embedded in the choice-decision space.

The computational concept of cost may be perceived from either the accounting calculus or the economic calculus of opportunity cost. The accounting concept of cost involves imprecise measurement while the economic concept of opportunity cost is characterized by vagueness and subjectivity of measurement. Both perceptives reveal various elements of fuzziness which are further complicated by subjective assessment of the burden of cost. Irrespective of the computational concept of cost adopted, we must deal with the problems of vagueness and subjectivity which must be simultaneously handled with the cost computations. Furthermore, the aggregate measure of cost, given various, different and real cost characteristics, depends on the common unit of cost measurement. In a monetary unit, the aggregate value depends on the prices used. But these prices of different cost characteristics are measured in fuzzy domain space bringing with them an increased imprecision. In all assessments and estimates of costs, therefore, one must take account of the presence of fuzzy characteristics and approximations that are intrinsically inherent in any humanistic system (for discussions on costs, see [129,131,150,180,188,189] and on fuzziness see [694, 735, 756, 770, 780]). These are the major justifications for the cost comprehensive framework for social costing that is developed here in addition to the fact that one cannot speak of (market) prices of units of inputs and outputs when they are not identified. Decision making is a cost-benefit process in relative standing. All optimizations can be viewed in cost-benefit terms.

3.4 The Cost Components and Decision

Every decision has a set of corresponding costs irrespective of what the benefits are. We thus begin the cost analysis by partitioning the decision space into decisions that are *private* and those that are *social*. The costs associated with each decision are made up of *explicit* and *implicit* costs irrespective of whether we are dealing with either a private or social decision. The private decisions and choices are associated with implicit private costs, C_{I}, and explicit private costs, C_{E}. The total costs of private decision, C_{P}, to an individual is thus

$$C_P = C_E + C_I.$$

(3.4.1)

The total social explicit cost, C_{SE} of a private decision is composed of total private costs in addition to real costs, C_n, to society incurred over and above the total private costs, C_P. Thus

$$C_{SE} = C_P + C_n$$

(3.4.2)

Such real costs, C_n, may be due to consumption and production externalities which are not accounted for in the market imputations. The total social costs, C_{ST}, of either private or social decision in the sum of explicit and implicit social costs where the social implicit costs, *C,* is the net benefit of the next best alternative of society's use of its resources and time. (See [29, 134, 135,151,153] for more discussions on these points). Thus,

$$C_{ST} = \underbrace{C_E + C_I + C_n}_{\downarrow} + C_{SI}$$
$$= C_{SE} + C_{SI}$$

(3.4.3

From Eq. (3.4.3) we may write

$$C_{ST} = \begin{cases} C_{SE} \;; & \text{in private decision } (C_{ST} = 0) \\ C_E + C_I & \begin{cases} \text{in private costing of private decision} \\ \text{based on market prices } (C_n = 0) \end{cases} \\ & \end{cases}$$

(3.4.4)

We now advance a framework for computing the total social cost that takes into account the burden of cost on an individual or a class.

In private economic decisions $C_{SI} = 0$ since it is not considered in the choice process by the individual. In private costing based on market process $C_n = 0$ since such costs are not accounted for by the market outcomes in terms of information flows. Under a perfectly competitive market organization for economic decisions where externalities and uncertainties are assumed away, $C_{SI} = C_n = 0$ and hence total social costs tends to coincide with total private costs in terms of private opportunities foregone. In other words, the total social costs of private decision becomes the market valuation of the alternative choice that may be made by some rational process of recourse allocation into other uses other than that observed. This variation which is calculated from market prices for all resource units used becomes simply measured by prospective money market outlay.

3.5 Cost, Choice and Decision

There are two views on the objective of cost analysis that one may consider in the choice-decision process. One view considers costs in a process of explaining and predicting rational 'socioeconomic' behavior. Costing in this sense is simply an organized recording of historic information. The other view looks at cost as a measure of some important information on the basis of which a decision is to be made in the process of setting the *potential* inherent in nature against the *actual* as observed. The former view of costs objectifies costs, in the sense that they are more or less independent of the analyst. The latter view, however, subjectifies costs in the sense that they exist in the mind of the decision maker. Costing in this sense is simply an organized process where such important information and its measurement are provided. The former concept is an ex-post one while the latter is an ex-ante concept.

The meaning that cost assumes in cost-benefit analysis depends on whether the theory of cost-benefit analysis is viewed as part of explanatory or prescriptive science. If the objective of the cognitive activity of either cost-benefit analysis or cost-effective analysis is to make a 'wise' decision (i.e., to decide on the alternative best preferred) then cost-benefit analysis may be viewed as an algorithm for realizing best decisions in some specific sense. Under such process of choice-decision analysis, costs may be rightly considered as subjective to the decision maker in the sense that they are not yet realized and exist in his or her mind and as such they are only potentials that may be actualized. Viewed in this light, real costs in decision are those elements which a decision maker thinks he or she may sacrifice in others to realize his or her best alternative. They, therefore, consist of his or her own subjective evaluation of foregone opportunities (benefits) due to his or her decision. Costs are sacrificed benefits and benefits are costs induced. This is the cost-benefit duality. As subjective valuations, costs are either completely fuzzy or contains fuzzy elements that must be properly dealt with in costing. The fuzzy elements of cost become more complex as we move from individual costs to social costs and social costing.

This view of costing in decision carries with it some important implications.

1a. Private costs of individual decisions are borne by the decision maker while their social costs of the same decision are borne by the society irrespective of the social character of the decision maker.

1b. Total costs of social decisions are borne by the society irrespective of whether the decision is collectively and unanimously made, imposed by majority will or detectorally imposed.

1c. While it is not possible to always impose private costs of individual decision on others, it is invariably always the case to impose social costs of private decision on the members of the society in other to create private benefit.

2. Costs in decision are based on expectation (or fuzzy anticipations); as such they are future oriented.
3. From the viewpoint of opportunity cost in terms of decision, cost is not realizable since the next alternative is always not chosen. It may be estimated.
4. Private costs of individual decision are assessed by the decision maker while their social costs and the social cost of social decision can only be assessed not necessarily by those who make decision, but by those who will be affected by the outcome of the decision; in other words, costs have utility dimensions.
5. Costs may be stock or flow, however, they may have time dimension relative to the moment of decision. (For some important aspects of cost characteristics see [129] [134] [155].

The concept of cost in the calculus of choice and decision is a very disturbing one. The sense in which the concept of cost is meaningful to the logic of choice must always be made explicit if we are to develop acceptable algorithms of choice and decision and avoid some confusion in choice theory. Viewed from the utility space, opportunity costs may be divided into *choice-influencing cost* (cost-influenced choice) and *choice- influenced cost* (cost-influencing choice in a case of choice) as provided in Buchanan [129]. The choice-influenced cost is the net utility foregone consequent on a choice of an alternative. It results from decision or choice that have been made. It is said to be choice produced. Thus it is not the cause of decision and may even be borne not by the decision or choosing agent. The choice-influencing cost, on the other hand, is that cost that directly or indirectly inhibits decision and choice. It is a genue immediate constraint on choice. The choice is said to be cost-influenced. In the individual decisions, the explicit and implicit private costs may be viewed as choice-influencing costs; that is, they constrained choice, while the social costs of individual decisions may be viewed as choice-influenced costs; that is, individual decisions produce social costs. Given proper social costing, social explicit costs will be choice-influencing while the social implicit costs will be choice-influenced in decision. The choice-influencing cost is an ex-ante decision cost; that is, that opportunity costs which are evaluated before choice. The choice-influenced costs is an ex-post decision-choice costs.

3.6 The Cost Space and Identification Matrix

Let us now look at the structure of costs of decision making and ascertain the required logic to define the cost characteristics and their relation to identification matrices that we have discussed in Chapter two. The idea is to bring the logic embodied in the theory of computable cost-benefit identification matrices to facilitate the development of cost database required for decision making.

3.6.1 The Cost Space Defined

The objective of social costing is to identify and estimate costs for decision analysis where the social costing should provide cost information on alternatives for choice. We shall thus turn our attention to the process through which our understanding of cost may be translated into cost identification and measurement that may be associated with a decision and choice. We shall utilize the analytical framework of the theory of computable cost-benefit identification matrices that we have developed in Chapter 2 in addition to the conceptual structure of decision provided in Chapter 1. The objective of cost identification and measurement is to assign costs to various relevant decision alternatives or programs. This requires an establishment of decision alternatives (programs) relative to certain socioeconomic objectives defined in Ω. As such, we shall consider a set of programs, Π, with a representative element, π, for a given set of social objectives also with a representative element $\omega \in \Omega$. The set Π is the same as decision alternatives, $\mathbb{A}$, used in Chapter 1. It is a set of vehicles to reach $\omega \in \mathbb{V} \subset \Omega$ where $\mathbb{V}$ is the attainable set of objectives. Let $\mathbb{C}$ be a set of cost measures whose generic element is σ for any given objective and benefit.

Definition 3.6.1.1

A cost space, Γ, of a decision consists of a triplet, $\{\mathbb{C}, \Pi, \Omega\}$, of sets of cost measures, programs and objectives respectively. Thus

$$\Gamma = \mathbb{C} \otimes \Pi \otimes \Omega = \{ (\sigma, \pi, \omega) \mid \sigma \in \mathbb{C}, \pi \in \Pi \text{ and } \omega \in \Omega \}$$

where $\otimes$ is the Cartesian product of sets.

Definition 3.6.1.2

A cost relation, R, is a rule that assigns to each program $\pi \in \Pi$ a cost measure, $\sigma \in \mathbb{C}$, given the set of objectives, Ω. That is,

$$R|_{\Omega} = \{ (\sigma, \pi) \mid {}_{\sigma}R_{\pi}, \sigma \in \mathbb{C} \text{ and } \pi \in \Pi \}$$

It is a mapping from the space of projects into the cost space $R|_{\Omega} : \Pi \to \mathbb{C}$

Definition 3.6.1.1 and 3.6.1.2 assume the existence of an aggregate value, $\sigma \in \mathbb{C}$. We now must find a process through which σ is computed given any $\pi \in \Pi$. First, we observe that $\sigma \in \mathbb{C}$ is an aggregate value. For such an aggregate value to exist the space of $\mathbb{C}$ must be measurable and computable. As such, we need to construct a σ-measurable and computable space. We shall require the σ-measurable space, given the same objectives and benefits, to exhibit some important properties of comparability and ranking.

Definition 3.6.1.3

A cost space, $\Gamma = \mathbb{C} \otimes \Pi \otimes \Omega$ is said to be value comparable if it is equipped with a σ-measurable difference function $\rho(\sigma_i \sigma_j) = \sigma_j - \sigma_i$ or a ratio function $\hat{\phi}(\sigma_i, \sigma_j) = (\sigma_j / \sigma_i)$ where $i, j \in \mathbb{I}$, an index set of cost characteristics such that only one of the following conditions holds at any time:

$$\rho(\sigma_i, \sigma_j) \begin{Bmatrix} > 0 \\ \text{or} \\ = 0 \\ \text{or} \\ < 0 \end{Bmatrix} \Rightarrow \hat{\phi}(\sigma_i, \sigma_j) \begin{cases} > 1 \text{ and } \sigma_j > \sigma_i \\ = 1 \text{ and } \sigma_j = \sigma_i \\ < 1 \text{ and } \sigma_j = \sigma_i \end{cases}, \quad i, j \in \mathbb{I}$$

where σ_i and $\sigma_j \in \mathbb{C}$.

Definition 3.6.1.4

A cost space, Γ, is said to be computable if there exist σ_i and σ_j for all *i*'s and *j*'s belonging to $\mathbb{I}$ such that the difference and ratio functions, $\rho(\cdot)$ and $\hat{\phi}(\cdot)$ are both quantitatively definable.

Proposition 3.6.1.1

A cost space Γ *is* σ *-computable if and only if it is* σ *-measurable and if it is* σ *-computable it is also value-comparable for any given data.*

Postulate 3.6.1.1 *Project comparability in the cost space.*

Projects with equal benefit and objectives in a cost space, Γ, are cost-comparable if the σ -measurable difference function $\rho(\cdot)$ exhibits

$$\rho(\sigma_i, \sigma_j) \begin{cases} > 0 \Rightarrow \pi_i \succ \pi_j \\ = 0 \Rightarrow \pi_i \sim \pi_j, \quad i \neq j \in \mathbb{I} \\ < 0 \Rightarrow \pi_j \succ \pi_i \end{cases}$$

where $\pi_i, \pi_j \in \Pi$, with corresponding cost measures σ_i and $\sigma_j \in \mathbb{C}$; and only one of the stated relations must hold for any $\pi_i, \pi_j \in \Pi$ with corresponding σ_i, $\sigma_j \in \mathbb{C}$, where $\succ$ and $\sim$ refer to superior and indifferent to, respectively.

Postulate 3.6.1.2 *Asymmetry of project comparability in the cost space*

If $\rho(\sigma_i, \sigma_j) > 0 \Rightarrow \pi_i \succ \pi_j$ then $\rho(\sigma_j, \sigma_i) \not> 0 \Rightarrow \pi_j \not\succ \pi_i$;
$\sigma_i\ \sigma_j \in \mathbb{C}$ and $\pi_i, \pi_j \in \Pi$ with $i, j \in \mathbb{I}$

Postulate 3.6.1.3 Transitivity of project comparability in the cost space

If $\rho(\sigma_i, \sigma_j) \begin{cases} >0 \Rightarrow \pi_i \succ \pi_j \\ =0 \Rightarrow \pi_i \sim \pi_j \\ <0 \Rightarrow \pi_i \prec \pi_j \end{cases}$ and $\rho(\sigma_j, \sigma_l) \begin{cases} >0 \Rightarrow \pi_j \succ \pi_l \\ =0 \Rightarrow \pi_j \sim \pi_l \\ <0 \Rightarrow \pi_j \prec \pi_l \end{cases}$

Then

$$\rho(\sigma_i, \sigma_l) \begin{cases} >0 \Rightarrow \sigma_i \succ \sigma_l \\ =0 \Rightarrow \sigma_i \sim \sigma_l \\ <0 \Rightarrow \sigma_i \prec \sigma_l \end{cases} \text{ for all } \pi \in \Pi \text{ and } \sigma \in \mathbb{C} \text{ where } i \neq j \neq l \in \mathbb{I}$$

Postulate 3.6. 4. *Reflexivity of project comparability in the cost space*

$$\rho(\sigma_i, \sigma_j) = 0 \Rightarrow \pi_i \succsim \pi_i \text{ for all } \sigma_i \in \mathbb{C} \quad \text{and} \quad \pi_i \in \Pi \quad .$$

The postulates of project comparability can be abstractly stated by using the cost relation, $R|_{\Omega}$, as stated in Definition 3.6.1.2. in addition to Definition 3.6.1.3. We now turn to the problems of measure and computability. The postulates of project comparability in the cost space, Γ, establish a cost-effective analysis. In this case the decision alternatives are ranked by costs as benefits are known and fixed for a given social objective.

3.6.2 Cost Identification Matrix for Costing

Our objective now is to develop a framework for computing $\sigma \in \mathbb{C}$ for any program alternative $\pi \in \Pi$. The value σ is a cost aggregate which is obtained from the real cost characteristics associated with $\pi \in \Pi$. Let the types of *real cost characteristics* associated with $\pi \in \Pi$, be a set, $\mathbb{X}^{C}$ whose generic element is *x*. The set $\mathbb{X}^{C}$ is called real cost characteristic set. It reveals the cost attributed of a project $\pi \in \Pi$. We shall assume that each $x \in \mathbb{X}^{C}$ is a distinct real cost characteristic and that $\mathbb{X}^{C}$ is of size T. The impact of each cost characteristic is distributed over the members of the society. However, such impact may differ from an individual to another. As a next step in the social costing process, the society is partitioned into social groups through appropriate *cluster analysis and stratification.* The criteria for cluster analysis and social group formation would depend on the problem and will vary from project to project. The criteria must be specified. Let $\mathbb{G}$ be a set of the social groups with generic element $G \in \mathbb{G}$ and where $\mathbb{G}$ is of size N relative to the decision alternative, $\pi \in \Pi$. Thus associated with each identified real cost characteristics $x \in \mathbb{X}^{C}$ and each social group (class), $G \in \mathbb{G}$, is a real cost value, c_{ij} of the ith characteristic given the jth group measured in real units. The $\mathbb{G} \otimes \mathbb{X}^{C}$ forms a matrix for identifying all costs at least in theory. The identified real cost measures c_{ij} are cohort-specific and may be represented as in Table 3.6.1 as real cost-identification matrix.

Table 3.6.3.1 Cost-identification matrix of decision alternative, $\pi \in \Pi$

X / G	G_1	G_2	$\cdots$	G_j	$\cdots$	G_{N-1}	G_N	Totals
x_1	c_{11}	c_{12}	$\ddots$			$c_{1,N-1}$	c_N	$C^x_{1\bullet}$
x_2	c_{21}			$\vdots$			c_{2N}	$C^x_{2\bullet}$
$\vdots$								$\vdots$
x_i		$\cdots$		c_{ij}	$\cdots$			$C^x_{i\bullet}$
$\vdots$								$\vdots$
x_{T-1}	$c_{T-1,2}$			$\vdots$			$c_{T-1,N}$	
x_T	c_{T1}	c_{T2}	$\ddots$	c_{Tj}	$\ddots$	$c_{T,N-1}$	c_{TN}	$C^x_{T\bullet}$
Totals	$C^G_{\bullet 1}$	$C^G_{\bullet 2}$	$\cdots$	$C^G_{\bullet j}$	$\cdots$		$C^G_{\bullet T}$	$C_{\bullet\bullet}$

Note:

$$c_{ij} \begin{cases} = 1, & \text{if the real cost is physically incommeasurable} \\ \geq 1, & \text{if the real cost is physically measurable} \\ = 0, & \text{if the cohort is unaffected} \end{cases}$$

Care must be exercised in assigning meanings and interpretations to the values in the cost-identification matrix. If $c_{ij} > 1$ for any fixed *i*, then c_{ij} is summable over all j's and $C^x_{i\bullet}$'s are single values and the cost characteristic, *i*, is quantitatively measurable. If $c_{ij} = 1$ for a fixed *i* and over all *j*'s then characteristic, *i*, may not be quantitatively measurable and hence $C^x_{i\bullet}$ has no quantitative meaning. In this case our only cost knowledge is that given an ith cost characteristic, a cohort may either be affected or not. When we consider the vertical aggregate value such as $C_{\bullet j}^{G}$ we immediately observe that they are in composite units in that, their sums are taken over costs measures that are defined on different non-summable units. Thus

$$C_{\bullet j}^{G} = \sum_{i=1}^{T} c_{ij}$$

(3.6.2.1)

$$C_{i\bullet}^{x} = \sum_{j=1}^{N} c_{ij}$$

(3.6.2.2)

And given a common measure over all cohorts the total cost C. is

$$C = \sum_{j=1}^{N} C_{\bullet j}^{G} = \sum_{i=1}^{T} C_{i\bullet}^{x} = \sum_{i=1}^{T}\sum_{j=1}^{N} c_{ij}$$

(3.6.2.3)

The values of C and $C_{\bullet j}^{G}$ cannot be computed in their current forms and hence require further analysis for conditions of computability.

If $C_{\bullet j}^{G} = 0$ then $c_{ij} = 0$ for any fixed jth class and for all i's. The implication here is that the jth social class is cost un-affected by the project or the choice alternative and hence $C_{\bullet j}^{G} \geq 0$ for all j. On the other hand, $C_{i\bullet}^{x} = 0$ implies that the identified cost characteristic does not affect any of the social classes and hence it is decision irrelevant. Thus it is also established that $C_{i\bullet}^{x} \geq 0$. The, C_{ij}'s ≥ 0 are the identified and estimated real costs for a given social class and cost characteristic. The identification and estimation of these real cost characteristics may be done with the help of experts and by the method of *information elicitation* from the members of the society and the experts [38, 57, 68b, 72,182, 498, 926, 928,]. Corresponding to each decision alternative is a cost identification matrix on the basis which, given the set of objectives and benefits, the associated total cost measured in a monetary unit is computed.

The units of measurement in which the real cost characteristics are recorded in the identification matrix will be determined by experts. The experts in collaboration with the social classes (groups) will assess the real values of the cohort entries. These values may be viewed in terms of economic products or by-products. Given that the units of measurement of real cost characteristics have been established and that the cost-identification matrix has been constructed, we must find a logical process to form the cost aggregate, $\sigma \in \mathbb{C}$ that corresponds to $\pi \in \Pi$. The cost-identification matrix reveals a number of important information. It shows the concentration of cost-characteristic impact, what cost-characteristic has been taken into account and whose cost has been considered in the project evaluation process. It does reveal comprehensive costing information about the decision alternative. Thus with cost information from the identification matrix we know what real cost characteristic and what social group has been cut off as unimportant in the costing process. The cost identification matrix will force us, by choice-decision rationality, to ask questions about the process of elimination and inclusion of the real cost characteristics and social groups. This is because not every real cost characteristic and every group will be in the identification matrix entries.

Given the identification matrix the next steps in the social costing process is 1) establish the *burden* of *real unit cost* characteristic that may be associated with each cost characteristic and each social group, and 2) find a general *common unit* of measure that will allow aggregation of the values in the cohorts. This common unit constitutes monetary unit that find expression in *prices.* The burden of real cost finds its expression in social weight attached to each cost cohort. It does require computation of social *distributive weights* that will allow an important transformation of the real cost characteristics where *grades* of suffering due to unit impact of real cost as socially assessed are incorporated into the real cost values before aggregation is performed. The social distributive weights must not be arbitrary selected. They must be optimally constructed to reflect the evaluative preferences of the social groups in addition to the preferences of decision makers and supported by the assessments of experts. The comprehensive framework of social costing that is being advanced here requires collaborative efforts on the part of the society for democratic decision making, economists and other experts for scientific computing that good decision based on democratic ideals requires. The collaborative process requires the development of a logical structure for combining different information obtained from social groups, experts and decision makers [11,33,40,41,68b, 69,71,77, 737]. In Chapter 4 we shall tighten up the definition of the cost characteristics in order to distinguish them from benefit characteristics.

3.6.3 The Burden of Social Cost and Cost Distributional Weights

We begin with the notion that costs of decision and choice exist in the minds of those who will be affected by such costs and hence are utility-based. The true burden of a unit of real cost can only be satisfactorily assessed by those who are or will be affected by the costs of either private or social decision. The individual assessments of the cost burden are interpersonally and intertemporally relative which may be either underestimation or overestimation. Thus any process of obtaining an aggregate real cost measure for any cost characteristic that accounts for the burden of real cost encounters the problems of interpersonal, interclass, intertemporal and intergenerational comparisons. This is due to the notion that the concept of cost burden is imprecisely measured and is also subjectively and vaguely defined in terms of ordinal scale in time and over time. [126,135,146,159,161].

To overcome such problems of interpersonal, interclass and intergenerational comparisons one may elicit information about individual assessments of the burden of cost of decision for each cohort. Here the problems of under valuation and overvaluation are encountered. Given the elicited information an attempt is then made to obtain a "median" for each cohort thus allowing one to assign the same distributional weight to the members of the same class. This must be done for each cost characteristic. The median in this sense is that value of distributional weight that minimizes the absolute sum of a well-defined Hamming distance. This method of constructing cost-distributional weights for each social class and each cost characteristic seems perhaps the most accurate from the viewpoint of subjective valuations. From the position of general social calculations this may be very costly and time-consuming.

A way out of these assessment and evaluation problems is for either the decision maker or the analyst to assign his or her own subjective weights to each social class given the cost characteristic. This process of assigning cost-distributional weights for each cost-characteristic and for each social class may produce some important distortions in the costing. The degree of distortions will depend on the decision maker's knowledge about the real cost characteristics, and the structure and form of each social class. The degree of distortion of cost calculation may be made smaller as much as possible by appealing to expert judgments and information elicitation from the members of the society. This implies a development of a computational process that will allow us to construct subjectively relative real cost-burden distributional weights from the information provided by experts and selected members of the society relative to the real cost characteristic and the social classes. Approaches in constructing acteristic and the social classes. Approaches in constructing distributional weights in cost-benefit analysis may be found in [40, 683, 927, 932,].

No matter what method is sued, the construction of class weights of the burden of real cost characteristics requires extreme care. Of all the approaches, it seems that the method of expert judgement is preferred. To proceed in this fashion we consider an index set, $\mathbb{L}$, of selected experts. Furthermore, let $\mathbb{J}$ be an index set of social classes and $\mathbb{I}$, an index set for the cost characteristics. For any cost characteristic, x_i , $i \in \mathbb{I}$, we consider an index set $\mathbb{L}_i$, $i \in \mathbb{I}$ *of* experts who will provide us with subjective *value weights* of the burden of a unit real cost for each social class, G_j , $j \in \mathbb{J}$, given that the class is affected by the *i*th cost characteristic.

The value weight assigned to each social class is viewed as belonging to a fuzzy set. Alternatively, it may be viewed in terms of measures on ordinal scale [8], [66]. The problem then is to develop a fuzzy measures of the distributional value weights. Suppose we have collected the subjective value weights, α of the expert for each x_i , $i \in \mathbb{I}$. Let such weights be a set, $\mathbb{D}_i$ that may be written as

$$\mathbb{D}_i = \left\{ \alpha_{ijl} \mid j \in \mathbb{J} , l \in_i \mathbb{L} , i \text{ fixed in } \mathbb{I} \text{ and } \alpha > 0 \right\}$$

(3.6.3.1)

From Eq. (3.6.3.1) we must construct unit value weights $\alpha_{ij}, j \in \mathbb{J}$ for any fixed $i \in \mathbb{I}$. The value may be computed as:

$$\min_{\alpha_{ij} \in \mathbb{D}} \sum_{l \in \mathbb{L}_i} d\left(\alpha_{ij}, \alpha_{ijl} \right) \rightarrow \alpha_{ij}$$

(3.6.3.2)

s.t. eqn. (3.6.3.1)

where $d(\cdot)$) defines a Hamming distance. The system of weighted social costing can be improved by introducing an element, λ_l, that allows one to capture the degree of competence of each expert. In other words, λ_l is some *competence weight* of the experts [33,926, 1153b]. In this respect we may write:

$$\hat{\alpha}_{ij} = \min_{\alpha_{ij}} \sum_{l \in \mathbb{L}} \lambda_l d\left(\alpha_{ij} , \alpha_{ijl} \right)$$

(3.6.3.3)

s.t. eqn. (3.6.3.1)

We may note that while Eq. (3.6.3.1) is a matrix, Eqs. (3.6.3.2) and (3.6.3.3) yield a vector for each fixed $i \in \mathbb{I}$. Let such a vector be a set, $\hat{\mathbb{D}}$ of the form

$$\hat{\mathbb{D}} = \left\{ \begin{array}{l} \alpha_{ij} \mid \alpha_{ij} \text{ is as computed in} \\ \text{eqn.(3.6.3.2) or } (3.6.3.3 \text{ and } j \in \mathbb{J}) \end{array} \right\}$$

(3.6.3.4)

and hence

$$\mathbb{D} = \left\{ \hat{\mathbb{D}}_i \mid \hat{\mathbb{D}}_i \text{ is as defined in eqn. (3.6.3.4)} \right\}$$

(3.6.3.5)

defines a matrix of estimated unit *cost-burden weights*. This allows one to write the weighted real cost-burden matrix as in Table 3.6.2.

Table 3.6.3.2 The cost-identification matrix with value weights of burden of real cost

$X \setminus G$	G_1	G_2	$\cdots$	G_j	$\cdots$	Totals
x_1	$\alpha_{11}c_{11}$	$\alpha_{12}c_{12}$		$\alpha_{1j}c_{1j}$		$\alpha^*_{1\bullet}c^*_{1\bullet}$
x_2				$\alpha_{2j}c_{2j}$		
$\vdots$	$\vdots$		$\vdots$			$\vdots$
x_i	$\alpha_{i1}c_{i1}$	$\cdots$		$\alpha_{ij}c_{ij}$	$\cdots$	$\alpha^*_{i\bullet}c^*_{i\bullet}$
$\vdots$	$\vdots$			$\vdots$		$\vdots$
Total	$\alpha^G_{\bullet 1}C^G_{\bullet 1}$	$\cdots$	$\cdots$	$\alpha^G_{\bullet j}C^G_{\bullet j}$	$\cdots$	αC

The real cost as represented in Table 3.6.2 may be represented in a compact form in the process of our social costing. Define block diagonal matrices of the forms

$$\mathbf{A} = \begin{bmatrix} A_{11} & 0 & \cdots & 0 \\ 0 & A_{22} & & \vdots \\ \vdots & & & \\ 0 & \cdots & \cdots & A_{ii} \end{bmatrix}, i \in \mathbb{I}$$

(3.6.3.6)

where

$$A_{ii} = \begin{bmatrix} c_{i1} & \cdots & \cdots & 0 \\ & c_{i2} & & \\ 0 & & & \\ 0 & \cdots & & c_{ij} \end{bmatrix}, \quad i \in \mathbb{I}, \ j \in \mathbb{J}$$

(3.6.3.7)

Define another block diagonal matrix, B of the form:

$$\mathbf{B} = \begin{bmatrix} B_{11} & \cdots & \cdots & 0 \\ 0 & B_{22} & & \vdots \\ \vdots & & \ddots & \\ 0 & \cdots & \cdots & B_{ii} \end{bmatrix}, \quad i \in \mathbb{I}.$$

(3.6.3.8)

where

$$\mathrm{B}_{ij} = \begin{bmatrix} \alpha_{i1} & & & \\ & \alpha_{i2} & & \\ & & \ddots & \\ & & & \alpha_{ij} \end{bmatrix}, \ i \in \mathbb{I}, \ j \in \mathbb{J}$$

(3.6.3.9)

From Eqs. (3.6.3.6) - (3.6.3.9), the real cost matrix, **C**, including the burden of cost may be written as

$$\mathrm{C} = \sum_{i \in \mathbb{I}} A_{ii} B_{ii}$$

(3.6.3.10)

and

$$Z_{ii} = A_{ii} B_{ii}, \ i \in \mathbb{I}$$

(3.6.3.11)

The total real cost, C_i^x, associated with each *i*-cost characteristic may be obtained as

$$C_i^x = \sum_{j \in \mathbb{J}} \alpha_{ij} c_{ij} = Z_{ij} \text{ , for any fixed } i \text{ in } \mathbb{I}$$

(3.6.3.12)

3.6.4 Constructing the Distributional Weights

The parameter α_{ij} is the cohort weight that is socially attached to corresponding real cost, c_{ij}. It reflects the society's evaluation of the intensity of impact of ith cost characteristic on social group, *j*. It has to be constructed by some logical process. To design an algorithm to realize α_{ij}, we must understand the character of α_{ij} and how `the need arises. The value α_{ij} contains elements of subjectivity, social judgement, and vagueness that generate fuzzy characteristics. The cohort cost values also contains elements of measurement imprecision and some cases vague concept of cost and perception of cost. The introduction of expert judgment and the level of competence also carry with them the problem of fuzzy characteristics. Given the structure of the computational problem of α_{ij}, there are few approaches that may be suggested for computing the α_{ij}, socially ethical weight. There are two that we like to discuss here. The first one is the optimal multi-utility measurement of α_{ij} based on similar group preferences to produce *Pareto weights*. The other approach is the fuzzy decision approach to generate *fuzzy equilibrium weights* based on information from different segments of the society and the method of fuzzy aggregation [2, 5, 10, 17, 22, 77, 78, 81].

3.6.4.1 Utility Measurement of Distributional Weights

From the basic economic theory of social decision, a decision alternative is selected from the set of alternatives to maximize the welfare (net benefit) of the society. The decision structure is equivalent to selecting a cost minimizing alternative when the social benefits are given or known. From the social welfare optimum problem as outlined in Chapter 1 section (1.5) if a social state is changed by selecting a project so also is the welfare level of the society. Let us suppose that various combinations of commodities x and y constitute a set of decision alternative. From this set a selection is to be made to achieve the welfare optimum of society composed of individuals A and B whose utilities are defined as $u^A(x^A, y^A)$ and $u^B(x^B, y^B)$. The society's welfare $W(\cdot)$ depends on the welfare of all its members. Thus

$$W = W\left(u^A(\bullet), u^B(\bullet)\right)$$

(3.6 .4.1)

Various alternative combinations of x, and y produce changes in individual net benefits and hence changes in the social welfare. Any change in social welfare may be represented from eqn. (3.6.4.1) as

$$dW = \frac{\partial W}{\partial u^A} du^A + \frac{\partial W}{\partial u^B} du^B .$$

(3.6.4.2)

with supporting changes in individual utilities in the form

$$du^A = \frac{\partial u^A}{\partial x^A} dx^A + \frac{\partial u^A}{\partial y^A} dy^A$$

(3.6.4.3)

$$du^B = \frac{\partial u^B}{\partial x^B} dx^B + \frac{\partial u^B}{\partial y^B} dy^B .$$

(3.6.4.4)

From equation (3.6.4.2) we know that (du's) are changes in individual net benefit and $(\partial W/\partial u)$'s are the importance that the society attaches to the individual benefit change or at least we can interpret it as such. The importance represents the social judgement weight as seen by society. Since the units and the functional forms of individual and social utilities are not known the numerical computation of the weights are not readily accessible. To make the distributional weights operationally computable 1) the individual welfare changes are transformed from the utility space into the commodity space where the utilities are measured in units of selected commodity called the numeraire good; and 2) distributional weights are attached to individual welfare changes on the basis of interpersonal comparisons of the individual welfare changes due to changes in the numeraire good accruing to different members of the society.

Let the numeraire good be y and divide and multiplier the right hand side of eqn. (3.6.4.2) by the corresponding individual marginal utility of numeraire good to obtain

$$dW = \frac{\partial W}{\partial u^A} u_y^A \frac{du^A}{u_y^A} + \frac{\partial W}{\partial u^B} u_y^B \frac{du^B}{u_y^B}$$

(3.6.4.5)

where u_y is the marginal utility of the numeraire good. Each term in the right hand side of eq. (3.6.4.5) may be broken down into two important components as:

1. (du/u_y) is the rate of utility gain (loss) measured in terms of the numeraire good, y. It approximates the number of units of y that each member is willing to give up in order to bring about the decision alternative associated with dW
2. $(\partial W/\partial u)u_y$ measures the social evaluation of extra unit of numeraire good y that may go to A (or B) given the choice of decision alternative. It is thus the social weight attached to each member's utility change.

In operationalizing the above we set $(du/u_y) = dy$ and $(\partial W/u_y) = \alpha$ hence eqn. (3.6.4.5) may then be written as

$$dW = \alpha_A dy^A + \alpha_B dy^B$$

(3.6.4.6)

If the distributional weights are to be assigned to leave the social welfare unchanged then

$$\alpha_A = \alpha_B \left| \frac{dy^B}{dy^A} \right|$$

(3.6.4.7)

If we assign the weight α_B then knowing the changes in dy's we can find α_A. For example, if dy^A = -100, dy^B = 400 and α_B = 1, then α_A = (1)(4) = 4. The α_B may be viewed as the value of central tendency for the group around which other weights are constructed and used. The value α_B is determined by the society and values dy^B, and dy^A are determined by the outcome of choice of an alternative. Generally if $W = W(u_i)$, $i \in \mathbb{J}$ an index set of the members of the society then

$$dW = \sum_{j \in \mathbb{J}} \alpha_j dy^j$$

(3.6.4.8)

The distributional weights α_i may be viewed as social shadow price that the society is willing to pay for the individual contribution, dy^j to the overall social welfare change. In a real cost analysis α_j is the intensity of cost burden that may be associated with a particular cost characteristic. If there are multiproducts (real cost characteristics) with and index set $\mathbb{I}$ and j is interpreted as a group then

$$dW = \sum_{i \in \mathbb{I}} \sum_{j \in \mathbb{J}} \alpha_{ij} dy^{ij}$$

(3.6.4.9)

as it is obtained in eq. (3.6.3.10). For other alternative approaches to the problem of distributional weight see [40] [41].

3.6.4.2 Fuzzy Computing of Distributional Weights, α_{ij}, for Burden of Real Cost.

There are a number of theoretical and algorithmic problems of the multi-utility measurement approach to compute the distributional weights. The important characteristics of the utility and welfare measures are vagueness, subjectivity and impression in the computational process. The method allows an arbitrary value to be assigned to $\alpha_{ij} = \left(\partial W / \partial u^{ij}\right) u^{ij}_{y}$ depending on who is assigning the values. It is this reason that has lead to a number of questions about the use of distributional weights. As it stands the weights are not optimally obtained. The multi-utility measurement approach does not allow the use of simultaneous assessment by policy makers, experts and the social groups. Even if one uses information elicitation, there is no channels for aggregating different information.

Given all these theoretical and algorithmic difficulties and the fact that there is the presence of fuzziness in assessment of the weights, a new computational approach is required. The introduction of expert judgement assessment; and the level of expert competence in section (3.4) bring into the computational process aggregation of differential fuzzy information. To get around the problems of fuzziness we view the competent experts as fuzzy set and the degree to which an expert belongs to the set of competent experts as its competence weight for the value, α_{ij}, that is assigned. We define a fuzzy set $\tilde{A}$ as

$$\tilde{A}_i = \{x_l \mid l \in \mathbb{L}_l, \mu_{\tilde{A}}(x_l) = \lambda_l \in [0,1], i \in \mathbb{I}\}$$

(3.6.4.2.1)

and a fixed level set

$$A_{\alpha_i^0} = \{x_l \mid \mu_{\tilde{A}_i}(x_l) = \lambda_l \geq \alpha_i^0 \in [0,1], \ i \in \mathbb{I}, \ l \in \mathbb{L}_i \}$$

(3.6.4.2.2)

where α_i^0 is the minimum acceptable level of expert competence weight, $\mathbb{I}$ is an index set of cost characteristics and $\mathbb{L}_i$, is an index set of experts for the *i*th cost characteristic. Thus $A_{\alpha_i^0}$ is a set of experts for the *i*th cost characteristic with confidence level greater than or equal to α_i^0.

Consider each cohort as a fuzzy ball of points and each evaluation, α_{ijl}, as a fuzzy number for $i \in \mathbb{I}, j \in \mathbb{J}$ and $l \in \mathbb{L}_i$, where $\mathbb{J}$ is an index set of social groups. Now define a fuzzy set as:

$$\mathbb{B}_{ijl} = \{ \alpha_{ijl} \mid \mu_{\mathbb{B}ijl}(\alpha_{ijl}) \in [0,1], \ i \in \mathbb{I}, \ j \in \mathbb{J}, l \in \mathbb{L}_i \}$$

(3.6.4.2.3)

that describes individual experts evaluation regarding the accuracy of α_{ijl} reflecting the true cost for the cohort. The problem of computing α_{ij} may be stated as an optimization of fuzzy accuracy index for the burden of real cost of decision for any given cohort subject to the α_i^0-level of competency set of experts as described by eqn. (3.6.4.2.2).

We may view the problem as a fuzzy decision problem, *D*, that may be specified as

$$D = \bigwedge_{l \in \mathbb{L}} \mu_{\mathbb{B}_{ijl}}(\alpha_{ijl})$$

(3.6.4.2.4)

for each fixed $i \in \mathbb{I}$ and $j \in \mathbb{J}$. It may be noted that the fuzzy decision problem, *D*, may be null for some *I*, $i \in \mathbb{I}$ if evaluations by experts are not needed. In this case the index set $\mathbb{L}_i = \varnothing$ for a particular $i \in \mathbb{I}$. Given the fuzzy decision as described by eqn. (3.6.4.2.4) we may state the problem of obtaining the optimal α_{ij}^{*} as a fuzzy mathematical programming problem where we optimize the membership function of the fuzzy decision subject to an appropriate constraint. The problem is simply

$$\alpha_{ij}^{*} = \begin{cases} \sup\limits_{\alpha_{ij}} \bigwedge\limits_{l \in \mathbb{L}_i} \mu_{\mathbb{B}_{ijl}}\left(\alpha_{ijl}\right) \\ \text{s.t.} \quad \text{Inf d}\left(\tilde{A}_i\ A_{\alpha_i^0}\right) \end{cases}$$

(3.6.4.2.5)

where $d\left(\tilde{A}_i\ A_{\alpha_i^0}\right)$ is a Hamming metric. The constraint inf. $d\left(\tilde{A}_i\ A_{\alpha_i^0}\right)$) may be viewed as a measure of maximum degree of consensus regarding the competency of experts. The algorithms required to solve problem of eqn. (3.6.4.2.5) or a class of such problems may be found in [694, 1059, 1072,1075,1085,1101, 1091,]. See also Chapters 5 and 6 of this book for quick reference. Equations (3.6.3.2), (3.6.3.3) and (3.6.4.2.4) allow one to obtain a *fuzzy equilibrium weights* as measures of cost-burden distributional weights, α_{ij}, given the set of competent experts as selected by the decision process. The experts are generated by the institutions of the social formation. It must be pointed out that α_{ij} as obtained by the calculus of fuzzy decisions can be adjusted to reflect social conditions as seen by politicians and planners while the cost calculation can be extended over time periods. The inclusion of social conditions will require extra fuzzy constraints that reflect assessments of interest groups, experts and policy makers. The multi-period extension allows the weights to be adjusted in a flexible manner.

From the cost-burden matrix we can write the total real composite cost, C^{AB} as

$$C^{AB} = \sum_{j \in \mathbb{J}} \sum_{i \in \mathbb{I}} \alpha_{ij} c_{ij}$$

(3.6.4.2.6)

This total cost concept is not useful for calculation since the sum cannot be taken. The composite total real cost, C^{AB} contains different cost characteristics. As such a single measure of total cost is not directly available to us. To get around this problem we assume that we can locate price P_i , $i \in \mathbb{I}$ for each ith cost characteristic and that such a price is uniform across social classes for all $j \in \mathbb{J}$. The weighted total social cost, C_{ST}, in monetary terms may then be written as

$$C_{ST} = \sum_{j \in \mathbb{J}} \sum_{i \in \mathbb{I}} P_i \alpha_{ij} c_{ij}$$

(3.6.4.2.7)

The price P_i *must* be found before Eq. (3.6.4.2.7) can be computed in full. If prices are differentiated across classes then we must use P_{ij}, as the imputed prices and the total social cost becomes

$$C^{ST} = \sum_{i \in \mathbb{I}} \sum_{j \in \mathbb{J}} P_{ij} \alpha_{ij} c_{ij}$$

(3.6.4.2.8)

Whether the aggregate, C_{ST}, can be found or not, depends essentially on the existence of P_{ij} which allow all or some of the cost characteristics to be reduced to one common unit of measure. This is the problem of the theory of efficient prices. There are a number of approaches to construct the P_{ij} given the cost characteristic $i \in \mathbb{I}$. Such approaches include a) marginal cost pricing [319] [304] [256] [276] [322], [348], hedonic pricing [288b] [300] [267] and optimal fuzzy prices [927] [929]. It may be noted from eqn. (3.6.4.2.8) and the cost identification matrix that $c_{ij}'s$ may be tangibles and intangibles. Given the tangibles, it is possible that prices may not exist for some $i \in \mathbb{I}$ and hence it neither might nor be possible to compute direct pecuniary measures. For example, when the tangible real cost is physically measured in terms of human lives where prices do not exist to monetize their values.

3.6.5 Tangibles, Intangibles, Prices and Cost Aggregates

The structure of cost aggregate as specified in Eq. (3.6.4.2.8) requires some analysis. It may be noted from Eq. (3.6.4.2.8), and the cost-identification matrix that $c_{ij}'s$ are either tangibles or intangibles. Prices may not exist for some of the tangibles and hence may not have pecuniary measures to be aggregated into C_{ST}. In most cases the tangibles have no directly recognizable physical measures. The cost characteristics, therefore, may be grouped into two sets of price measurables and price-non-measurables. Let the index sets for the two groups be $\mathbb{I}^{\mathbb{P}}$ and $\mathbb{I}^{\mathbb{N}}$, respectively. The price-non-measurable set of cost characteristics may further be divided into a set of cost characteristics that are both quantitatively and pecuniarily non-measurable. Let the index sets of these two groups of cost characteristics be $\mathbb{K}$ and Δ, respectively, such that $\mathbb{I}^{\mathbb{N}} = \mathbb{K} \cup \Delta$.

This comprehensive social costing in cost-benefit analysis of choice alternatives provides the decision maker with rich information set on the basis of which a selection can be made. Three important information sets must be processed and combined for the selection decision. They are

(a) total monetary cost, $C_{\mathrm{ST}}^{\mathrm{M}}$.

$$C_{\mathrm{ST}}^{\mathrm{M}} = \sum_{i \in \mathbb{I}^{\mathbb{P}}} \sum_{j \in \mathbb{J}} P_{ij} \alpha_{ij} c_{ij} \quad , \quad \mathbb{I}^{\mathbb{P}} \subseteq \mathbb{I}$$

(3.6.5.1)

where P_{ij} are to be constructed or ascertained.

(b) a composite quantitatively measurable sum, $C_{\mathrm{ST}}^{\mathrm{Q}}$,

$$C_{\mathrm{ST}}^{\mathrm{Q}} = \left[\alpha_{i1} c_{i1} , \cdots , \alpha_{ij} c_{ij}\right], \qquad i \in \mathbb{I},$$

(3.6.5.2)

where $\mathbb{K} \cap \mathbb{I}^{\mathbb{P}} = \varnothing$, $\mathbb{K} \subset \mathbb{I}$ and $\mathbb{K} \bigcup \mathbb{I}^{\mathbb{P}} \subset \mathbb{I}$ and

(c) Identified cost characteristics that are non-measurable in either quantitative or monetary terms, $C_{\mathrm{ST}}^{\mathrm{N}}$, where

$$C_{\mathrm{ST}}^{\mathrm{N}} = \lfloor \alpha_{i1} c_{i1} , \cdots , \alpha_{ij} c_{ij} \rfloor , \quad i \in \Delta$$

(3.6.4.3)

where the $\alpha_{ij}'s$ are constructed as distributional weights for the cost burden and $\Delta \cap \mathbb{K} = \emptyset = \mathbb{K} \cap \mathbb{I}^{\mathbb{P}} \cap \Delta$, $\mathbb{K} \cup \mathbb{I}^{\mathbb{P}} \cup \Delta = \mathbb{I}$. The real cost characteristics and the corresponding distributional weights of the burden of the costs are presented in Figure 3.6.4.1 in a Venn diagram for the index sets.

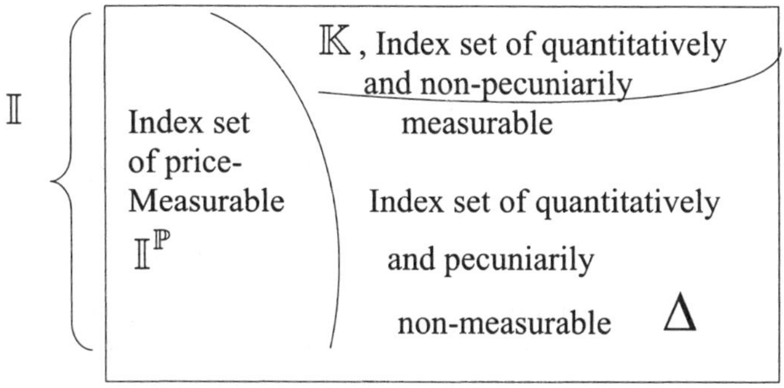

Fig. 3.6.4.1, Index set of Real Cost Characteristics

3.6.6 The Effect of Time on Cost Computing

The framework of social casting that has been presented so far is static in the sense that cost effects are calculated for only one period. In social cost-benefit analysis cost-shifting is not only possible interpersonally and across different classes but also cost-shifting is an intertemporal or intergenerational possibility when the cost effects of decision alternatives spread over time. This is also true of the distribution for the burden of real cost characteristics. The implication here is that the cost computation must take place at each point of time. Dynamic cost-identification matrix must then be constructed to provide the cost information for each point in time [358], [364] [369] [384] [927].

If $\mathbb{T}$ is a time set then each cost characteristic will be c_{ijt}; the corresponding cost burden weight, α_{ijt} and the cost information may then be presented for each t fixed in $\mathbb{T}$ as

$$\text{(a)}\quad C^{\mathbb{M}}_{\mathrm{ST}_t} = \left[\sum_{i\in\mathbb{I}^{\mathbb{P}}}\sum_{j\in\mathbb{J}}\alpha_{ijt}c\right]_{t\in\mathbb{T}}, \quad \mathbb{I}^{\mathbb{P}}\subseteq\mathbb{I}$$

(3.6.6.1)

$$\text{(b)}\quad C^{\mathbb{Q}}_{\mathrm{ST}_t} = \left[\alpha_{i1t}c_{i1t}, \ldots ,\alpha_{ijt}c_{ijt}\right]_{t\in\mathbb{T}}, \quad i\in\mathbb{K}\subset\mathbb{I}$$

(3.6.6.2)

and

$$\text{(c)}\quad C^{\mathbb{N}}_{\mathrm{ST}} = \left[\alpha_{i1t}c_{i1t}, \ldots , \alpha_{ijt}c_{ijt}\right]_{t\in\mathbb{T}}, \quad i\in\Delta\subset\mathbb{I}.$$

(3.6.6.3)

Equations (3.6.6.1 - 3.5.6.3) provide time series cost information for the project or decision alternative in question. Equation (3.6.5.1) is a static matrix aggregate over cost characteristics and across social classes. To obtain the total social cost of projects over time, current value equivalence must be computed for future cost values. This leads to intertemporal aggregation which is done through discounting. Such a discounting process may be done in the traditional way [201,203a, 203b, 204, 206, 239] or in a non-traditional way [927, 739, 909, 932, 1153,] after the time series of cost information has been obtained (see an extensive treatment on fuzzy and non-fuzzy discounting process in companion volume devoted to theory of fuzzy value computation [931b].

3.7 Example of Social Costing: The Social Cost of Mental Illness

Let us now sketch an example of social costing by using the framework that has been advanced. We consider the mentally ill where our objective is to compute the cost to society irrespective of the treatment alternative. The first task is to identify the real cost characteristics and then divide the section of the population that may be affected by the cost into or groups.

The *real cost characteristics* of mental illness irrespective of alternative methods of treatment must be identified. The identifications may be done with expert, policymakers and community advice and input.

3.7.1 Identification of Real Cost Characteristics

The real cost characteristics are identified as:

1. Mortality	x_1
2. Morbidity - illness and disability	x_2
3. In-patient hospital treatment	x_3
4. Out-patient hospital treatment	x_4
5. Rehabilitation	x_5
6. Social services	x_6
7. Community monitoring services	x_7
8. Police and jail services	x_8
9. Court Services	x_9
10.Probation and parole	x_{10}
11.Housing services	x_{11}
12.Food	x_{12}
13.Property damage	x_{13}
14.Human physical injury	x_{14}
15.Loss of income not attributed to death	x_{15}
16.Anxiety (psychic loss, pain and suffering)	x_{16}

These identified real cost characteristics are sufficient to illustrate the matrix identification approach to comprehensive costing of mental illness.

3.7.2 Social Groups (Classes)

The identified real cost characteristics have differential impact on social groups. The social groups need to be identified in order to assess the cost for the group in this case of mental illness. The required groups are identified as:

1. Mentally ill person
   (a) Young
      (i) Females G_1

(ii) Males G_2

(b) Old

(i) Females G_3

(ii) Males G_4

2. Families G_5
3. Workers G_6
4. Neighbors G_7
5. Public philanthropists G_8
6. Tax payers
   a) Local government G_9
   b) State government G_{10}
   c) Federal government G_{11}

3.7.3 Forming the Computable Cost-Identification Matrix

Given the identified real cost characteristics and social groups we construct the cost identification matrix for the mental illness that is to provide a comprehensive accounting of cost. The required identification matrix is shown in Table 3.6.3.1. It may also be noted that G_9 , G_{10} and G_{11} may have different names depending on society's organization and institutions of Government and policy making.

From the matrix we know that $c_{ij} \geq 1$ and the blank spaces in the cohorts suggest that the class is unaffected by the cost characteristics. Care must be taken *not* to *double count.* For example the services provided by local, State and Federal governments must be distinguished and separated. Similarly, the costs of hospital services assigned to the mentally ill individuals must not be included in those attributable to either the family or a government agency or a philanthropic organization. Furthermore, care must be taken in identifying inter-class cost shifting as well as inter-temporal cost shifting when the costing is done over time.

The matrix may be partitioned into (a) price measurable, (b) quantitatively but non-pecuniarily measurable and (c) quantitatively and pecuniarily non-measurable. The cost characteristics,

x_3 , x_4 , x_5 , x_6 , x_7 , x_8 , x_9 , x_{10} , x_{11} x_{12} , x_{13} , x_{14} and x_{15}

are price and pecuniarily measurable while the characteristic, x_{16}, is quantitatively and pecuniarily non-measurable.

Table 3.7.3.1 The Real Cost-identification Matrix

$x \backslash G$	G_1	G_2	G_3	G_4	G_5	G_6	G_7	G_8	G_9	G_{10}	G_{11}	Total
x_1	c_{11}	c_{12}	c_{13}	c_{14}	c_{15}	c_{16}	c_{17}					C_1^x
x_2	c_{21}	c_{22}	c_{23}	c_{24}	c_{25}	c_{26}	c_{27}					C_2^x
x_3	c_{31}	c_{32}	c_{33}	c_{34}								
x_4	c_{41}	c_{42}	c_{43}	c_{44}								
x_5	c_{51}	c_{52}	c_{53}	c_{54}								
x_6								c_{68}	c_{69}	$c_{6,10}$	$c_{6,11}$	
x_7								c_{78}	c_{79}	$c_{7,10}$	$c_{7,11}$	•
x_8									c_{89}	$c_{8,10}$	$c_{8.11}$	•
x_9									c_{99}	$c_{9,10}$	$c_{9,11}$	•
x_{10}									$c_{10,9}$	$c_{10,10}$	$c_{10,11}$	
x_{11}					$c_{11.5}$				$c_{11,9}$			
x_{12}					$c_{12,5}$				$c_{12,9}$			
x_{13}					$c_{13,5}$	$c_{13,6}$	$c_{13,7}$					
x_{14}	$c_{14,1}$	$c_{14,2}$	$c_{14,3}$	$c_{14,4}$	$c_{14,5}$	$c_{14,6}$	$c_{14,7}$					
x_{15}	$c_{15,1}$	$c_{15,2}$	$c_{15,3}$	$c_{15,4}$	$c_{15,5}$	$c_{15,6}$	$c_{15,7}$					C_{15}^x
x_{16}	$c_{16,1}$	$c_{16,1}$	$c_{16,3}$	$c_{16,4}$	$c_{16,5}$	$c_{16,6}$	$c_{16,7}$					C_{16}^x
Total	C_1^G	C_2^G				•	•	•				C

As a hypothetical situation let us consider a small community in and industrial society with (140) one-hundred and forty mentally ill persons. From these persons, community and families a cost-identification frequency matrix is developed in Table 3.7.3.2.

Table 3.7.3.2
Real cost-identification matrix for computing social cost of mental illness

	G_1	G_2	G_3	G_4	G_5	G_6	G_7	G_8	G_9	G_{10}	G_{11}	Total
x_1		2	1	4	3	1	3					14
x_2	10	10	49	64	2	1	2					138
x_3	5	6	20	34								65
x_4	5	4	29	30								68
x_5	10	10	49	64								133
x_6								10	10	46	64	133
x_7								6	5	35	40	86
x_8									100	30	3	133
x_9									100	30	3	133
x_{10}									100	30	3	133
x_{11}					70				63			133
x_{12}					70				63			133
x_{13}					25	4	3					32
x_{14}	3	2	10	16	15	3	10					59
x_{15}	10	10	49	64	48	1	8					190
x_{16}	10	10	49	64	93	12	6					244
Total					•	•	•					C

One must be careful in interpreting the cost aggregates at the end of each real cost characteristic. For example, $C_1^x = 14$ implies that 14 people died during the period due to mental illness in the community while $C_{14}^X = 59$ is interpreted as 59 people experienced human physical injury due to mental illness in the community. Similar meaningful interpretation cannot be found for column aggregate, $C_1^G = 53$. The matrix of Table 3.7.3.2 is an identification construct for one period. It will form part of initial data to estimate the future cost given a set of institutional and non-institutional assumptions

3.7.4 Computing the Aggregate Cost Data

To obtain an aggregate cost for the period we utilize, as a matter of an illustrated example, the unit cost computations and other cost information in Weisbroad [602,603]. The cost information in [603] is restructured to be meaningful and consistent with the theoretical framework that is developed in this chapter. Since we have not yet developed in this book the theory of optimal prices we shall take advantage of the cost information in [603]. To arrive at the aggregate cost data we need to construct a) a data partitioning into price measurables, quantitatively and non-price measurables and quantitatively and pecuniarily non-measurables, b) prices and c) distributional weights.

3.7.4.1 Prices and Cost-Data Partitioning

From cost information in Table 3.7.3.2 and Weisbrod [603] we construct matrices $\mathbb{C}_1$,$\mathbb{C}_2$, $\mathbb{C}_3$ and $\mathbb{C}_4$ with supporting price vectors $\mathbb{P}_1$, $\mathbb{P}_2$ and $\mathbb{P}_3$ as in eqns. (3.7.4.1.1) to (3.7.4.1.8). From the general cost information in [603], the following price vector in dollar unit is developed

$$\mathbb{P} = [P_3 , P_4 , P_5 , P_6 , P_7 , P_8 , P_9 , P_{10} , P_{11} , P_{12} , P_{13} , P_{15}]$$
$$= [3096, 42, 1744, 185, 122, 203, 17, 189, 950, 537, 250, 200]$$

(3.7.4.1.1)

These prices are constructed from information about the cost characteristics for an individual over a period of one year. Each price is a unit cost per year for each person affected. For example, x_3, is in-patient hospital treatment and P_3 is the average unit cost for each patient per year. From the real cost-identification matrix of Table (3.7.3.2) the following submatrices may be constructed

$$\mathbb{C}_1 = \begin{bmatrix} 5 & 6 & 20 & 34 \\ 5 & 4 & 29 & 30 \\ 10 & 10 & 49 & 64 \\ 10 & 10 & 49 & 64 \end{bmatrix} = \begin{bmatrix} c_{31} & c_{32} & c_{33} & c_{34} \\ c_{41} & c_{42} & c_{43} & c_{44} \\ c_{51} & c_{52} & c_{53} & c_{54} \\ c_{15,1} & c_{15,2} & c_{15,3} & c_{15,4} \end{bmatrix} = \left[c_{ij}\right]$$

(3.7.4.1.2)

with a corresponding price vector.

$$\mathbb{P}_1 = \left[P_3 , P_4 ,P_5 , P_{15}\right] = [3096, 42, 1744, 200] = \vec{\mathrm{P}}_j .$$

(3.7.4.1.3)

Similarly, we may write

$$\mathbb{C}_2 = \begin{bmatrix} 10 & 10 & 49 & 64 \\ 6 & 5 & 35 & 40 \\ 0 & 100 & 30 & 3 \\ 0 & 100 & 30 & 3 \\ 0 & 100 & 30 & 3 \end{bmatrix} = \begin{bmatrix} c_{68} & c_{69} & c_{6,10} & c_{6,11} \\ c_{78} & c_{79} & c_{7,10} & c_{7,11} \\ c_{88} & c_{89} & c_{8,10} & c_{8,11} \\ c_{98} & c_{99} & c_{9,10} & c_{9,11} \\ c_{10,8} & c_{10,9} & c_{10,10} & c_{10.11} \end{bmatrix} = \left[c_{lk}\right]$$

(3.7.4.1.4)

with a corresponding price vector

$$\mathbb{P}_2 = \left[P_{6,}, P_7, P_8, P_9, P_{10}\right] = [185, 122, 203, 17, 1898] = \vec{\mathrm{P}}_{\mathrm{k}}$$

(3.7.4.1.5)

Similarly, we may construct, $\mathbb{C}_3$ as

$$\mathbb{C}_3 = \begin{bmatrix} c_{11,5} & c_{11,6} & c_{11,7} & c_{11,8} & c_{11,9} \\ c_{12,5} & c_{12,6} & c_{12,7} & c_{12,8} & c_{12,9} \\ c_{13,5} & c_{13,6} & c_{13,7} & c_{13,8} & c_{13,9} \\ c_{15,5} & c_{15,6} & c_{15,7} & c_{15,8} & c_{15,9} \end{bmatrix} = \begin{bmatrix} 70 & 0 & 0 & 0 & 63 \\ 70 & 0 & 0 & 0 & 63 \\ 25 & 4 & 3 & 0 & 0 \\ 48 & 1 & 8 & 0 & 0 \end{bmatrix} = \left[c_{rs}\right]$$

(3.7.4.1.6)

with a corresponding price vector

$$\mathbb{P}_3 = \left[P_{11}, P_{12}, P_{13}, P_{15}\right] = [950, 537, 250, 200] = \vec{\mathrm{P}}.$$

(3.7.4.1.7)

The final sub-matrix, $\mathbb{C}_4$ may be written as

$$\mathbb{C}_4 = \begin{bmatrix} c_{11} & c_{12} & c_{13} & c_{14} & c_{15} & c_{16} & c_{17} \\ c_{21} & c_{22} & c_{23} & c_{24} & c_{25} & c_{26} & c_{27} \\ c_{14,1} & c_{14,2} & c_{14,3} & c_{14,5} & c_{14,6} & c_{14,7} & c_{14,7} \\ c_{16,1} & c_{16,2} & c_{16,3} & c_{16,4} & c_{16,5} & c_{16,6} & c_{16,7} \end{bmatrix} = \begin{bmatrix} 0 & 2 & 1 & 4 & 3 & 1 & 3 \\ 10 & 10 & 49 & 64 & 2 & 1 & 2 \\ 3 & 2 & 10 & 16 & 15 & 3 & 10 \\ 10 & 10 & 49 & 64 & 93 & 12 & 6 \end{bmatrix} = \left[c_{hq}\right]$$

(3.7.4.1.8)

The sub-matrix, $\mathbb{C}_4$, has no corresponding price vector since it is both price and quantitatively non-measurable. Given this cost partitioning, the next step is to construct distributional weights that reflect the cost burden of each social group.

3.7.4.2 Construction of Distributional Weights on the Basis of Fuzzy Decision

The degree of the real cost burden may vary from social group to the other. This is the social intensity of cost impact. Each weight may be viewed as the shadow price that society attaches to the unit of the real cost characteristic. The social evaluation of such burden of the real costs may be obtained by information elicitation from the relevant classes, the decision-or policy-making unit and a set of experts. The information is then processed to obtain aggregate values as weights for the various classes. The aggregation can be done by a method of fuzzy decision [11-14, 74-78] or as has been presented in section (3.6.4.2). Alternatively, we may obtain the distributional intensity weights of burden of real cost by the method of Section (3.6.4.1). Since our emphasis is more on fuzzy computing of costs and benefits than on other techniques for realizing distributional weights we shall concentrate on the development of fuzzy equilibrium weights.

We shall now consider the value characterization of the burden of real cost characteristic. To develop the index of burden of cost we consider the value of $1.00 for each cost characteristic over social groups. Furthermore, we consider the social intensity of the burden of cost over different real cost characteristics. These two valuations will be combined to develop the matrix of burden of costs over social groups and different cost characteristics. We consider two experts and their evaluational information on the basis of pairwise preference matrices. Since we are dealing with heterogeneous social groups and cost characteristics we also ascertain the degree of expertness of the two experts in providing evaluations of the burden of costs for the social classes and cost characteristics. The degrees of expertness define the levels of confidence that the society attaches to the valuations of the experts (the number of evaluators may be increased).

The following hypothetical data is constructed to illustrate the fuzzy computational steps and the process developed in Section (3.6.4.2). We define the following sets for the fuzzy computing.

$\mathbb{E}$ = a set of valuations of burden of one dollar cost over social classes.

$\mathbb{M}$ = a set of measures of degree of expertness associated with social confidence in the evaluator's ability to evaluate the classes.

$\mathbb{F}$ = a set of evaluations of social burden of real cost characteristics.

$\mathbb{H}$ = a set of measures of degree of expertness associated with social confidence in the evaluator's ability to evaluate the social burden of real cost characteristics.

The index set of the evaluators contains only two elements ($\#\mathbb{L} = 2$). For each evaluator we specify two fuzzy sets of the form

$$\mathbb{Q} = \mathbb{E}_l \otimes \mathbb{M}_l = \{ (x, \mu(x)) \mid x \in \mathbb{E}_l \ \mu \in \mathbb{M}_l \text{ and } l \in \mathbb{L} \}$$

(3.7.4.2.1)

$$\mathbb{Z}_l = \mathbb{F}_l \otimes \mathbb{H}_l = \{ (y , \mu(y)) \mid y \in \mathbb{F}_l, \ \mu \in \mathbb{H}_l \text{ and } l \in \mathbb{L} \}$$

(3.7.4.2.2)

whose values for this example are shown in Tables (3.7.4.1) and (3.7.4.2).

Table 3.7.4.1Intensity Distribution over Social Classes

		G_1	G_2	G_3	G_4	G_5	G_6	G_7	G_8	G_9	G_{10}	G_{11}
$\mathbb{Q}_1$	$\mathbb{E}_1$	3.0	2.0	1.0	1.0	1.0	2.0	2.0	1.0	0.75	0.5	0.5
	$\mathbb{M}_1$	0.8	0.6	0.8	0.7	0.5	0.6	0.7	0.9	0.9	1.0	1.0
$\mathbb{Q}_2$	$\mathbb{E}_2$	3.0	4.0	2.0	1.0	3.0	2.0	3.0	0.75	0.75	0.75	0.5
	$\mathbb{M}_2$	0.6	0.5	0.5	0.7	0.6	1.0	1.0	0.9	1.0	1.0	1.0

Table 3.7.4.2 Intensity distribution over cost characteristics

		x_1	x_2	x_3	x_4	x_5	x_6	x_7	x_8	x_9	x_{10}	x_{11}	x_{12}	x_{13}	x_{14}	x_{15}	x_{16}
$\mathbb{Z}_1$	$\mathbb{F}_1$	3.0	2.0	1.0	1.0	1.0	1.0	2.0	1.0	2.0	1.0	0.75	0.5	1.0	2.0	2.0	2.0
	$\mathbb{H}_1$	0.7	0.6	1.0	1.0	0.8	0.6	0.5	0.4	0.4	0.6	0.9	1.0	0.8	0.7	0.5	1.0
$\mathbb{Z}_2$	$\mathbb{F}_2$	3.0	3.0	1.0	1.0	2.0	2.0	1.0	1.0	2.0	1.0	0.5	0.5	2.0	3.0	3.0	2.0
	$\mathbb{H}_2$	0.5	0.6	0.7	0.5	0.9	0.8	1.0	1.0	0.7	0.6	1.0	1.0	1.0	0.7	1.0	1.0

The fuzzy decision logic and its mathematics that are being utilized here may be obtained from Chapter 5 and 6 as well as in [661] [756] [806] [694].

From Tables (3.7.4.1) and (3.7.4.2) we construct a matrix of values and corresponding matrix of membership values through fuzzy algebraic product composition for each evaluator as

$$\mathbb{E}_l \odot \mathbb{F}_l = \mathbb{R}_l = \{x \cdot y \mid x \in \mathbb{E}_l , y \in \mathbb{F}_l \text{ and } l \in \mathbb{L}\}$$

(3.7.4.2.3)

where $\mathbb{R}_l$ is a relational matrix of evaluations distributed over the cohorts by taking into account the cost burdens over social groups and cost characteristics. Corresponding to $\mathbb{R}_l$ we construct the corresponding membership values, $\mu_{\mathbb{R}_l}(x, y)$ by an algebraic product composition as

$$\mu_{\mathbb{HML}}(x,y) = \{(\mu_{\mathbb{E}_l}(x) \cdot \mu_{\mathbb{F}_l}(y)) \mid \mu_{\mathbb{E}_l} \in \mathbb{M}, \mu_{\mathbb{F}_l} \in \mathbb{H}_l \text{ and } l \in \mathbb{L}\}$$

(3.7.4.2.4)

By using Equations (3.7.4.2.1) and (3.7.4.2.2) and the data in Tables (3.7.4.1) and (3.7.4.2) four matrices may be constructed for the two evaluators. For the first evaluator we obtain:

$$\mathbb{R}_1 = \begin{bmatrix} 9.0 & 6.0 & 3.0 & 3.0 & 3.0 & 6.0 & 6.0 & 3.0 & 2.25 & 1.5 & 1.5 \\ 6.0 & 4.0 & 2.0 & 2.0 & 2.0 & 4.0 & 4.0 & 2.0 & 1.5 & 1.0 & 1.0 \\ 3.0 & 1.0 & 1.0 & 1.0 & 1.0 & 2.0 & 2.0 & 1,0 & 0.75 & 0.5 & 0.5 \\ 3.0 & 1.0 & 1.0 & 1.0 & 1.0 & 2.0 & 2.0 & 1.0 & 0.75 & 0.5 & 0.5 \\ 3.0 & 1.0 & 1.0 & 1.0 & 1.0 & 2.0 & 2.0 & 1.0 & 0.75 & 0.5 & 0.5 \\ 3.0 & 1.0 & 1.0 & 1.0 & 1.0 & 2.0 & 2.0 & 1.0 & 0.75 & 0.5 & 0.5 \\ 6.0 & 2.0 & 2.0 & 2.0 & 2.0 & 4.0 & 4.0 & 2.0 & 1.5 & 1.0 & 1.0 \\ 3.0 & 1.0 & 1.0 & 1.0 & 1.0 & 2.0 & 2.0 & 1.0 & 0.75 & 0.5 & 0.5 \\ 6.0 & 2.0 & 2.0 & 2.0 & 2.0 & 4.0 & 4.0 & 2.0 & 1.5 & 1.0 & 1.0 \\ 3.0 & 1.0 & 1.0 & 1.0 & 1.0 & 2.0 & 2.0 & 1.0 & 0.75 & 0.5 & 0.5 \\ 2.25 & 1.5 & 0.75 & 0.75 & 0.75 & 1.5 & 1.5 & .75 & 0.56 & 0.5 & 0.5 \\ & 0.5 & 0.5 & 0,5 & 0.5 & 1.0 & 1.0 & 1.0 & 0.38 & 0.25 & 0.25 \\ 3.0 & 1.0 & 1.0 & 1.0 & 1.0 & 2.0 & 2.0 & 2.0 & 0.75 & 0.5 & 0.5 \\ 6.0 & 2.0 & 2.0 & 2.0 & 2.0 & 4.0 & 4.0 & 2.0 & 1.5 & 1.0 & 1.0 \\ 6.0 & 2.0 & 2.0 & 2.0 & 2.0 & 4.0 & 4.0 & 2.0 & 1.5 & 1.0 & 1.0 \\ 6.0 & 4.0 & 2.0 & 2.0 & 2.0 & 4.0 & 4.0 & 2.0 & 1.5 & 1.0 & 1.0 \end{bmatrix} = \lfloor R_{ij1} \rfloor$$

(3.7.4.2.5)

$$\mu_{\mathbb{E}_1 \cdot \mathbb{F}_1}(x,y) = \begin{bmatrix} 0.56 & 0.42 & 0.56 & 0.49 & 0.35 & 0.42 & 0.49 & 0.63 & 0.63 & 0.70 & 0.70 \\ 0.48 & 0.36 & 0.48 & 0.42 & 0.42 & 0.36 & 0.42 & 0.54 & 0.54 & 0.60 & 0.60 \\ 0.80 & 0.60 & 0.80 & 0.70 & 0.50 & 0.60 & 0.70 & 0.90 & 0.90 & 1.00 & 1.00 \\ 0.80 & 0.60 & 0.80 & 0.70 & 0.50 & 0.60 & 0.70 & 0.90 & 0.90 & 1.00 & 1.00 \\ 0.64 & 0.48 & 0.64 & 0.56 & 0.40 & 0.48 & 0.56 & 0.72 & 0.72 & 0.80 & 0.80 \\ 0.48 & 0.36 & 0.48 & 0.42 & 0.30 & 0.36 & 0.42 & 0.54 & 0.54 & 0.60 & 0.60 \\ 0.40 & 0.30 & 0.40 & 0.35 & 0.25 & 0.30 & 0.35 & 0.45 & 0.45 & 0.50 & 0.50 \\ 0.32 & 0.24 & 0.32 & 0.28 & 0.20 & 0.24 & 0.28 & 0.36 & 0.36 & 0.40 & 0.40 \\ 0.32 & 0.24 & 0.32 & 0.28 & 0.20 & 0.24 & 0.28 & 0.36 & 0.36 & 0.40 & 0.40 \\ 0.48 & 0.36 & 0.48 & 0.42 & 0.30 & 0.36 & 0.42 & 0.54 & 0.54 & 0.60 & 0.60 \\ 0.72 & 0.54 & 0.72 & 0.63 & 0.45 & 0.54 & 0.63 & 0.81 & 0.81 & 0.90 & 0.90 \\ 0.80 & 0.60 & 0.80 & 0.70 & 0.50 & 0.60 & 0.70 & 0.90 & 0.90 & 1.00 & 1.00 \\ 0.64 & 0.48 & 0.64 & 0.56 & 0.40 & 0.48 & 0.56 & 0.72 & 0.72 & 0.80 & 0.80 \\ 0.56 & 0.42 & 0.56 & 0.49 & 0.35 & 0.42 & 0.49 & 0.63 & 0.63 & 0.70 & 0.70 \\ 0.40 & 0.30 & 0.40 & 0.35 & 0.25 & 0.30 & 0.35 & 0.45 & 0.45 & 0.50 & 0.50 \\ 0.80 & 0.60 & 0.80 & 0.70 & 0.50 & 0.60 & 0.70 & 0.90 & 0.90 & 1.00 & 1.00 \end{bmatrix}$$

$$= \mu_{\mathbb{R}_1}(x,y)$$

(3.7.4.2.6)

Similarly, we shall define two matrices for the second evaluator. These matrices have particular structures such that they complement each other and together provide us with the needed information as input to decision making. Thus,

$$\mathbb{R}_2 = \mathbb{E}_2 \odot \mathbb{F}_2 = \begin{bmatrix} 9.0 & 12.0 & 6.0 & 3.0 & 9.0 & 6.0 & 9.0 & 2.25 & 2.25 & 2.25 & 1.50 \\ 12.0 & 16.0 & 8.0 & 4.0 & 12.0 & 8.0 & 12.0 & 3.0 & 3.0 & 3.0 & 2.0 \\ 3.0 & 4.0 & 2.0 & 1.0 & 3.0 & 2.0 & 3.0 & 0.75 & 0.75 & 0.75 & 0.5 \\ 3.0 & 4.0 & 2.0 & 1.0 & 3.0 & 2.0 & 3.0 & 0.75 & 0.75 & 0.75 & 0.5 \\ 6.0 & 8.0 & 4.0 & 2.0 & 6.0 & 4.0 & 6.0 & 1.50 & 1.50 & 1.50 & 1.0 \\ 6.0 & 8.0 & 4.0 & 2.0 & 6.0 & 4.0 & 6.0 & 1.50 & 0.75 & 1.50 & 1.0 \\ 3.0 & 4.0 & 2.0 & 1.0 & 3.0 & 2.0 & 3.0 & 0.75 & 0.75 & 0.75 & 0.5 \\ 3.0 & 4.0 & 2.0 & 1.0 & 3.0 & 2.0 & 3.0 & 0.75 & 0.75 & 0.74 & 0.5 \\ 3.0 & 8.0 & 4.0 & 2.0 & 6.0 & 4.0 & 6.0 & 1.50 & 1.50 & 1.50 & 1.0 \\ 6.0 & 4.0 & 2.0 & 1.0 & 3.0 & 2.0 & 3.0 & 0.75 & 0.75 & 0.75 & 0.5 \\ 3.0 & 2.0 & 1.0 & 0.5 & 1.5 & 1.0 & 1.50 & 0.38 & 0.38 & 0.38 & 0.25 \\ 1.5 & 2.0 & 1.0 & 0.5 & 1.50 & 1.0 & 1.50 & 0.38 & 0.38 & 0.38 & 0.25 \\ 6.0 & 8.0 & 4.0 & 2.0 & 6.0 & 4.0 & 6.0 & 1.50 & 1.50 & 1.50 & 1.0 \\ 9.0 & 12.0 & 6.0 & 3.0 & 9.0 & 6.0 & 9.0 & 2.25 & 2.25 & 2.25 & 1.50 \\ 9.0 & 12.0 & 6.0 & 3.0 & 9.0 & 6.0 & 9.0 & 2.25 & 2.25 & 2.25 & 1.50 \\ 6.0 & 8.0 & 4.0 & 2.0 & 6.0 & 2.0 & 6.0 & 1.50 & 1.50 & 1.50 & 1.00 \end{bmatrix}$$

$$= R_{ij2}$$

(3.7.4.2.7)

where $\lfloor R_{ij1} \rfloor$ and $\lfloor R_{ij2} \rfloor$ are the matrices of evaluations which are specified by their column vectors,
R_{ijl} , $i \in \mathbb{I}$, $j \in \mathbb{J}$, $l \in \mathbb{L}$ and $\#\mathbb{L}=2$; also

$$\mu_{\mathbb{R}_2}(x,y) = \begin{bmatrix} 0.30 & 0.25 & 0.25 & 0.35 & 0.30 & 0.45 & 0.50 & 0.45 & 0.50 & 0.50 & 0.50 \\ 0.36 & 0.30 & 0.30 & 0.42 & 0.36 & 0.54 & 0.60 & 0.54 & 0.60 & 0.60 & 0.60 \\ 0.42 & 0.35 & 0.35 & 0.49 & 0.42 & 0.63 & 0.70 & 0.63 & 0.70 & 0.70 & 0.70 \\ 0.30 & 0.25 & 0.25 & 0.35 & 0.30 & 0.45 & 0.50 & 0.45 & 0.50 & 0.50 & 0.50 \\ 0.54 & 0.45 & 0.45 & 0.63 & 0.54 & 0.81 & 0.90 & 0.81 & 0.90 & 0.90 & 0.90 \\ 0.48 & 0.40 & 0.40 & 0.56 & 0.48 & 0.72 & 0.80 & 0.72 & 0.80 & 0.80 & 0.80 \\ 0.60 & 0.50 & 0.50 & 0.70 & 0.60 & 0.90 & 1.00 & 0.90 & 1.00 & 1.00 & 1.00 \\ 0.60 & 0.50 & 0.50 & 0.70 & 0.60 & 0.90 & 1.00 & 0.90 & 1.00 & 1.00 & 1.00 \\ 0.42 & 0.35 & 0.35 & 0.49 & 0.42 & 0.63 & 0.70 & 0.63 & 0.70 & 0.70 & 0.70 \\ 0.36 & 0.30 & 0.30 & 0.42 & 0.36 & 0.54 & 0.60 & 0.54 & 0.60 & 0.60 & 0.60 \\ 0.60 & 0.50 & 0.50 & 0.70 & 0.60 & 0.90 & 1.00 & 0.90 & 1.00 & 1.00 & 1.00 \\ 0.60 & 0.50 & 0.50 & 0.70 & 0.60 & 0.90 & 1.00 & 0.90 & 1.00 & 1.00 & 1.00 \\ 0.60 & 0.50 & 0.50 & 0.70 & 0.60 & 0.90 & 1.00 & 0.90 & 1.00 & 1.00 & 1.00 \\ 0.42 & 0.35 & 0.35 & 0.49 & 0.42 & 0.63 & 0.70 & 0.63 & 0.70 & 0.70 & 0.70 \\ 0.60 & 0.50 & 0.50 & 0.70 & 0.60 & 0.90 & 1.00 & 0.90 & 1.00 & 1.00 & 1.00 \\ 0.60 & 0.50 & 0.50 & 0.70 & 0.60 & 0.90 & 1.00 & 0.90 & 1.00 & 1.000. & 1.00 \end{bmatrix}$$

(3.7.4.2.8)

We now combine the values of $\mathbb{R}_l$ and $\mu_{\mathbb{R}_l}(\cdot)$ to abstract the optimal values of the burden of cost for each cohort. The process is to select those elicited values whose evaluators have the highest degree of expertness. The following algorithmic process is designed to arrive at the final matrix of values for the burden of cost. Let $\mu_{\mathbb{R}_l}(\cdot)$'s be specified in terms of their column vectors, $\hat{P}_{lj}$, $l \in \mathbb{L}$ and $j \in \mathbb{J}$. Thus,

$$\mu_{\mathbb{R}_l}(\cdot) = \left[\hat{P}_{l1}\hat{P}_{l2}\hat{P}_{l3}\cdots\hat{P}_{lj}\cdots\right].$$

(3.7.4.2.9)

From Eq. (3.6.4.13) we construct a new matrix of optimal degree of expertness and corresponding valuations by *first projection* defined over the matrix of valuations for each social class as (see also [810] and Chapter 5 of this volume

$$\mu_{\mathbb{R}}(\cdot) = \bigvee_{l \in \mathbb{L}} \mu_{\mathbb{R}}(\cdot) = \left[\hat{P}_{\cdot j} = \bigvee_{l \in \mathbb{L}} \hat{P}_{lj} \mid l \in \mathbb{L} \text{ and } j \in \mathbb{J}\right]$$

(3.7.4.2.10)

Utilizing the information of the example at hand where $\#\mathbb{L} = 2$, $\#\mathbb{J} = 11$ and for $j = 1$ we obtain the column vectors $\hat{P}_{11}^{\mathrm{T}}$, $\hat{P}_{21}^{\mathrm{T}}$ with the first projection, $\hat{P}_{\cdot 1}^{\mathrm{T}}$ *as* shown in Table 3.7.4.2.1, where $\hat{R}_{11}^{\mathrm{T}}$, $\hat{R}_{21}^{\mathrm{T}}$ and $\hat{R}_{\bullet 1}^{\mathrm{T}}$ are the corresponding values of evaluations from $\mathbb{R}_1$ and $\mathbb{R}_2$, and $R_{\bullet 1}^{\mathrm{T}}$ is a column vector of values corresponding to the first projection. By the same first projections of the column vectors of the individual evaluators we can construct the fuzzy matrix as defined in Eq. (3.7.4.2.6) with the corresponding values of the cost burden $W = \left[W_{\bullet 1}\ W_{\bullet 2}, \cdots, W_{\bullet j} \right]$ where $W_{\bullet j}$ are column vectors, $j \in \mathbb{J}$

$$
\mu_{\mathbb{R}_2}(\cdot) = \begin{bmatrix}
P_{\cdot 1} & P_{\cdot 2} & P_{\cdot 3} & P_{\cdot 4} & P_{\cdot 5} & P_{\cdot 6} & P_{\cdot 7} & P_{\cdot 8} & P_{\cdot 9} & P_{\cdot 10} & P_{\cdot 11} \\
0.56 & 0.42 & 0.56 & 0.49 & 0.35 & 0.45 & 0.50 & 0.63 & 0.63 & 0.70 & 0.70 \\
0.48 & 0.36 & 0.48 & 0.42 & 0.42 & 0.54 & 0.60 & 0.54 & 0.60 & 0.60 & 0.60 \\
0.80 & 0.60 & 0.80 & 0.70 & 0.50 & 0.63 & 0.70 & 0.90 & 0.90 & 1.00 & 1.00 \\
0.80 & 0.60 & 0.80 & 0.70 & 0.50 & 0.60 & 0.70 & 0.90 & 0.90 & 1.00 & 1.00 \\
0.64 & 0.48 & 0.64 & 0.63 & 0.54 & 0.81 & 0.90 & 0.81 & 0.90 & 0.90 & 0.90 \\
0.48 & 0.40 & 0.48 & 0.56 & 0.48 & 0.72 & 0.80 & 0.72 & 0.80 & 0.80 & 0.80 \\
0.60 & 0.50 & 0.50 & 0.70 & 0.60 & 0.90 & 1.00 & 0.90 & 1.00 & 1.00 & 1.00 \\
0.60 & 0.50 & 0.50 & 0.70 & 0.60 & 0.90 & 1.00 & 0.90 & 1.00 & 1.00 & 1.00 \\
0.42 & 0.35 & 0.35 & 0.49 & 0.42 & 0.63 & 0.70 & 0.63 & 0.70 & 0.70 & 0.70 \\
0.48 & 0.36 & 0.48 & 0.42 & 0.36 & 0.54 & 0.60 & 0.54 & 0.60 & 0.60 & 0.60 \\
0.72 & 0.54 & 0.72 & 0.70 & 0.60 & 0.90 & 1.00 & 0.90 & 1.00 & 1.00 & 1.00 \\
0.80 & 0.60 & 0.80 & 0.70 & 0.60 & 0.90 & 1.00 & 0.90 & 1.00 & 1.00 & 1.00 \\
0.64 & 0.50 & 0.64 & 0.70 & 0.70 & 0.90 & 1.00 & 0.90 & 1.00 & 1.00 & 1.00 \\
0.56 & 0.42 & 0.56 & 0.49 & 0.42 & 0.63 & 0.70 & 0.72 & 0.72 & 0.80 & 0.80 \\
0.60 & 0.50 & 0.50 & 0.70 & 0.60 & 0.90 & 1.00 & 0.90 & 1.00 & 1.00 & 1.00 \\
0.80 & 0.60 & 0.70 & 0.70 & 0.600 & 0.90 & 1.00 & 0.90 & 1.00 & 1.00 & 1.00
\end{bmatrix}
= \bigvee_{l \in \mathbb{L}} \mu_{\mathbb{R}_l}(\cdot)
$$

(3.7.4.2.11)

$$W = \begin{bmatrix} W_{.1} & W_{.2} & W_{.3} & W_{.4} & W_{.5} & W_{.6} & W_{.7} & W_{.8} & W_{.9} & W_{.10} & W_{.11} \\ 9.0 & 6.0 & 3.0 & 3.0 & 3.0 & 6.0 & 9.0 & 3.0 & 2.25 & 1.50 & 0.50 \\ 6.0 & 4.0 & 2.0 & 4.0 & 2.0 & 8.0 & 12.0 & 3.0 & 3.0 & 3.0 & 2.0 \\ 3.0 & 2.0 & 1.0 & 1.0 & 1.0 & 2.0 & 3.0 & 1.0 & 0.75 & 0.5 & 0.50 \\ 3.0 & 2.0 & 1.0 & 1.0 & 1.0 & 2.0 & 2.0 & 1.0 & 0.75 & 0.5 & 0.50 \\ 3.0 & 2.0 & 1.0 & 2.0 & 6.0 & 4.0 & 6.0 & 1.5 & 1.50 & 1.50 & 1.0 \\ 6.0 & 8.0 & 1.0 & 2.0 & 6.0 & 4.0 & 6.0 & 1.5 & 1.50 & 1.50 & 1.0 \\ 3.0 & 4.0 & 2.0 & 1.0 & 3.0 & 2.0 & 3.0 & 0.75 & 0.75 & 0.75 & 0.50 \\ 3.0 & 4.0 & 2.0 & 1.0 & 3.0 & 2.0 & 3.0 & 0.75 & 0.75 & 0.75 & 0.50 \\ 3.0 & 8.0 & 4.0 & 2.0 & 6.0 & 4.0 & 6.0 & 1.50 & 1.50 & 1.50 & 1.0 \\ 3.0 & 2.0 & 1.0 & 1.0 & 3.0 & 2.0 & 3.0 & 1.0 & 0.75 & 0.75 & 0.50 \\ 2.25 & 1.5 & 0.75 & 0.5 & 1.5 & 2.0 & 1.5 & 0.38 & 0.13 & 0.38 & 0.25 \\ 1.5 & 1.0 & 0.5 & 0.5 & 1.5 & 1.0 & 1.5 & 0.50 & 0.13 & 0.38 & 0.25 \\ 3.0 & 8.0 & 1.0 & 2.0 & 6.0 & 4.0 & 6.0 & 1.50 & 1.50 & 1.50 & 1.0 \\ 6.0 & 4.0 & 2.0 & 3.0 & 9.0 & 6.0 & 9.0 & 2.0 & 1.50 & 1.0 & 1.0 \\ 9.0 & 12.0 & 6.0 & 3.0 & 9.0 & 6.0 & 9.0 & 2.25 & 2.25 & 2.25 & 1.50 \\ 6.0 & 4.0 & 2.0 & 2.0 & 6.0 & 2.0 & 6.0 & 2.0 & 1.50 & 1.50 & 1.0 \end{bmatrix}$$

(3.7.4.2.12)

Table 3.6.4.3
A computational example using eqns. (3.7.4.2.9) and (3.7.4.2.10

$\hat{P}_{11}^{T}$	0.56	0.48	0.80	0.80	0.64	0.48	0.40	0.32	0.32	0.48	0.72	0.80	0.64	0.56	0.40	0.80
$\hat{R}_{11}^{T}$	9.0	6.0	3.0	3.0	3.0	3.0	6.0	3.0	6.0	3.0	2.24	1.5	3.0	6.0	6.0	6.0
$\hat{P}_{21}^{T}$	0.30	0.36	0.42	0.30	0.54	0.48	0.60	0.60	0.42	0.36	0.60	0.60	0.60	0.42	0.60	0.60
R_{21}^{T}	9.0	12.0	3.0	3.0	6.0	6.0	3.0	3.0	3.0	6.0	3.0	1.50	6.0	9.0	9.0	6.0
$\hat{P}_{.1}^{T}$	0.56	0.48	0.80	0.80	0.64	0.48	0.60	0.63	0.42	0.48	0.72	0.80	0.64	0.56	0.60	0.80
$\hat{R}_{.1}^{T}$	9.0	6.0	3.0	3.0	3.0	6.0	3.0	3.0	3.0	3.0	2.25	1.50	3.0	6.0	9.0	6.0

Where W-matrix is constructed on the process that if $\mu_{\mathbb{R}_2}(a_{112}) > \mu_{\mathbb{R}_1}(a_{111})$ and $a_{112} \in R_{112,}$, $a_{111} \in R_{111}$ then in general $\mu_{\mathbb{R}_l}(a_{11l}) > \mu_{\mathbb{R}_k}(a_{11k})$ and

$a_{112} \in W_{\bullet 1}$ $[a_{11l} \in \mathbb{R}_{11l}, a_{11k} \in \mathbb{R}_{11k}$ then $a_{11l} \in W_{\bullet 1}$

From Eqns. (3.7.2.11) and (3.7.2.12) construct sub-matrices $\mathbb{D}_k$ and $\mathbb{N}_k$ of valuations and weights of burden of cost that correspond to the $\mathbb{C}_k$ sub-matrices specified in Eqns. (3.7.4 .1.2) (3.7.4.1.4), (3.7.4.1.6) and (3.7.4.1.8). Thus,

$$\mathbb{D}_1 = \begin{bmatrix} \alpha_{31} & \alpha_{32} & \alpha_{33} & \alpha_{34} \\ \alpha_{41} & \alpha_{42} & \alpha_{43} & \alpha_{44} \\ \alpha_{51} & \alpha_{52} & \alpha_{53} & \alpha_{54} \\ \alpha_{15,1} & \alpha_{15,2} & \alpha_{15,3} & \alpha_{15,4} \end{bmatrix} = \begin{bmatrix} 3.0 & 2.0 & 1.0 & 1.0 \\ 3.0 & 2.0 & 1.0 & 1.0 \\ 3.0 & 2.0 & 1.0 & 2.0 \\ 9.0 & 12.O & 6.0 & 3.0 \end{bmatrix} = \left[\alpha_{ij}\right]$$

(3.7.4.2.13)

$$\mathbb{N}_1 = \begin{bmatrix} \beta_{31} & \beta_{32} & \beta_{33} & \beta_{34} \\ \beta_{41} & \beta_{42} & \beta_{43} & \beta_{44} \\ \beta_{51} & \beta_{52} & \beta_{53} & \beta_{54} \\ \beta_{15,1} & \beta_{15,2} & \beta_{15,3} & \beta_{15,4} \end{bmatrix} = \begin{bmatrix} 0.80 & 0.60 & 0.80 & 0.70 \\ 0.80 & 0.60 & 0.80 & 0.70 \\ 0.64 & 0.48 & 0.64 & 0.63 \\ 0.60 & 0.50 & 0.50 & 0.70 \end{bmatrix} = \left[\beta_{ij}\right]$$

(3.7.4.2.14)

$$\mathbb{D}_2 = \begin{bmatrix} \alpha_{68} & \alpha_{69} & \alpha_{6,10} & \alpha_{6,11} \\ \alpha_{78} & \alpha_{79} & \alpha_{7,10} & \alpha_{7,11} \\ \alpha_{88} & \alpha_{89} & \alpha_{8,10} & \alpha_{8,11} \\ \alpha_{98} & \alpha_{99} & \alpha_{9,10} & \alpha_{9,11} \\ \alpha_{10,8} & \alpha_{10,9} & \alpha_{10,10} & \alpha_{10,11} \end{bmatrix} = \begin{bmatrix} 1.50 & 1.50 & 1.50 & 1.0 \\ 0.75 & 0.75 & 0.75 & 0.50 \\ 0.75 & 0.75 & 0.75 & 0.50 \\ 1.50 & 1.50 & 1.50 & 1.00 \\ 1.00 & 0.75 & 0.75 & 0.50 \end{bmatrix} = \left[\alpha_{lk}\right]$$

(3.7.4.2.15)

$$\mathbb{N}_2 = \begin{bmatrix} \beta_{68} & \beta_{69} & \beta_{6,10} & \beta_{6,11} \\ \beta_{78} & \beta_{79} & \beta_{7,10} & \beta_{7,11} \\ \beta_{88} & \beta_{89} & \beta_{8,10} & \beta_{8,11} \\ \beta_{98} & \beta_{99} & \beta_{9,10} & \beta_{9,11} \\ \beta_{10,8} & \beta_{10,9} & \beta_{10,10} & \beta_{10,11} \end{bmatrix} = \begin{bmatrix} 0.72 & 0.80 & 0.80 & 0.90 \\ 0.90 & 1.00 & 1.00 & 0.80 \\ 0.90 & 1.00 & 1.00 & 1.00 \\ 0.63 & 0.70 & 0.70 & 1.00 \\ 0.54 & 0.60 & 0.60 & 0.70 \end{bmatrix} = \left[\beta_{lk}\right]$$

(3.7.4.2.16)

$$\mathbb{D}_3 = \begin{bmatrix} \alpha_{11,5} & \alpha_{11,6} & \alpha_{11,7} & \alpha_{11,8} & \alpha_{11,9} \\ \alpha_{12,5} & \alpha_{12,6} & \alpha_{12,7} & \alpha_{12,8} & \alpha_{12,9} \\ \alpha_{13,5} & \alpha_{13,6} & \alpha_{13,7} & \alpha_{13,8} & \alpha_{13,9} \\ \alpha_{15,5} & \alpha_{15,6} & \alpha_{15,7} & \alpha_{15,8} & \alpha_{15,9} \end{bmatrix} = \begin{bmatrix} 1.50 & 1.00 & 1.50 & 0.38 & 0.38 \\ 1.50 & 1.00 & 6.00 & 0.50 & 0.38 \\ 6.0 & 4.00 & 9.00 & 1.50 & 1.50 \\ 9.0 & 6.00 & 9.00 & 2.25 & 2.25 \end{bmatrix} = \left[\alpha_{rs}\right]$$

(3.7.4.2.17)

$$\mathbb{N}_3 = \begin{bmatrix} \beta_{11,5} & \beta_{11,6} & \beta_{11,7} & \beta_{11,8} & \beta_{11,9} \\ \beta_{12,5} & \beta_{12,6} & \beta_{12,7} & \beta_{12,8} & \beta_{12,9} \\ \beta_{13,5} & \beta_{13,6} & \beta_{13,7} & \beta_{13,8} & \beta_{13,9} \\ \beta_{15,5} & \beta_{15,6} & \beta_{15,7} & \beta_{15,8} & \beta_{15,9} \end{bmatrix} = \begin{bmatrix} 0.60 & 0.90 & 1.00 & 0.90 & 1.00 \\ 0.60 & 0.90 & 1.00 & 0.90 & 1.00 \\ 0.70 & 0.90 & 1.00 & 0.90 & 1.00 \\ 0.60 & 0.90 & 0.70 & 0.90 & 0.72 \end{bmatrix} = [\beta_{rs}]$$

(3.7.4.2.18)

$$\mathbb{D}_4 = \begin{bmatrix} \alpha_{11,5} & \alpha_{12} & \alpha_{13} & \alpha_{14} & \alpha_{15} & \alpha_{16} & \alpha_{17} \\ \alpha_{21} & \alpha_{22} & \alpha_{23} & \alpha_{24} & \alpha_{25} & \alpha_{26} & \alpha_{27} \\ \alpha_{14,1} & \alpha_{14,2} & \alpha_{14,3} & \alpha_{14,4} & \alpha_{14,5} & \alpha_{14,6} & \alpha_{14,7} \\ \alpha_{16,1} & \alpha_{16.6} & \alpha_{16.3} & \alpha_{16,4} & \alpha_{16,5} & \alpha_{16,6} & \alpha_{16,7} \end{bmatrix} = \left[\alpha_{hq}\right]$$

(3.7.4.2.19)

$$\mathbb{N}_4 = \begin{bmatrix} \beta_{11} & \beta_{12} & \beta_{13} & \beta_{14} & \beta_{15} & \beta_{16} & \beta_{17} \\ \beta_{21} & \beta_{22} & \beta_{23} & \beta_{24} & \beta_{25} & \beta_{26} & \beta_{27} \\ \beta_{14,1} & \beta_{14,2} & \beta_{14,3} & \beta_{14,4} & \beta_{14,5} & \beta_{14,6} & \beta_{14,7} \\ \beta_{16,1} & \beta_{16,2} & \beta_{16,3} & \beta_{16,4} & \beta_{16,5} & \beta_{16,6} & \beta_{16,7} \end{bmatrix} = \left[\beta_{hq}\right]$$

(3.7.4.2.20)

By combining the distributional weights for the burden of real cost and the degree of expertness we can compute the aggregate monetary cost, C, for mental illness. The computation is through the use of the values in the matrices. From Eqs. (3.7.4.1.3), (3.7.4.1.5) and (3.7.4.1.7) define matrices, $\hat{\mathbb{Q}}_1$, $\hat{\mathbb{Q}}_2$ and $\hat{\mathbb{Q}}_3$ with the same column elements. Thus

$$\hat{\mathbb{Q}}_1 = \left[P_{ij}\right], \quad \hat{\mathbb{Q}}_2 = \left[P_{ij}\right] \text{ and } \hat{\mathbb{Q}}_3 = \left[P_{ij}\right]$$

where for any fixed row the numerical values are the same. Thus,

$$C = \sum_i \sum_j \beta_{ij} \alpha_{ij} c_{ij} P_{ij} + \sum_l \sum_k \beta_{lk} \alpha_{lk} c_{lk} P_{lk} + \sum_r \sum_s \beta_{rs} \alpha_{rs} c_{rs} P_{rs} \quad (3.6.4.29)$$

$$= 517291.60 + 167352.50 + 214384.04 \quad = \$899028.14$$

The total cost of mental illness without taking account of social assessment of the burden of cost may easily be computed as

$$\hat{C} = 462648 + 79831 + 217171 = \$759650.$$

The difference is easily assessed to be

$$\Delta C = C - \hat{C} = \$139378.14.$$

The one period monetary cost of mental illness of our hypothetical community by taking into account the burden of cost and the degree of expertness of the evaluators is \$899028.14. Without accounting for the burden of cost, the monetary cost is \$759650 and the difference due to the burden of cost is \$139378.14. Additionally, there are other real costs represented by (3.7.4.1.8) that must be considered when the monetary cost is to be used in a decision making. Such costs must be computed for each period covering the time frame for the decision. The cost values of each period over the time frame must be discounted to obtain the present value of cost if such costs are to be used in current decision. The true cost of cost characteristics expressed in Eq. (3.7.4.1.9) may be ascertained by taking into account the corresponding weights as expressed in matrices $\mathbb{D}_4$ and $\mathbb{N}_4$ when the decision is contemplated.

3.8 Conclusion

A method of comprehensive social costing is required if decision errors due to limited information space are to be minimized in applications of cost-benefit analysis. This has been the objective of this chapter. To achieve this objective, an emphasis has been placed on accounting for all relevant variables even if their quantitative and pecuniary valuations are practically impossible. The comprehensive approach is through the use of computable *cost-identification matrix.* The cost-identification matrix assists us to answer the questions of (a) what are the cost characteristics whose measures are sought?, (b) How are these cost characteristics to be measured and in what units? (c) What are the fair distributional weights that may be attached to the burden of real cost? (d) What are the interpersonal and intertemporal cost shiftings?, and (e) Whose cost would be included?

The framework also allows us to group the cost characteristics into (a) those that have monetary measures and hence summable, and (b) those that have real quantitative measures but in different units without monetary measures and hence not summable and (c) those that have no quantitative and monetary measures. The framework, thus allows an explicit identification of latent costs, non-monetary costs and cost shifting over cohorts. It brings into focus the differential distribution of socially *ethical weights* that may be assigned to cohorts due to the social evaluation of the cost burden. Furthermore, it makes explicit those costs that may be omitted in the decision-making process because of measurement difficulties. Such a deliberately explicit omission would indicate the direction of possible bias in the social costing and its possible effects on the decision and choice alternatives.

Finally, the static cost identification matrix is extended to generate a time series of cost information that includes intertemporal distributional weights. Intertemporal aggregation of cost requires constructing present value equivalencies which in turn demands a discounting process by utilizing intertemporal distributional weights. All these involve soft computing with judicious uses of theories of computable cost-benefit identification matrices, fuzzy decisions and economic logic. The steps of costing are provided in section 3.9.

3.9 The Steps in Social Costing

This approach is used to bring together a set of rules that will aid one in a practical social costing based on the comprehensive framework that we have advanced. The set of rules provides logical steps constituting the set of conceptual algorithms for obtaining good cost information on explicit and implicit cost characteristics. The required steps may be written as:

1. Ascertain the structure of the population and divide the population into social classes based on some chosen criterion (such as territorial or climatological location, income, age, etc.)
2. Ascertain the real cost characteristics by either direct or indirect method.
3. Construct a cost identification matrix and assign to each cohort a real cost $C_{ij} \geq 1$ if the cohort is affected by the specific cost characteristic and $C_{ij} = 0$ if the cohort is not affected.
4. Determine the burden of each cost characteristic relative to a given social class by assigning *ethical weights* say $\alpha_{ij} > 0$ that reflects the assessment of the degree of suffering due to the project's costs. Such ethical weights may be obtained through the use of information from expert judgments, a well-designed survey of the society and the decision makers.

5. Process the information of the expert judgments, sample surveys of the society, class and the decision makers through the use of fuzzy mathematics and decision or other methods of combination of information from differential sources to obtain equilibrium distributional weights for the burden of the real cost characteristics in each cohort.
6. For each cost characteristic check whether market price exists for the real cost and make proper corrections for the price. If market price does not exist construct a shadow price with proper corrective adjustment if possible. Check to see whether there are distortions in the market prices and make corrections. The efficient prices may also be constructed by the method of fuzzy logic to give us approximate prices (see a companion volume on fuzzy value theory [931b].
7. Partition the cost-identification matrix into three, (a) cost characteristics that are price-measurable, (b) cost characteristics that are quantitatively but non-pecuniarily measurable and (c) cost-characteristics that are non-qualifiable in any form.
8. Multiply each vector of price-measurable real cost characteristics by the corresponding price vectors that have been adjusted for distortions and sum them up for any given time point to obtain total of social costs.
9. Construct present value equivalence of future cost values through a discounting [206,220,228] process such as fuzzy discounting (see chapter 5 of [931b] as a companion volume to this volume in addition to [1153b]) or traditional method of discounting.

In identifying the cost characteristics, it is useful to find the net inputs of either the project or the decision alternative. Care must be taken to identify those inputs that go to reduce the total supply of the economy's inputs and those that go to absorb resources in order to keep the input supplies constant through the expanding production in the economy. In the case of those inputs that reduce total supply, estimate the "willingness to pay" through proper corrections. Such corrections are:

(a) corrections that are related to the value of rationed input at some stage or another;
(b) corrections that are related to the presence of monopoly power in either buying or selling at the current or future market;

(c) corrections that are related to the size of the reduction in the inputs and their impact on market prices if market prices are used in the aggregation process. The presence of real impact on the market price will call for some estimation of the demand conditions in order to correct for the "willingness to pay" as representing an efficient price. The process must be carried to later stages in production involving the products made with those inputs. These corrections can be avoided by constructing approximate prices through the use of fuzzy preferences in the price space (see [929]).

If those inputs that go to absorb producer goods in the project are to be compensated by expanding the supply of these resources from other sources such as increased domestic production then the actual costs involved in the expansion must be computed in either monetary or real values. If some of the absorbing resources are either imported or secured at the expense of potential or actual export then the loss of foreign exchange must be calculated and corrected by the shadow prices of foreign exchange. Corrections for both labor and land needed must be in terms of reduction of supply since they cannot be met by production expansion. Furthermore, it may be pointed out that the economist's concept of "willingness to pay" may be inapplicable in some cost calculations even when the cost characteristics are price-measurable. The structure and the nature of appropriate prices must be analyzed before cost aggregates are computed in monetary terms. The alternative approach is to develop a set of *fuzzy prices* through the methods of information elicitations and fuzzy mathematical programming or any approach for combining evidence as it is discussed and presented in fuzzy value theory [931b].

4 The Theory of Social Benefits and Benefit Accounting

In Chapter 3 we pointed out that there are two information sides to any decision alternative. They are cost and benefit as were suggested. The analytical structure of decision as viewed in framework of cost-benefit analysis is that every decision alternative is cost-benefit supported; every cost has its benefit support and every benefit has its cost support. The cost and benefit constitute a duality with a conceptual separation and unity in practice. The problem of cost has been considered in Chapter 3. As such the basic objective of this chapter is to consider the problem of benefits that may be associated with each feasible alternative when a decision to choose is to be made given the associated costs and objectives.

4.1 A Rationale for the Basic Framework

Some reasons were given for the analytical approach of cost-benefit framework being perused in this volume. Other reasons may be added to epistemologically tighten the validity of the approach. The first extra reason of the approach is found in the increasing importance and policy acceptance of broad use of cost-benefit analysis in social regulatory systems [101, 102, 120, 548, and 562], such as environment, product safety, technology, law, social infrastructure and many others. In all these cases, the involved input-output values may have no markets whose outcomes can be of help to access the aggregate values of important elements including risk reductions in natural, social and technological disasters. Here, risk reduction must be viewed as a benefit no matter how and in what units risk is measured [90, 111].

The second reason is derived from the relative effectiveness of direct and indirect approaches to ascertain both benefit and cost information required to develop the cost-benefit data. The direct approach is through an information elicitation and has come to be known in benefit-cost analysis as the *contingent valuation method* (CVM) where the information on *willingness-to-pay* (WTP) and *willingness-to-accept* (WTA) is elicited.

The indirect approach is where the benefit and cost information is inferred from observed market behavior. It has come to be known as the revealed preference approach (RPA). For logical arguments in support of CVM see [548, 550, 562]; for critique and critical reflection see [545, 566] for responses to the critique see [554, 558, 579] and for other comparative analysis of the two approaches see [554, 562, 571, 579].

The third reason in support of the general approach taken in this volume is abstracted from the problem induced by interpersonal and intertemporal aggregation of differential benefit and cost information that may require the development of logically valid method for combining evidence from differential expert and non-expert valuations, judgments and multi-attribute information where some weighting may be required. Such weights, as have been indicated in the previous chapter, reflect ethical judgments, relative social importance and the degree of social confidence attached to the information source and credibility of the transmission process.

The fourth reason comes from the general observation that there are numerous social decisions where the indirect method (market revealed preferences) fails in the sense that the market mechanism is incapable of generating benefit and cost data that can support a reasonable analysis of such social decisions in a democratic setup where preferences of the residents must count [544a, 546, 547, 594].

The fifth reason comes from the basic principle that links different aspects of our research agenda in broadening the application's framework of cost-benefit analysis into non-traditional areas of decision making. The principle is simply that the proper measure of individual preferences, in a democratic social organization and decision system, is the choice behavior among competing alternatives, and interest groups, and that in a sizable subspace of the general decision space, pertinent to participatory social choice, the individual preferences are best obtained by expressed preferences through non-market instrument of information elicitation rather than through abstraction by revealed preference as presented by market prices. This is particularly important in cost and benefit assessments of projects and decision alternatives that are required for productions of non-marketables as well as those for the production of tradables where the market organization is plagued by market failures due to disinformation on the part of the private sector, government-private sector coalitions, misleading advertisements and erroneous policy propaganda (such as debates on government down-sizing, regulatory overkill, privatization, the size of government, the government's role in social production, rent-seeking activities and others ([see Chapter 6 of volume on fuzzy value theory[931b] as a companion to this volume]) that destroy and reshape preferences.

Other reasons are given in [927, 929] while discussions on the relative strengths on different approaches to benefit and cost computing are given in [544a, 590, 592, 596, 603].

The theoretical framework for benefit accounting presented in this chapter may be used to complement the direct and indirect methods irrespective of their weaknesses and strengths. The structure of the computable cost-benefit identification matrices has particularly an added advantage in providing a process of a comprehensive accountability of benefits (costs) for assessing the relative efficiency of different regimes in social regulatory systems where benefits are seen in terms of *risk reduction,* time saving, health improvement, preventive activities or damage reduction in a situation of possible human and environmental disasters with irreversible damages. The above advantage of comprehensive accounting framework is amplified by the fact that many different opinions, assessments and judgments of experts, policy makers, politicians, social interest groups and other citizens can be assembled for aggregation leading to a collective decision and social choice on the principles of democratic institution. It may also be added that the information assessment about prices either by direct (CVM) or indirect (RPA) method is not possible until the real benefit and cost characteristics are identified. Here the real cost characteristics may be viewed as inputs while real benefit characteristics may be viewed as output in a special way.

4.2 A Reflection on the Concept of Benefit in a Choice-Decision Process

The concept of benefit is either ex ante or ex post relative to a social event. We mostly concern ourselves with the ex post concept of benefit when explanatory and predictive processes are sought, or when structural adjustment decisions of an on-going project is being considered. In this respect, the required analytical construct is not only driven by historical net-benefit data but it takes claim to objectivity in the sense that the benefit data is independent of subjective evaluation of the analyst. Alternatively, we concern ourselves with the ex ante concept of benefit when a prescriptive theory is to be constructed to guide a choice process.

In this case, the required benefit data for processing is future oriented and the concept of benefit is subjective in the sense that it exists in the mind of either the decision maker or the analyst or the beneficiary of the decision. The true values, from the viewpoint of valuation, exist in the minds of the persons who would be affected by the decision given the relevant information support of the assessments. The benefit values are those that either the decision maker or analyst or beneficiary would consider as the fruits (satisfaction) that may be realized from either a commitment of resources or undertaking a cost. Every (aggregate) benefit has its cost support and every cost has its benefit support in a choice-decision process. The values of these ex ante benefits and costs in decision analysis are evaluated and valued in the utility space before the decision alternative is selected.

In valuing the benefits in choice-decision space two types of difficulties arise from the valuation differences between the decision maker and the beneficiary of the results of the decision. There are two cases that demand analytical attention. The first case is a situation where the decision maker and the beneficiary are one and the same agent. This case has two sides where (a) the decision maker is the sole beneficiary and (b) the decision maker is a partial beneficiary. The second case involves a situation where the decision maker and the beneficiary are two different agents of the society. As viewed relative to cost, this situation is divisible into (a) the decision maker is not a beneficiary but bears all or some of the costs, for example, public policy maker, and (b) the decision maker is neither a beneficiary nor a bearer of the cost (e.g. a consultant.)

Both exp post and ex ante concepts of benefits like the cost may be viewed from *accounting* or *economic* perspective. The two perspectives are epistemically interdependent. While in general the two perspectives and their corresponding measurement are different, they may be identical in some cases or in a certain class of problems of social choice. Let us examine the two perspectives.

4.2.1 Accounting Concept and Measurement of Benefits

The accounting concept and measurement of benefit characteristics refer to special benefits that accrue to a society or a section of it given a socio-economic activity such as a project of a change of state. Such benefits are simply and usually but not necessarily monetary value of output from the activity. In other words, it is the money value that is received from carrying on a social activity or must be received if such a socially productive activity is to be carried on. In light of this, the concept and benefits of social choice may be divided into *historical benefits* and *predicted benefits*.

The historical benefits are the proceeds from the previous periods due to a selection of a similar or same activity or project. They summarize the historical information about benefits on the benefit-cost balance sheet due to outcomes of a socioeconomic process. The benefits on the benefit-cost balance sheet for social accounting may include real values in addition to recorded monetary values.

Predicted benefits, on the other hand, are measures of proceeds of expected future activities when they are carried on. Such predicted benefits may be based either on historical benefit data or on a class of information bits that allows expectations to be formed regarding the proceeds of a future activity or potential change of a socio-economic state.

Such concept and measurements of benefits from a choice of socio-economic activity are mostly internal to a particular enterprise on the basis of which present and future decisions and choices are made. In all cases they do not include other benefits that may accrue to society. The systematic process of providing such benefit data for any given decision unit will be referred to as *private benefit accounting*. Such benefit accounting is based solely on *market prices* and hence fails to account for other important benefits whose values are not registered in the market as seen from the broader scope of interest of society. Because of this difficulty we shall consider the economic concept and measurement of benefits.

4.2.2 The Economic Concept and Measurement of Benefits

In an attempt to find broader and appropriate measures for social benefits that may be associated with a change of a socio-economic state due to either a social choice or project selection we shall look at the structure of social benefits. We may begin the analysis of the structure by classifying the general benefits into direct (explicit) and indirect (implicit) benefits. The same benefit characteristics may also be grouped into internal benefits and external benefits. The benefit characteristics may also be classified into private and public depending on their nature, ownership structure, choice and the benefit recipients.

Each of these classifications will place the problems and difficulties in the concept and measurement of benefit characteristics in their proper perspective not only within the cost-benefit analysis but also in the whole structure of decision theory. We shall now define some of these terms used for the classifications.

Definition 4.2.2.1

Internal benefits of a change of a social state due to social choice or decision are the real benefit characteristics that may directly or indirectly accrue to the entity on behalf of which the decision or the choice is made.

Example 4.2.2.1

An increase in output flow due to a choice of an additional investment project in an organization will be considered as direct internal benefit while the increased market power or increased administrative productivity due to an increased size of the organization that allows for specialization and division of administrative labor will be considered as indirect internal benefits.

Definition 4.2.2.2

External benefits are said to occur from a change of social state if some real benefit characteristics tend to directly or indirectly accrue involuntarily, without payment, to other individuals or groups of individuals besides the individuals or groups of individuals on behalf of which the decision is made.

Example 4.2.2.2

Suppose a social decision to build a dam for residence in a location A is made and the dam is built. Furthermore, suppose that such a dam serves to prevent flood for the people in location B at down stream as well as provide irrigational facility to people in location C at upstream. The prevention of the real physical destruction and the creation of the irrigational facility that result from the construction of the dam will both be considered as real external direct benefit to people in location B and C. If social and productive activities increase in location B and C due to the absence of flood and creation of irrigation facility then such benefits will be referred to as indirect benefit externality.

Definition 4.2.2.3

Direct benefits are those output (benefit) characteristics that either flow or will flow from either a project when it is completed or from a decision when it is made and implemented. Usually, their values depend on market prices when market exists. The project or the decision is the primary source of the benefits. Direct benefits are sometimes referred to as explicit benefits.

Example 4.2.2.3

Refer to example 4.2.2.2 and suppose that the dam was built for the purpose of generating electric power. The electric current generated through such a dam will be direct benefits.

Definition 4.2.2.4

Indirect benefits are extra output flows over and above those that may be directly associated with a project or a choice of a socio-economic state. Their values may or may not depend on market prices. Indirect benefits are sometimes referred to as *implicit benefits*.

Example 4.2.2.4

Refer to example 4.2.2.3 where the dam is built for the purpose of generating electric power. If building such a dam leads to flood control in the down stream, then the prevention of physical and other destructions that usually accompany flooding will be considered as indirect (implicit) benefits.

Definition 4.2.2.5

Private benefit characteristics are the real output flows that accrue to the individuals or group of individuals because of their own socio-economic activities (or choices). In real terms, they are the sum of direct (explicit) and indirect (implicit) internal and external benefit characteristics.

Example 4.2.2.5

Suppose an individual makes a decision to build a well-designed shopping mall in a suburban area and that such decision is implemented. The rental value of the mall will be considered as private benefit.

Definition 4.2.2.6

Public benefit characteristics of a project consist of all those benefit characteristics that the society receives but not captured by the private decision maker. It is the sum of direct and indirect external benefit characteristics reduced to some common unit.

Example 4.2.2.6

Refer to example 4.2.2.5. The reduction in time for shopping, cost of transportation and the improvement in the areas of the shopping mall may all be considered as public benefits.

Connected to these distinctions are the concepts of social and private benefits and the difficulties they present in benefit identification and measurement. When the social agent is both the decision maker and sole beneficiary of all aspects of direct and indirect benefits of choice, the agent's conceptual framework is well defined in all relevant spaces of decision, choice, and output. In this case, the concept of benefit is less *ambiguous* and its measurement less difficult because the agent has the same perception over the general utility space. For any framework of benefit accounting to be comprehensive, it must take these problems and difficulties into account by developing an analytical process not only for accessing benefits but for aggregation and combination of differential evidence or assessments in order to create benefit data associated with alternatives.

4.3 The Structure of Benefit Space and Decision Alternatives

In developing the *benefit data* we consider a set of programs (or projects), Π, with corresponding sets of benefit characteristics, $\mathbb{X}^{B}$, and cost characteristics, $\mathbb{X}^{C}$. Corresponding to the sets of benefit and cost characteristics are sets of benefit and cost measures, $\mathbb{B}$ and $\mathbb{C}$, whose generic values are projections from $\mathbb{X}^{B}$ and $\mathbb{X}^{C}$ respectively.

Definition 4.3.1

A set of decision alternatives, Π, with a generic element, π, is called a set of programs for a choice action. It is non-trivial if $\# \Pi > 1$, null if $\# \Pi = 0$ and trivial if $\# \Pi = 1$ given a set of objectives, Ω, that a decision agent may pursue.

Definition 4.3.2

The general set of elements from which each $\pi \in \Pi$ may be uniquely identified for choice is called a characteristic set, $\mathbb{X}$, such that $\pi \in \Pi$ is identifiable.

Definition 4.3.3

A general utility space, $\mathbb{U}$ is a set of measures, consistent with a given set of preferences, such that there exists a utility function, $u: \mathbb{X} \rightarrow \mathbb{U}$; i.e.,

$$\mathbb{U} = \left\{ u_x \mid u(x) = u_x \text{ and } x \in \mathbb{X} \right\}$$

where $\mathbb{U} \subset \mathbb{R}$, the set of real numbers.

Note: $u(.)$ is assumed to be regular in that it is monotonic, quasi-concave and continuously differentiable.

Definition 4.3.4

A set, $\mathbb{X}_{\mathbb{B}^{\pi}} \subset \mathbb{X}$ is called a set of real benefit characteristics associated with π, and with generic elements, $B_i \in \mathbb{X}_{\mathbb{B}^{\pi}}$, $i \in \mathbb{I}$, an index set for benefit characteristics if

$$\mathbb{X}_{\mathbb{B}^{\pi}} = \left\{ x \mid x \in \mathbb{X} \text{ and } u(x) > 0 \text{ , } \pi \in \Pi \right\}$$

Similarly, $\mathbb{X}_{\mathbb{C}^{\pi}} \subset \mathbb{X}$ is a set of real cost characteristics associated with, π and with generic elements, $c_j \subset \mathbb{C}, j \in \mathbb{J}$, an index set for the cost characteristics if

$$\mathbb{X}_{\mathbb{C}^{\pi}} = \left\{ x | x \in \mathbb{X} \text{ and } u(x) < 0, \text{for } \pi \in \Pi \right\}$$

where $\mathbb{X}_{\mathbb{B}^{\pi}} \cap \mathbb{X}_{\mathbb{C}^{\pi}} = \varnothing$ and $\mathbb{X}_{\mathbb{B}^{\pi}} \cup \mathbb{X}_{\mathbb{C}^{\pi}} = \mathbb{X}$.

Definition 4.3.5

If $\mathbb{X}_{\pi}$ is a set of defining characteristics for $\pi \in \Pi$ then $\mathbb{B}$ and $\mathbb{C}$ are called sets of benefit and cost measures relative to $\pi \in \Pi$ if there is an aggregation function, A , such that

$$\mathbb{B} = \left\{ b \mid b = A\left(\mathbb{X}_{\mathbb{B}^{\pi}}\right), \mathbb{X}_{\mathbb{B}^{\pi}} \subset \mathbb{X} \text{ and } b > 0 \right\}$$

and

$$\mathbb{C} = \left\{ c \mid c = A\left(\mathbb{X}_{\mathbb{C}^{\pi}}\right), \ \mathbb{X}_{\mathbb{C}^{\pi}} \subset \mathbb{X} \text{ , and } c > 0 \right\}$$

For each $\mathbb{X}_{\mathbb{B}^{\pi}} \subset \mathbb{X}$, $A\left(\mathbb{X}_{\mathbb{B}^{\pi}}\right) = b$ is called the benefit support of decision on $\pi \in \Pi$. Alternatively, if $\mathbb{X}_{\mathbb{C}^{\pi}} \subset \mathbb{X}$ then $A\left(\mathbb{X}_{\mathbb{C}^{\pi}}\right) = c$ is called the *cost support* of decision on $\pi \in \Pi$. The condition $u(x) > 0$ defines positive utility and hence a benefit while $u(x) < 0$ specifies negative utility and hence a cost where the axiomatic foundation of the preferences on the basis of which the utility function is constructed is of neoclassical type.

We consider a general decision space, Θ, that is composed of a set of programs, Π , a set of objectives, Ω, a set of benefit measures, $\mathbb{B}$, and a set of cost measures, $\mathbb{C}$, with generic elements, $\theta \in \Theta$, $\pi \in \Pi$, $\omega \in \Omega$, $b \in \mathbb{B}$ and $c \in \mathbb{C}$ respectively. Thus $\Theta = \Pi \otimes \Omega \otimes \mathbb{B} \otimes \mathbb{C}$ where $\otimes$ is used for Cartesian product.

Definition 4.3.6

A benefit space Φ, of a decision with a generic element, $\varphi \in \Phi$ is composed of a triplet of sets of benefit measures, $\mathbb{B}$, programs, Π, and objectives, Ω, that may be specified as:

$$\Phi = \mathbb{B} \otimes \Pi \otimes \Omega = \{\varphi = (b, \pi, \omega), b \in \mathbb{B}, \pi \in \Pi \text{ and } \omega \in \Omega\}$$

Definition 4.3.7

A benefit relation, R, is a rule that assigns to each program, $\pi \in \Pi$, a unique benefit measures, $b \in \mathbb{B}$, conditional on a set of objectives, Ω, and is specified as: $R|_{\Omega} = \{(b, \pi) \mid {}_b R_{\pi}, b \in \mathbb{B} \text{ and } \pi \in \Pi\}$.
It is a mapping from the set of projects into the set of benefits (i.e. $R|_{\Omega} : \Pi \to \mathbb{B}$)

It is assumed in Definitions 4.3.6 and 4.3.7 that the set, $\mathbb{B}$, exists and its generic elements are identifiable and measurable in some physical or abstract unit. For these definitions to be operational, we must find an analytical process where each element, $b \in \mathbb{B}$, can be identified and measured relative to $\pi \in \Pi$ and then be related to the decision space, Θ. Since each program, $\pi \in \Pi$, may produce different benefit characteristics (attributes) $\mathbb{X}_{\mathbb{B}^{\pi}} \subset \mathbb{X}$, the benefit value corresponding to any $\pi \in \Pi$ is an aggregate, $b \in \mathbb{B}$. The existence of these aggregate values requires that the benefit space should be equipped with a *unit measure* that allows heterogeneous real benefit attributes (characteristics) $B_i \in \mathbb{X}_{\mathbb{B}}$ to be summable over all $i \in \mathbb{I}$. In other words, there must be a summable function, $A: \mathbb{X}_{\mathbb{B}^{\pi}} \to \mathbb{B}$. Additionally, the benefit measurable space must satisfy some important properties of comparability of project alternatives given their cost supports and objective.

Definition 4.3.8

A decision space, Θ, is said to be cost and benefit separable if there exists a cost space, $\Gamma = \mathbb{C} \otimes \Pi \otimes \Omega$, and a benefit space, $\Phi = \mathbb{B} \otimes \Pi \otimes \Omega$, such that

$$\Phi, \Gamma \subset \Theta$$
$$\Gamma \cap \Phi = \Pi \otimes \Omega$$
$$\Gamma \cup \Phi = \Theta$$
$$\Gamma \neq \varnothing \text{ and } \Phi \neq \varnothing$$
$$\mathbb{C} \cap \mathbb{B} = \varnothing$$

For the purpose of identification, and benefit accounting the benefit set $\mathbb{B}$ and its index set, Λ may be partitioned into public, π, private, p, external, E and internal, I as have been defined in section (4.2.2) and then into explicit (direct), D and implicit (indirect), C. We consider partitions of $\mathbb{B}$ and Λ with generic elements b and λ respectively into $\mathbb{B}^{E}$, external benefits with an index set, Λ^{E}; $\mathbb{B}^{I}$ internal benefits with an index set, Λ^{I}; $\mathbb{B}^{D}$, direct benefits with an index set, Λ^{D}; $\mathbb{B}^{C}$ indirect benefits with an index set, Λ^{C}; $\mathbb{B}^{P}$, private benefits with an index set, Λ^{P}; $\mathbb{B}^{\pi}$ public benefits with an index set, Λ^{π}. Thus

$$\mathbb{B}^{E} = \left\{ b_{\lambda} \mid b \in \mathbb{B},\ \lambda \in \Lambda \right\} \qquad \text{(External)}$$

(4.3.1)

$$\mathbb{B}^{I} = \left\{ b_{\lambda} \mid b \in \mathbb{B}, \lambda \in \Lambda^{I} \right\} \qquad \text{(Internal)}$$

(4.3.2)

$$\mathbb{B}^{D} = \left\{ b_{\lambda} \mid b \in \mathbb{B},\ \lambda \in \Lambda^{D} \right\} \qquad \text{(Explicit or Direct)}$$

(4.3.3)

$$\mathbb{B}^{C} = \left\{ b_{\lambda} \mid b \in \mathbb{B},\ \lambda \in \Lambda^{C} \right\} \qquad \text{(Implicit or Indirect)}$$

(4.3.4)

$$\mathbb{B}^{P} = \left\{ b_{\lambda} \mid b \in \mathbb{B},\ \lambda \in \Lambda^{P} \right\} \qquad \text{(Private)}$$

(4.3.5)

$$\mathbb{B}^{\pi} = \left\{ b_{\lambda} \mid b \in \mathbb{B},\ \lambda \in \Lambda^{\pi} \right\} \qquad \text{(Public)}$$

(4.3.6)

where these sets are such that

$$\Lambda = \Lambda^{E} \cup \Lambda^{I} = \Lambda^{D} \cup \Lambda^{C} = \Lambda^{P} \cup \Lambda^{\pi}$$
$$\Lambda^{E} \cap \Lambda^{I} = \Lambda^{D} \cap \Lambda^{C} = \Lambda^{P} \cap \Lambda^{\pi} = \varnothing$$
$$\Lambda^{P} = \Lambda^{I} \cap \Lambda^{D} \cup \Lambda^{I} \cap \Lambda^{C}$$
$$\Lambda^{\pi} = \Lambda^{E} \cap \Lambda^{D} \cup \Lambda^{E} \cap \Lambda^{C}$$
$$\Lambda = \Lambda^{E} \cap \Lambda^{D} \cup \Lambda^{E} \cap \Lambda^{C} \cup \Lambda^{I} \cap \Lambda^{D} \cup \Lambda^{I} \cap \Lambda^{C}$$
$$\mathbb{B}^{P} = \mathbb{B}^{I} \cap \mathbb{B}^{D} \cup \mathbb{B}^{I} \cap \mathbb{B}^{C}$$
$$\mathbb{B}^{\pi} = \mathbb{B}^{E} \cap \mathbb{B}^{D} \cup \mathbb{B}^{E} \cap \mathbb{B}^{C}$$

These set relations in terms of union and intersection provide us with an organic view of the benefit space. This view brings into focus the problem of aggregation of heterogeneous benefit real values.

Let us suppose that a measure can be defined over the set of real benefit characteristics such that their values are summable. Let hats denote their corresponding values. The following aggregate values may be written

$$\hat{\mathbb{B}}^{\mathrm{P}} = \mu\left(\mathbb{B}^{\mathrm{I}} \cap \mathbb{B}^{\mathrm{D}}\right) + \mu\left(\mathbb{B}^{\mathrm{I}} \cap \mathbb{B}^{\mathrm{C}}\right)$$

(4.3.7)

where μ is a measure on the set of benefit characteristics. Similarly,

$$\hat{\mathbb{B}}^{\pi} = \mu\left(\mathbb{B}^{\mathrm{E}} \cap \mathbb{B}^{\mathrm{D}}\right) + \mu\left(\mathbb{B}^{\mathrm{E}} \cap \mathbb{B}^{\mathrm{C}}\right)$$

(4.3.8)

Definition 4.3.9

The set of social benefit characteristics, $\mathbb{B}^{\mathrm{s}}$ with an index set, Λ^{s}, is

$$\mathbb{B}^{\mathrm{s}} = \left\{ b_{\lambda} \mid b \in \mathbb{B}^{\mathrm{P}} \cup \mathbb{B}^{\pi} \text{ and } \lambda \in \Lambda^{\pi} \cup \Lambda^{\mathrm{P}} \right\}$$

The total social benefit, $\mathbb{B}^{\mathrm{s}}$ given a measure on the set of benefit characteristics may be written as the sum of private and public implicit and explicit benefits. Thus

$$\hat{\mathbb{B}}^{\mathrm{s}} = \hat{\mathbb{B}}^{\mathrm{P}} + \hat{\mathbb{B}}^{\pi}$$

(4.3.9)

In private socio-economic decisions, the implicit and explicit public benefits are not included in the decision-making process in the sense that $\hat{\mathbb{B}}^{\pi} = 0$ for $\mu \neq 0$ by construct. Production-consumption decisions under competitive market conditions where externalities and uncertainties are absent will rationalize this situation. Under such ideal perfectly competitive market conditions the total social benefits tends to coincide with the total private benefits. Thus $\hat{\mathbb{B}} = \hat{\mathbb{B}}^{\mathrm{P}}$.

The total social benefits in value terms under such specification conditions, becomes the market value of the implicit and explicit output flows due to a choice of a state that in turn may be due to some rational process for allocating resources into other uses other than the observed. The value of the total social benefits is then calculated from market prices for all units of real output flow. The social benefits are valued, therefore, by the *principle of willingness to pay* in order to acquire the right to enjoy the benefits as we have outlined in Chapter 1 section (1.5). From the benefit data obtained from eqns. (4.3.7 - 4.3.10), and definitions (4.3.1 - 4.3.8), we can examine the comparability principle in the benefit space.

Definition 4.3.10

A benefit space, Φ , is said to be project comparable if it is equipped with a b-measurable difference function $\xi(b_i\ b_j) = b_j - b_i$ or a ratio function, $\zeta(b_i\ b_j) = b_j / b_i$ such that at any point of time one of the following conditions holds:

$$\xi(b_i\ b_j)\ \begin{Bmatrix} >0 \\ =0 \\ <0 \end{Bmatrix} \Rightarrow \zeta(b_i\ b_j)\ \begin{cases} >1 & b_j > b_i \\ =1 \quad \text{if} & b_j = b_i \\ <1 & b_j < b_i \end{cases}$$

Definition 4.3.11

A benefit space Φ is computable if there exist, $b_i, b_j \in \mathbb{B}$ such that $\xi(\cdot)$ and $\zeta(\cdot)$ are definable and measurable in some unit of measure.

Proposition 4.3.1

If a benefit set, $\mathbb{B}$, *is measurable then the benefit space,* Φ, *is both computable and project comparable given the cost support and objective.*

We shall require the decision space, Θ, to have certain desirable properties in terms of project ordering if the space is benefit measurable and computable. These desirable properties are stated in terms of postulates.

Postulate 4.3.1 (*Project comparability in the benefit space*).

Projects with the same cost support and objectives in a measurable benefit space, Φ, are benefit comparable if the b-measurable ratio function $\zeta(\cdot) = b_j / b_i$, is such that whenever $\pi_i, \pi_j \in \Pi$ and $b_i, b_j \in \mathbb{B}$ then for all values of $0 < \xi(\cdot) < \infty$

$$\zeta(b_i\ , b_j)\ \begin{Bmatrix} >1 \\ =1 \\ <1 \end{Bmatrix} \Rightarrow \begin{cases} \pi_j \succ \pi_i \\ \pi_j \sim \pi_i \\ \pi_j \prec \pi_i \end{cases}$$

and only one relation must hold where and $\sim$ refer to superior and indifference respectively.

Postulate 4.3.2 (*Asymmetry of project comparability in the benefit space*). If $\pi_i, \pi_j \in \Pi$ and $b_i, b_j \in \mathbb{B}$ for a given cost support then

$$\zeta\left(b_i\ , b_j\right) > 1 \Rightarrow \pi_j \succ \pi_i \text{ and } \zeta\left(\mathrm{b}_j \mathrm{b}_i\right) > 1 \Rightarrow \pi_i \not\succ \pi_j$$

Postulate 4.3.3 (*Transitivity of project comparability in the benefit space*). If $\pi_i, \pi_j, \pi_l \in \Pi$ and $b_i, b_j, b_l \in \mathbb{B}$ for a given cost support, and

$$\zeta\left(b_i, b_j\right)\begin{Bmatrix} >1 \\ =1 \\ <1 \end{Bmatrix} \Rightarrow \begin{Bmatrix} \pi_j \succ \pi_i \\ \pi_j \sim \pi_i \\ \pi_j \prec \pi_i \end{Bmatrix} \text{,and} \quad \zeta\left(b_j, b_l\right) \begin{Bmatrix} >1 \\ =1 \\ >1 \end{Bmatrix} \Rightarrow \begin{Bmatrix} \pi_l \succ \pi_j \\ \pi_l \sim \pi_j \\ \pi_l \prec \pi_j \end{Bmatrix}$$

$$\text{then } \zeta\left(b_i, b_l\right)\begin{Bmatrix} >1 \\ =1 \\ <1 \end{Bmatrix} \Rightarrow \begin{Bmatrix} \pi_l \succ \pi_i \\ \pi_l \sim \pi_i \\ \pi_l \prec \pi_i \end{Bmatrix} \text{, for } j \neq l \neq i$$

Postulate 4.3.4 (*Reflexivity of project comparability in the benefit space*). If $\pi_i \in \Pi$ and $b_i \in \mathbb{B}$ for a given cost support then

$$\zeta\left(b_i\ b_i\right) = 1 \text{ and } \pi_i \succsim \pi_i$$

The postulates of project comparability simply assert that if the benefit space is measurable then all projects with the same cost support can be completely ranked according to their benefit values. Thus for any given cost the choice space is spun by the benefit basis around which a selection is made. The analytical procedure for the selection of preferred alternatives is said to be of *benefit-effectiveness analysis* if the choice space is completely spun by the benefit basis. It is said to be of *cost-effectiveness* (cost-benefit) analysis if the choice space is completely spun by the cost (relative cost-benefit) basis.

4.4 Benefit Identification Matrix for Benefit Accounting

We consider a project space, Π, where each project, π, provides multiple benefit characteristics that must be identified. A characteristic is considered to be a benefit if it produces a positive satisfaction or utility to the beneficiary. Let the total number of such distinct benefit characteristics associated with $\pi \in \Pi$ be $\mathrm{T} = \#\mathbb{X}_{\mathbb{B}^{\pi}}$ and each real benefit characteristics identified as $h_i \in \mathbb{X}_{\mathbb{B}^{\pi}}$. To produce a comprehensive benefit data, we partition the social space into *social classes,* G_j, where *M* is the total number of social classes. The criterion for partitioning the social space will depend on the problem for study and such a criterion must be explicitly specified. The combined interactions between the sets of benefit characteristics (attributes) and social classes induce a crisp partition on the space of benefit imputations where each cohort is identified with a real value b_{ij} of *i*-characteristic and *j*-social class. The resulting real benefit accounting is represented as a matrix of benefits in Table 4.4.1 where $h_i \in \mathbb{X}_{\mathbb{B}^{\pi}}, \forall i \in \mathbb{I}$.

From the structure of the benefit-identification matrix, if $B^{G}_{\bullet j} = 0$ for any *j*, then $b_{ij} = 0$ for all *i* and a fixed *j*. In this case, a particular social class does not experience benefit from the project or the decision and hence must be deleted. On the other hand, if $B^{R}_{i\bullet} = 0$ for any fixed *i*, then the *i*th benefit characteristic is not realized to any of the classes and hence must be deleted. It is also possible that $b_{ij} = 0$ for some *i-j* cohorts. In this case the *j*th class does not share in the *i*th benefit characteristic. The b_{ij}'s are the identified and estimated real value of

Table 4.4.1 Benefit Identification Matrix of Decision

	Social classes							
	G_1	G_2	$\cdots$	G_j	$\cdots$	G_{M-1}	G_M	Total
h_1	b_{11}	b_{12}		b_{1j}		$b_{1,M-1}$	b_{1M}	$B^R_{1.}$
h_2	b_{21}	b_{22}		b_{2j}		$b_{2,M-1}$	b_{2M}	$B^R_{2.}$
$\vdots$				$\vdots$				$\vdots$
h_i	b_{i1}	b_{i2}	$\ddots$	b_{ij}		$b_{i,M-1}$	b_{iM}	$B^R_{i.}$
$\vdots$	$\vdots$			$\vdots$			$\vdots$	$\vdots$
h_{T-1}	$b_{T-1,1}$	$b_{T-1,2}$		$b_{T-1,j}$		$b_{T-1,M-1}$	$b_{T-1,M}$	$B^R_{T-1.}$
h_T	$b_{T,1}$	$b_{T,2}$	$\cdots$	b_{Tj}		$b_{T,M-1}$	b_{TM}	$B^R_{T.}$
Total	$B^G_{.1}$	$G^G_{.2}$	$\cdots$	$B^G_{.j}$		$B^G_{.M-1}$	$B^G_{.M}$	$B_{..}$

The numerical conditions on the values in the table are such that

$$b_{ij}\begin{cases} = 0, & \text{if the benefit does not apply to the cohort} \\ = 1, & \text{if the real benefit is physically incommensurable} \\ \geq 1, & \text{if the real benefit is physically measurable} \end{cases}$$

$$B^G_{\bullet j} = \left\{B_{1j},\, B_{2j},\, \cdots,\, B_{Tj}\right\},\quad j = 1,2,\cdots,\mathrm{M}$$

$$B^R_{i\bullet} = \sum_{j=1}^{\mathrm{M}} b_{ij},\quad i = 1,2,\ \cdots,\ \mathrm{T}.$$

benefit characteristics for all the i - j cohorts. The process of identification and estimation of the real benefits may be designed with the help of different experts, and by the method of information elicitation from the members of the society and the policy makers. Similar but non-fuzzy approaches are pursued in [579, 582, 588.]. For applied and theoretical difficulties of information elicitation, see [67, 332,937]. The logical foundation of this approach to information construct is provided in Chapter 2.

4.5 Tangibles, Intangibles, Corresponding Prices and Benefit Aggregates

The real benefit characteristics in the benefit-identification matrix fall into two groups of tangibles and intangibles. The tangibles and intangibles may also be subdivided into those that are pecuniarily and non-pecuniarily measurable. In the case of intangibles, directly recognizable objective physical measure may not exist (e.g. reduction in pain and suffering). In most cases, pecuniary measures may not exist except perhaps through a construction of surrogate monetary measures. These differences induce a partition on the space of benefit characteristics with an index set, $\mathbb{I}$, into three mutually exclusive subsets of price measurable with an index set, $\mathbb{K}$; quantitatively but non-pecuniarily measurable whose index set is $\mathbb{Q}$ and "pecuniarily" and quantitatively non-measurable with an index set, $\mathbb{R}$. Structurally, $\mathbb{I} = \mathbb{Q} \cup \mathbb{R} \cup \mathbb{K}$, and $\mathbb{Q} \cap \mathbb{R} = \varnothing = \mathbb{R} \cap \mathbb{K} = \mathbb{Q} \cap \mathbb{K}$. This partition is superimposed on the social classes whose index set is, $\mathbb{N}$.

The comprehensive social accounting for benefits of decision from Table 1 provides the decision maker with rich information on the basis of which analysis and selection may be undertaken. Three data sets are available for processing. They are:

(a) Total non-weighted value of monetary benefit, B_{ST}^{M} where

$$B_{ST}^{M} = \sum_{I \in \mathbb{K}} \sum_{j \in \mathbb{N}} p_{ij} b_{ij} \ , \quad b_{ij} \in \mathbb{B}$$

(4.5.1)

where the p_{ij}'s are the corresponding prices that are to be constructed or ascertained either from the market imputations [22, 25, 51] or from a non-market rational construct [6,20].

A composite quantitatively and non-pecuniarily measurable vector sum of benefit characteristics, B_{ST}^{Q} in the form

$$B_{ST}^{Q} = \left\{ b_{i1}, b_{i2}, \ \cdots \ b_{ij} \cdots \right\} \ , \qquad i \in \mathbb{Q} \ , b_{ij} \in \mathbb{B}$$

(4.5.2)

where $\mathbb{K} \cap \mathbb{Q} - \varnothing$ and $\mathbb{Q} \subset \mathbb{I}$.

Non-measurable benefit data, B_{ST}^{R}, whose structure is of the form,

$$B_{ST}^{R} = \left\{ b_{i1}, \, b_{i2}, \ \cdots \ b_{ij}, \ \cdots \right\} \ , \quad i \in \mathbb{R}, \, b_{ij} \in \mathbb{B}$$

(4.5.2)

We shall illustrate the nature of the data set, B_{ST}^{Q} and B_{ST}^{R}. The following list belongs to B_{ST}^{Q} for $1 \in \mathbb{Q}$.

b_{11} = number of fatalities,

b_{12} = number of serious injuries,

b_{13} = reduction in air pollution as measured in terms of pollutant particles per million,

b_{14} = reduction in noise pollution as measured in decibels.

In the case of data set, B_{ST}^{R}, for $i \in \mathbb{R}$ we may list

b_{11} = reduction in pain

b_{12} = reduction in grief,

b_{13} = increase in inventiveness,

b_{14} = increase in happiness.

The following two definitions are necessary for the computation of eqn., (4.5.1).

Definition 4.5.1

A benefit subspace, $\mathbb{B}$, is said to be pecuniarily measurable if there exists a price, p_i, $i \in \mathbb{I}$, such that for any social class the total value of benefits, $B_{\bullet j}^{G}$, may be written as $B_{\bullet j}^{G} = \sum_{i \in \mathbb{I}} p_i b_{ij}$ for non-differentiated prices and $B_{\bullet j}^{G} = \sum p_{ij} b_{ij}$ for differentiated price and for a fixed $j \in \mathbb{N}$.

Lemma 4.5.1

If the benefit subspace is pecuniarily measurable,then it is also said to be price measurable and the total benefit, B_T^R, *may be computed as*

$$B_T^R = \sum_{i \in \mathbb{I}} \sum_{j \in \mathbb{N}} p_{ij} b_{ij}$$

If the benefit space is price measurable then the next problem is to find the vector of prices that would allow the benefit characteristics to be price measurable and hence summable. The task of obtaining the vector of best prices belongs to the theory of efficient prices. We now consider the problem of *benefit intensity* of different groups G_j and benefit characteristics, h_i.

4.6 The Concept of Benefit Intensity and Distributional Weights

The three types of benefit data sets provide raw data without socially relative judgment of benefit intensities of each cohort. In social decisions, interpersonal and intergenerational weighting of benefits and costs is the norm. The acceptance of the raw data for decision implies an implicit acceptance of equal weights where such weights are assigned values whose normalized sum equals one. The implicit assumption is that past decisions and the corresponding institutions are correct (in a particular sense) for current and future decisions. Every unit of either cost or benefit is always weighted in every social decision. Such weights may or may not be of numerical nature. They may be subjectively expressed as linguistic variables (such as very important, less important, equally important, and others ([see Chapters 5 and 6 of this volume for mathematical discussions on these linguistic variables]) or in any other form.

The selection of such weights that reflect social judgment of importance (intensity) in a democratic choice process must be rational for reasons of social efficiency and/or equity. For reasons of equity, the selection should not be left solely in the hands of politicians because of lack of technical skills for combining differential evidence and judgment from different social classes and experts. Additionally, rent-seeking environment and activities may distort the perceptions and social vision of the politicians that constitute the decision making core (see Chapter 6 of the companion volume [931b] for detailed discussions). The analyst is independent and less influenced by games in the political market, has the tools and must use them to provide the weights as well as the logic behind their constructs. We thus take a different position from others such as the one in [652]. Furthermore, the impact of each benefit characteristic of either a tangible or intangible on any given class may vary substantially depending on the nature of the social class, the benefit characteristics and the trends of national goals and objectives in addition to how these goals and objectives are formed..

The benefit intensity, holding a class constant, may vary over different benefit attributes. Additionally, certain social decisions tend to produce benefit attributes that are latent or primary and/or secondary and time-deferred. There are some benefit attributes that activate other inactive benefit characteristics depending on the environment of social production and the structure of social goals and objectives (in other words there are new benefit characteristic that act as complementary benefit attributes to some existing but inactive benefits). These essential elements of benefit accounting are too important for the social efficiency and equity of a social entity organized around the principles of democracy to be left to the dictatorship of policy makers and politicians whose weight assignments may be driven by their selfish interest to remain in power rather than by general social welfare of the social unit.

Democracy works best if assessment of social utility (benefit) allows evidential inputs from different segments of the society, including those of politicians, policy makers and other experts to be integrated into the benefit assessment and decision process. The needed ethical weights and their distribution are an aggregative process that requires a combination of differential evidence on the social intensity of benefit characteristic for each cohort. Such aggregative construct of benefit intensity like the burden of cost may be approached from a number of directions as have been discussed in Chapter 3, section 3.5.4 and also [17, 41]. We think, however, that fuzzy logical aggregation provides a superior process in combining differential information on social assessment of the degree of benefit intensity.

The fuzzy aggregation logic allows us to incorporate into the accounting process differential evidence from intra-class and inter-class assessments as well as those of the experts and policy makers. The process is to elicit evaluations from the representative members of each class, group of experts and policy makers as to the social importance (benefit intensity) that must be attached to a benefit characteristic of each class. Such valuations are subjective, vague, imprecise, ill-defined and approximation that require a non-classical approach to their analysis and computations. Such a non-classical approach is suggested to operate through fuzzy logical reasoning to which we turn our attention.

4.7 The Construction of Class Benefit Intensity Weights Through Fuzzy Aggregation

We consider a set of selected experts with an index set $\mathbb{L}$. Let $\mathbb{N}$ be an index set of affected social classes including policy makers, and $\mathbb{I}$ an index set of real benefit characteristics. For any benefit characteristic (attribute), h_i , $i \in \mathbb{I}$, we consider a partition on the set of experts with index set $\mathbb{L}_i$, $i \in \mathbb{I}$ where $\bigcup_{i\in\mathbb{I}} \mathbb{L}_i = \mathbb{L}$. Thus $\mathbb{L}_i$ is an index set of experts required to provide the informed subjective weights for the ith benefit characteristic. Such weights are viewed as fuzzy numbers of the form α_{lij}, $l \in \mathbb{L}_i$, $i \in \mathbb{I}$ and $j \in \mathbb{N}$, for each cohort. A set, $\mathbb{E}_i$, of elicited values from the experts may be specified as

$$\mathbb{E}_i = \left\{ \alpha_{lij} \geq 1 \mid l \in \mathbb{L}_i \ , j \in \mathbb{N} \text{ and } i \text{ is fixed in} \mathbb{I} \right\}$$

(4.7.1)

From each set, $\mathbb{E}_i$, $i \in \mathbb{I}$, we seek an aggregate value, $\alpha_{\bullet ij}$, over all selected experts that will provide the measure of *social intensity weight* of each benefit attribute, $i \in \mathbb{I}$, and for each class $j \in \mathbb{N}$. eqn. (4.7.1) generates a sub-matrix where each column vector is obtained by fixing j in $\mathbb{N}$ and varying $i \in \mathbb{I}$, and each row vector is obtained by fixing i in $\mathbb{I}$ and varying $j \in \mathbb{N}$. From eqn. (4.7.1) a set $\mathbb{E}$, a family of data sets may be constructed as

$$\mathbb{E} = \left\{ \mathbb{E}_i \mid i \in \mathbb{I} \text{ and } \mathbb{E}_i \text{ as defined in Eq. 4.7.1} \right\}$$

(4.7.2)

It may be observed that for all $i \in \mathbb{I}$, eqn. (4.7.2) provides a complete matrix that presents informed intensity weight distribution over all benefit characteristics as seen by all the experts.

The problem, given the relevant data set, is to construct an algorithm that will induce a combination of expert judgments so as to obtain equilibrium benefit intensity weights in the sense of most preferred. The constructed algorithm must incorporate in its structure the benefit-intensity assessments of various social classes and policy makers in addition to those of the experts. Such assessments may directly enter as constraints on the acceptance of the experts' assessments. In this respect, two possibilities are opened to us. The first possibility is to allow the society or social classes to select the experts and assign them with weights that capture the degree of their expertness and thus accept their assessments with their competence weights as it has been done in [927, 928]. Alternatively, inter-class and intra-class assessments of benefit-intensity weights may be elicited as fuzzy numbers in addition to those of the policy makers. These weights are then used as direct constraints on the fuzzy aggregation process. We select the former. For the case of the latter see [11, 69, 76, 78]. The concepts of social groups and social classes are interchangeably used. They define the same idea and their structures are decision specific.

To proceed we present the problem of obtaining the equilibrium weights as a fuzzy aggregation one. Let $\mathbb{E}$ be as specified in eqn. (4.7.2). Additionally, let $\widehat{\mathbb{D}}$ be a set of elicited measures of the degree of expertness associated with the evaluator's ability to assess the value of benefit attribute, b_{ij} , $i \in \mathbb{I}$ and $j \in \mathbb{N}$. $\mathbb{H}$ represents a set of elicited fuzzy measures for the degree of expertness associated with society's trust in the evaluators' abilities to correctly value the class and benefit intensities, α_{lij} , where the elicited values are normalized as fuzzy numbers.

To obtain these sets of information we first construct the *optimal* set of expert evaluators. Let $\mathbb{A}$ be the general set of experts. Let the benefit characteristics induce a partition on the set $\mathbb{A}$ such that

$$\mathbb{A}_i = \mathbb{A} \otimes [0,1] = \{(x_l, \omega_l) \mid l \in \mathbb{L}_i, x \in \mathbb{A},\ \mu_{\mathbb{A}_i}(x_l) = \omega_l \in [0,1], i \in \mathbb{I}\}$$

(4.7.3)

The set $\mathbb{A}_i$ is a fuzzy set of experts that can evaluate the *i*th benefit characteristic. From the fuzzy set, $\mathbb{A}$, we develop an efficient set of experts through the method of fixed-level cut that may be specified as

$$\mathbb{A}_{\omega_i^0} = \{(x_l, \omega_l^0) \mid \mu_{\mathbb{A}_i}(x_l) = \omega_l \geq \omega_l^0 \in [0,1],\ l \in \mathbb{L}_i, i \in \mathbb{I}, x \in \mathbb{A}\}$$

(4.7.4)

The optimal sets of experts as defined by Eq. (4.7.4) are used to develop the benefit database as presented in Table 4.9.1. Conceptually and algorithmically, the ω_i^0 , for the fixed-level cut, is obtained by fuzzy optimization of a fuzzy decision problem where an aggregate fuzzy set of the policy makers is optimized subject to an aggregate fuzzy set developed from the sample of members of social classes of the society. The fuzzy set of experts, $\mathbb{A}$, is thus decomposed by its fixed-level cuts whose structure and properties are stated as the theorem below.

Theorem 4.7.1

Let $\mathbb{M} = \bigcup_{i\in\mathbb{I}} \mathbb{A}_{\omega_i^0}$, $i \in \mathbb{I}$, *define a power set* $\boldsymbol{P}(\mathbb{M})$*; then*

(a) $\bigcup_{i\in\mathbb{I}} \mathbb{A}_{\omega_i^0} \subset \mathbb{A}, \mathbb{M} \in \boldsymbol{P}(\mathbb{M})$ and $\mathbb{M}' \notin \boldsymbol{P}(\mathbb{M})$,

(b) $\bigcap_{i\in\mathbb{I}} \mathbb{A}_{\omega_i^0} \subset \mathbb{A}$ and $\bigcap_{i\in\mathbb{I}} \mathbb{A}_{\omega_i^0} \in \boldsymbol{P}(\mathbb{M})$

(c) $\exists\ \mathbb{A}_{\omega_i^0}$ and $\mathbb{A}_{\omega_j^0} \ni \mathbb{A}_{\omega_i^0} \cap \mathbb{A}_{\omega_j^0} \neq \varnothing$ for $i \neq j \in \mathbb{I}$,

(d) $\exists\, \mathbb{A}_{\omega_i^0}$ and $\mathbb{A}_{\omega_k^0} \ni \mathbb{A}_{\omega_i^0} \cap \mathbb{A}_{\omega_k^0} = \varnothing,\ i \neq k \in \mathbb{I}.$

Proof

The proof of the theorem is straightforward from a general knowledge of set permutation.

The condition (c) implies that there is at least one evaluator that is qualified to evaluate more than one benefit characteristic. Condition (d) states that there are some benefit characteristics whose evaluators are not qualified to evaluate other benefit characteristics. Theorem 4.7.1 presents only the structure of the set of expert evaluators.

Let us now consider each cohort with real benefit aggregate value, b_{ij} as obtained from the experts. Furthermore, let the unit value of benefit intensity, α_{lij} as assessed by the l-expert for each (i- j) cohort be a fuzzy number that is expressed as linguistic or interval value. For each competent expert and for each (i - j) cohort we specify a fuzzy set $\mathbb{F}_{lij}$ as

$$\mathbb{F}_{lij} = \mathbb{E}_{lij} \otimes \mathbb{H}_{lij} = \left\{ \left(\alpha_{lij}, \mu_{\mathbb{F}_{lij}}\left(\alpha_{lij}\right)\right) \mid \alpha_{lij} \in \mathbb{H} \text{ and } \mu_{\mathbb{F}_{lij}} : \mathbb{E}_{lij} \to \mathbb{H}_{lij} \subset [0,1] \right\}$$
(4.7.5)

From eqn. (4.7.5) we can obtain an aggregate value for each cohort through a fuzzy decision construct where a fuzzy aggregate index of degree of benefit intensity is optimized subject to the ω_l^0-level set that will provide lowest deviations among the abilities of the experts regarding assessments. We shall refer to this approach as *indirect fuzzy aggregation.* For discussions on direct fuzzy aggregation see [22, 26, 77, 78, and 829]. From the angle of indirect fuzzy aggregation two structural problems may be developed for solution. One structure does not allow evaluative interactions among the experts while the other does. The first one is constructed through a fuzzy set intersection. The second one is through algebraic product of fuzzy sets. The structures lead to the following theorems for the required optimal fuzzy decision.

Theorem 4.7.2A

$$\alpha_{ij}^{*} = \begin{cases} \sup\limits_{\alpha_{lij}} \bigcap\limits_{l \in \mathbb{L}_i} \mathbb{F}_{lij} \\ \text{s.t. } \inf d\left(\mathbb{A}_l\ ,\ \mathbb{A}_{\omega_l^0}\right) \end{cases}$$

Theorem 4.7.2B

$$\hat{\alpha}_{ij}^{*} = \begin{cases} \sup_{\alpha_{lij}} \bigodot_{l \in \mathbb{L}_i} \mathbb{F}_{lij} \\ \text{s.t. } \inf d\left(\mathbb{A}_l , \mathbb{A}_{\omega_l^0}\right) \end{cases}$$

The symbol $\odot$ is an algebraic product operator on the fuzzy set, $d\left(\mathbb{A}_l , \mathbb{A}_{\omega_l^0}\right)$ is a Hamming distance and $\mathbb{A}_{\omega_l^0}$ is as specified in eqn. (4.7.4).

The α_{ij}^{*} and $\hat{\alpha}_{ij}^{*}$ provide the intensity weights for the (*i* - *j*) cohort based solely on the experts' judgments. Such judgments may contain some important evaluative distortions that may have their origins in the class and social background of the experts. As such, the optimal benefit intensity weights obtained from the assessments of the experts for some social classes may be either downward or upward biased. To reduce this possible biasedness we construct a fuzzy constraint based on the inter-class and intra-class assessments from a sample of each class. Let $\mathbb{K}_j, j \in \mathbb{N}$, be an index set of sample from the *j*th class. Furthermore, let α_{kij}, $k \in \mathbb{K}_j$, be the assessment of the *k*th evaluator for benefit intensity of the (*i* - *j*) cohort. We may then specify a fuzzy set of the form:

$$\mathbb{F}_{kij} = \left\{\alpha_{kij} \mid \mu_{\mathbb{F}_{kij}}\left(\alpha_{kij}\right) \in [0,1], k \in \mathbb{K}_j, j \in \mathbb{N}, \text{ and } i \in \mathbb{I}\right\}$$

(4.7.8)

From eqn. (4.7.8) we develop two types of aggregate constraints, $\bigcap_{k \in \mathbb{K}} \mathbb{F}_{kij}$ and $\bigodot_{k \in \mathbb{K}} \mathbb{F}_{kij}$ which are used to restrict the collective expert decision on the optimal benefit intensity weight, α_{lij}. By combining eqn. (4.7.8) with eqns. (4.7.6) and (4.7.7) we can obtain an improved benefit intensity weights. The solutions to the fuzzy decision problems are stated as Theorems 4.7.3A and B.

Theorem 4.7.3A

$$\alpha_{ij}^{**} = \begin{cases} \sup_{\alpha_{lij}} \bigwedge_{l \in \mathbb{L}_i} \mu_{\mathbb{F}_{lij}}\left(\alpha_{lij}\right), \; j \in \mathbb{N} \\ \quad \text{s.t} \left[\bigwedge_{l \in \mathbb{L}_i} \mu_{\mathbb{F}_{lij}}\left(\alpha_{lij}\right)\right] - \left[\bigwedge_{k \in \mathbb{K}_j} \mu_{\mathbb{F}_{kij}}\left(\alpha_{kij}\right)\right] \geq 0 \\ \quad \inf d\left(\mathbb{A}_i \, , \, \mathbb{A}_{\omega_l^0}\right) \end{cases}$$

Theorem 4.7.3B

$$\alpha_{ij}^{**} = \begin{cases} \sup_{\alpha_{lij}} \prod_{l \in \mathbb{L}_i} \mu_{\mathbb{F}_{lij}}\left(\alpha_{lij}\right) \\ \quad \text{s.t.} \left[\prod_{l \in \mathbb{L}_i} \mu_{\mathbb{F}_{lij}}\left(\alpha_{lij}\right)\right] = \left[\prod_{k \in \mathbb{K}_j} \mu_{\mathbb{F}_{kij}}\left(\alpha_{kij}\right)\right] \geq 0 \\ \quad \inf d\left(\mathbb{A}_i \, , \, \mathbb{A}_{\omega_l^0}\right) \end{cases}$$

The application of Theorems 4.7.3A and B yields a matrix of optimal distributional weights of the form:

$$\mathbb{D} = \begin{bmatrix} \alpha_{11} & \alpha_{12} & \cdots & \alpha_{1j} & \cdots \\ \alpha_{21} & \alpha_{22} & \cdots & \alpha_{2j} & \cdots \\ \vdots & & \cdots & & \vdots \\ \alpha_{i1} & \alpha_{i2} & \cdots & \alpha_{ij} & \cdots \\ & \cdot & \cdot & \cdot & \end{bmatrix}, \; i \in \mathbb{I} \, , j \in \mathbb{N}$$

(4.7.9)

From Table 4.4.1 we also obtain the assessment values of real benefit attributes of the form

$$\mathbb{B} = \begin{bmatrix} b_{11} & b_{12} & \cdots & b_{1j} & \cdots \\ b_{21} & b_{22} & \cdots & b_{2j} & \cdots \\ \vdots & & \vdots & & \vdots \\ b_{i1} & b_{i2} & \cdots & bij & \cdots \\ & \cdots & \cdots & & \vdots \end{bmatrix}, \; i \in \mathbb{I} \, , j \in \mathbb{N}$$

(4.7.10)

By combining eqns. (4.7.9) and (4.7.10) we obtain a vector of values of the form

$$\mathbb{G} = \mathbb{B}\mathbb{D}^{T}$$

(4.7.11)

By substituting eqn. (4.7.11) in eqn. (4.5.1) we can write the total monetary value of all relevant benefit characteristics as

$$\hat{\boldsymbol{B}}_{ST}^{M} = \sum_{i \in \mathbb{K}} \sum_{j \in \mathbb{N}} p_{ij} \alpha_{ij}^{*} b_{ij}$$

(4.7.12)

It must be noted that α_{ij}^{*} is a complicated distributional weight whose optimal value is obtained through a series of fuzzy algorithms. The net difference between the total value with and without distributional weights is obtained from eqns. (4.7.11) and (4.5.1) as:

$$\Delta B_{ST} = \sum_{i \in \mathbb{K}} \sum_{j \in \mathbb{N}} \left(\alpha_{ij} - 1\right) p_{ij} b_{ij}$$

(4.7.13)

4.8 Conclusion and Effect of Time on Benefit Data and Accounting

So far we have presented a benefit accounting for only one period. It is, thus, static. In social decisions and choices where comparative cost and benefit values are essential for decision, there is always the possibility of benefit-shifting across different classes and generations as the decision alternatives spread over time. Furthermore, the distributional weights may also shift along with the benefit-shifting as the structure of social values alters. Implied in these is that if the outcomes of the decision spread over years then the benefit accounting must take place at each point of essential decision time so as to provide us with a time series of benefit information [927]. When the benefit assessment is time-dependent we obtain vectors of vectors of three important information sets.

If $\hat{\mathbb{T}}$ is a time set, then each benefit characteristic will be represented in the form b_{ijt} with corresponding distributional weight α_{ijt} that is to be attached. The time series benefit information for cost-benefit analysis for any $t \in \hat{\mathbb{T}}$ and $j \in \mathbb{N}$ is

$$\hat{B}_{ST_t}^{M} = \left[\sum_{i} \sum_{j} \alpha_{ijt} p_{ijt} b_{ijt} \right]_{t \in \hat{\mathbb{T}}}, \quad i \in \mathbb{K}$$

(4.8.1)

$$B_{ST}^{Q} = \left[b_{i1t}, b_{i2t}, \cdots, b_{ijt} \cdots \right]_{t \in \hat{\mathbb{T}}}, \quad i \in \mathbb{Q}$$

(4.8.2)

$$B_{ST}^{R} = \left[b_{i1t}, b_{i2t}, \cdots, b_{ijt} \cdots \right]_{t \in \hat{\mathbb{T}}}, \quad i \in \mathbb{R}$$

(4.8.2)

Eqns. (4.8.1) - (4.8.3) provide us with a time series benefit information. To calculate the total social benefit of the project overtime, we must develop present value equivalence. This leads to intertemporal aggregation which is obtained by means of a discounting algorithm. Such a discounting algorithm may follow the known tradition [356, 363, 385] or non-traditional way [927, 931b, 1152b].

To conclude, we want to mention that the comprehensive framework developed here for benefit assessment is intended to provide a rich benefit data for decision. As such, emphasis has been placed on accounting for all relevant benefit values even if their quantitative and pecuniary valuations cannot be ascertained. The method of assessment is through benefit-identification matrices that help us to answer the following questions. (a) What are the benefit characteristics? (b) Who are the recipients of the benefits? (c) How are these benefit characteristics valued and in what units? (d) What are the appropriate distributional weights that will reflect social assessment of benefit intensities? (e) What is the structure of intertemporal, interpersonal and intergenerational benefit-shifting? The advantage of such an approach is to make explicit those benefits that may be omitted in the decision-making process because of measurement difficulties. Such deliberately explicit omissions will suggest directions of possible decision bias resulting from benefit accounting and its effect on choice alternatives. An illustrative example of how the theory and its developed algorithms work is presented as an example in Section (4.9) separately. The steps for developing the social benefit data with the required distributional weights are similar to those provided for cost in Chapter 3 by replacing cost with benefit. In some cases some minor modifications may be useful as well as required.

4.9 An Example of Social Benefits and Benefit Accounting: Treatment of Mental Illness

This section is an illustration of how to apply the theory and the system of algorithms developed in previous section. The selected area of illustration is mental illness and the community of interest is a hypothetical one. The example is to complement the comprehensive costing given in Chapter 3, section (3.6). The data to be used is created and the data structure reflects the counterpart of cost assessments undertaken in section (3.6) of Chapter 3. Some of the prices are abstracted from the insightful work of Weisbrod [603]. The resulting benefit data sets that will constitute the essential knowledge base for cost-benefit decision are obtained through a series of fuzzy optimization that allows the initial data to be transformed in a manner where benefit aggregation in the presence of differential assessments by experts, policy makers and residents is made possible. The initial data set is assumed to be elicited through some acceptable instrument such as instruments for contingent evaluation. The logic of fuzzy decision allows us to deal with the problem of combination of differential judgments and information regarding assessments and the degrees of intensity of benefits that reflect distributional weights over cohorts.

4.9.1 Developing the Required Initial Information Base

Let us return our attention to considering the problem of computing the social benefits of treating the mentally ill irrespective of the type of treatment regime. The approach requires us to identify the real benefit characteristics associated with this project and then divide the affected population into social classes. The real benefit characteristics associated with the treatment may be listed as:
Improved mental health (state) of patient:

(1) Improved mental health (state) of patient:
- (a) More satisfied with life h_1
- (b) Improved quality of life h_2
- (c) Leisure-time with others h_3

(2) Increase in productivity:
- (a) Paid work (measured in No. of hours) h_4
- (b) Unpaid house work (measured in hours) h_5

(c) Job stability and continuity (*# of absentees) h_6

(d) Job retention (fired or forced job change) h_7

(e) Fruitful job change (measured by wage differentials) h_8

(3) Improvements in decision-making:

(a) Financial management (increased savings) h_9

(b) Management of non-financial affairs h_{10}

(c) Management of present-future trade-offs h_{11}

(4) Reduction in risk:

(a) Reduction in mortality h_{12}

(b) Reduction in frequency of morbidity h_{13}

(c) Reduction in the frequency of physical injury h_{14}

(d) Reduction in anxiety (psychic loss, etc.) h_{15}

(5) Direct community benefit:

(a) Reduction in police and jail services h_{16}

(b) Reduction in court services h_{17}

(c) Reduction in probation and parole services h_{18}

(d) Reduction in property damage h_{19}

(e) Reduction in community monitoring services h_{20}

(f) Reduction in social services h_{21}

(g) Reduction in income loss not attributed to death h_{22}

These are the identified real benefit characteristics. To assess them we need to identify the classes of society that such benefit characteristics may affect as a result of the mental illness. The required classes may be identified as:
Mentally ill persons

(1) Mentally ill persons

(a) Young

(i) Females G_1

(ii) Males G_2

(b) Old

(i) Females G_3

(ii) Males G_4

(2) Families G_5

(3) Workers G_6

(4) Neighbors G_7

(5) Public philanthropists G_8

(6) Tax payers

(a) Local government G_9

(b) State government G_{10}

(c) Federal government G_{11}

With these real benefit characteristics and the social classes we can construct a benefit-identification matrix as in Table 4.9.11. It may also be noted that G_9, G_{10} and G_{11} may have different names depending on society's organization.

From the identification matrix of Table 4.9.1 we know that $b_{ij} \geq 1$ and the blank spaces in the (*i* - *j*) cohorts suggest that the *j*th class does not benefit from the *i*th characteristic. Care must also be taken to avoid *double counting*. For example, the reduction in the property damage attributed to the family should be distinguished from those attributed to workers and neighbors. Similarly, the reduction in services assigned to local, state and federal governments must be distinguished, and not to be included in those reductions attributable to public philanthropists. Special care must also be taken in identifying inter-class and inter-temporal shifting of benefits.

The benefit-identification matrix of Table 4.9.1 may be reorganized and partitioned into (a) price measurable, (b) quantitatively but non-pecuniarily measurable, (c) quantitatively and pecuniarily non-measurable. The following groups may be made. $h_1, h_2, h_9, h_{10}, h_{11}$, and h_{15} are non-price measurable. The characteristics $h_4, h_5, h_6, h_7, h_8, h_{16}, h_{17} h_{18}, h_{19}, h_{20}, h_{21}$ and h_{22} are either directly or indirectly price measurable. The characteristics, h_3, h_{12}, h_{13} and h_{14} , are quantitatively measurable as measured by frequency while the characteristic h_{15} is quantitatively and pecuniarily non-measurable as viewed from the affected.

Table 4.9.1 The Real Benefit-Identification Matrix

	G_1	G_2	G_3	G_4	G_5	G_6	G_7	G_8	G_9	G_{10}	G_{11}	Total
h_1	b_{11}	b_{12}	b_{13}	b_{14}								B_1^R
h_2	b_{21}	b_{22}	b_{23}	b_{24}								B_2^R
h_3	b_{31}	b_{32}	b_{32}	b_{34}								.
h_4	b_{41}	b_{42}	b_{43}	b_{44}								.
h_5	b_{51}	b_{52}	b_{53}	b_{54}								.
h_6	b_{61}	b_{62}	b_{63}	b_{64}								
h_7	b_{71}	b_{72}	b_{73}	b_{74}								B_i^R
h_8	b_{81}	b_{82}	b_{83}	b_{84}								
h_9	b_{91}	b_{92}	b_{93}	b_{94}								
h_{10}	$b_{10,1}$	$b_{10,2}$	$b_{10,3}$	$b_{10,4}$								
h_{11}	$b_{11,1}$	$b_{11,2}$	$b_{11,3}$	$b_{11,4}$	$b_{11,5}$							
h_{12}	$b_{12,1}$	$b_{12,2}$	$b_{12,3}$	$b_{12,4}$	$b_{12,5}$	$b_{12,6}$	$b_{12,7}$					
h_{13}	$b_{13,1}$	$b_{13,2}$	$b_{13,3}$	$b_{13,4}$	$b_{13,5}$	$b_{13,6}$	$b_{13,7}$					
h_{14}	$b_{14,1}$	$b_{14,2}$	$b_{14,3}$	$b_{14,4}$	$b_{14,5}$	$b_{14,6}$	$b_{14,7}$					
h_{15}	$b_{15,1}$	$b_{15,2}$	$b_{15,3}$	$b_{15,4}$	$b_{15,5}$	$b_{15,6}$	$b_{15,7}$					
h_{16}									$b_{16,9}$	$b_{16,10}$	$b_{16,11}$	.
h_{17}									$b_{17,9}$	$b_{17,10}$	$b_{17,11}$	.
h_{18}									$b_{18,9}$	$b_{18,10}$	$b_{18,11}$	.
h_{19}					$b_{19,5}$	$b_{19,6}$	$b_{19,7}$					
h_{20}					$b_{20,5}$			$b_{20,8}$	$b_{20,9}$	$b_{20,10}$	$b_{20,11}$	
h_{21}					$b_{21,5}$			$b_{21,8}$	$b_{21,9}$	$b_{21,10}$	$b_{21,11}$	
h_{22}	$b_{22,1}$	$b_{22,2}$	$b_{22,3}$	$b_{22,4}$	$b_{22,5}$	$b_{22,6}$	$b_{22,7}$					B_{22}^R
Total	B_1^G	B_2^G	.	.	.	B_j^G	.	.	.	B_{10}^G	B_{11}^G	B

Let us visit our hypothetical situation of a small community in an industrial society given in Chapter 3, Sec. 3.6. Again assume that there are one hundred and forty mentally ill persons. From these persons, community, families and society a benefit-identification frequency matrix is developed in Table 4.9.2.

We must exercise care in interpreting the frequency distribution in Table 4.9.2. The frequencies represent numbers of persons estimated either to benefit or as real units of benefit. For example, 5 young females who are mentally ill are estimated either to benefit or as real units of benefit. For example, 5 young females who are mentally ill are estimated to have more satisfaction with life due to treatment while 63 persons out of 140 are estimated to have satisfaction with life after they have gone through the treatment regime. The numbers corresponding to a reduction in services such as court services are weighted frequencies. They are obtained as a reduction in workforce. From Table 4.9.2 and conditions of measurability the following sub-matrices $\mathbb{C}_1$, $\mathbb{C}_2$, $\mathbb{C}_3$ and $\mathbb{C}_4$ may be constructed.

Table 4.9.2 Real Numerical Benefit-Identification Matrix

	G_1	G_2	G_3	G_4	G_5	G_6	G_7	G_8	G_9	G_{10}	G_{11}	Total
h_1	5	6	20	32								63
h_2	6	8	32	40								86
h_3	1	2	25	38								66
h_4	1	2	20	30								53
h_5	5	6	30	40								81
h_6	0	1	20	30								51
h_7	0	1	20	30								51
h_8	0	0	5	10								15
h_9	2	2	30	40								74
h_{10}	6	6	32	38								82
h_{11}	2	2	10	20								34
h_{12}	2	3	3	2	2	1	3					16
h_{13}	4	5	30	51	3	3	1					97
h_{14}	2	2	8	10	12	1	5					40
h_{15}	3	4	30	34	90	10	5					176
h_{16}									5	10	15	30
h_{17}									3	15	10	28
h_{18}									2	3	3	8
h_{19}						0	0	0				0
h_{20}								3	4	5	5	17
h_{21}								3	4	2	2	11
h_{22}	4	6	30	51	3	3	1					97
Total				•		•		•				

The non-price measurable matrix, $\mathbb{C}_1$, after some regrouping of the matrix elements may be written as:

$$\mathbb{C}_1 = \begin{bmatrix} b_{11} & b_{12} & b_{13} & b_{14} \\ b_{21} & b_{22} & b_{23} & b_{24} \\ b_{91} & b_{92} & b_{93} & b_{94} \\ b_{10,1} & b_{10,2} & b_{10,3} & b_{10,4} \\ b_{11,1} & b_{11,2} & b_{11,3} & b_{11,4} \end{bmatrix} = \begin{bmatrix} 5 & 6 & 20 & 32 \\ 6 & 8 & 32 & 40 \\ 2 & 2 & 30 & 40 \\ 6 & 6 & 32 & 38 \\ 2 & 2 & 10 & 20 \end{bmatrix}$$

(4.9.1)

The price-measurable characteristics generate a frequency distribution matrix, $\mathbb{C}_2$ after some rearrangements of the entries.

$$\mathbb{C}_2 = \begin{bmatrix} b_{31} & b_{32} & b_{33} & b_{34} \\ b_{41} & b_{43} & b_{43} & b_{44} \\ b_{51} & b_{52} & b_{53} & b_{54} \\ b_{16,8} & b_{16,9} & b_{16,10} & b_{16,11} \\ b_{17,8} & b_{17,9} & b_{17,10} & b_{17,11} \\ b_{18,8} & b_{18,9} & b_{18,10} & b_{18,11} \\ b_{20,8} & b_{20,9} & b_{20,10} & b_{20,11} \\ b_{21,8} & b_{21,9} & b_{21,10} & b_{21,11} \\ b_{22,1} & b_{22,2} & b_{22,3} & b_{22,4} \\ b_{22,5} & b_{22,6} & b_{22,7} & b_{22,8} \end{bmatrix} = \begin{bmatrix} 1 & 2 & 25 & 38 \\ 1 & 2 & 32 & 40 \\ 5 & 6 & 30 & 40 \\ 0 & 5 & 10 & 15 \\ 0 & 3 & 15 & 10 \\ 0 & 2 & 3 & 3 \\ 3 & 4 & 5 & 5 \\ 3 & 4 & 2 & 2 \\ 4 & 5 & 30 & 51 \\ 3 & 3 & 1 & 0 \end{bmatrix}$$

(4.9.2)

Similarly, quantitatively measurable characteristics generate a frequency distribution matrix, $\mathbb{C}_3$ while quantitatively and pecuniarily non-measurable characteristics generate a frequency distribution matrix, $\mathbb{C}_4$ where $\mathbb{C}_3^{\mathrm{T}}$ is the transpose of $\mathbb{C}_3$.

$$\mathbb{C}_3^{\mathrm{T}} = \begin{bmatrix} b_{81} & b_{12,1} & b_{13,1} & b_{14,1} & b_{19,5} \\ b_{82} & b_{12,2} & b_{13,2} & b_{14,2} & b_{19,6} \\ b_{83} & b_{12,3} & b_{13,3} & b_{14,3} & b_{19,7} \\ b_{84} & b_{12,4} & b_{13,4} & b_{14,4} & b_{19,8} \\ 0 & b_{12,5} & b_{13,5} & b_{14,5} & 0 \\ 0 & b_{12,6} & b_{13,6} & b_{14,6} & 0 \\ 0 & b_{12,7} & b_{13,7} & b_{14,7} & 0 \end{bmatrix} = \begin{bmatrix} 0 & 2 & 4 & 2 & 10 \\ 0 & 3 & 5 & 2 & 8 \\ 5 & 3 & 30 & 8 & 6 \\ 10 & 2 & 51 & 10 & 0 \\ 0 & 1 & 3 & 1 & 0 \\ 0 & 1 & 3 & 1 & 0 \\ 0 & 3 & 1 & 5 & 0 \end{bmatrix} = [B_{rs}]$$

(4.9.3)

$$\mathbb{C}_4 = \begin{bmatrix} b_{61} & b_{62} & b_{63} & b_{64} & 0 & 0 & 0 \\ b_{71} & b_{72} & b_{73} & b_{74} & 0 & 0 & 0 \\ b_{15,1} & b_{15,2} & b_{15,3} & b_{15,4} & b_{15,5} & b_{15,6} & b_{15,7} \end{bmatrix} = \begin{bmatrix} 0 & 1 & 20 & 30 & 0 & 0 & 0 \\ 0 & 1 & 20 & 30 & 0 & 0 & 0 \\ 3 & 4 & 30 & 34 & 90 & 10 & 5 \end{bmatrix} = \left[b_{ij} \right]$$

(4.9.4)

These hypothetical values are constructed for only one period. The values of the rest of decision relevant periods must be estimated likewise. To estimate the monetary component of the benefit characteristics we consider the matrix, $\mathbb{C}_2$. We then estimate the average man-hours of savings time per week that correspond to increases in productivity for the corresponding cohorts. The distribution of weekly hourly savings per worker is given in eqn. (4.9.5) matrix, T_2.

$$T_2 = \begin{bmatrix} T_{31} & T_{32} & T_{33} & T_{34} \\ T_{41} & T_{42} & T_{43} & T_{44} \\ T_{51} & T_{52} & T_{53} & T_{54} \\ T_{16,8} & T_{16,9} & T_{16,10} & T_{16,11} \\ T_{17,8} & T_{17,9} & T_{17,10} & T_{17,11} \\ T_{18,8} & T_{18,9} & T_{18,10} & T_{18,11} \\ T_{20,8} & T_{20,9} & T_{20,10} & T_{20,11} \\ T_{21,8} & T_{21,9} & T_{21,10} & T_{21,11} \\ T_{22,1} & T_{22,2} & T_{22,3} & T_{22,4} \\ T_{22,5} & T_{22,6} & T_{22,7} & T_{22,8} \end{bmatrix} = \begin{bmatrix} 2 & 2 & 3 & 2 \\ 15 & 20 & 30 & 30 \\ 5 & 4 & 8 & 8 \\ 40 & 40 & 40 & 40 \\ 40 & 40 & 40 & 40 \\ 40 & 40 & 40 & 40 \\ 40 & 40 & 40 & 40 \\ 40 & 40 & 40 & 40 \\ 15 & 20 & 30 & 30 \\ 40 & 40 & 40 & 40 \end{bmatrix} = \left[T_{ij} \right]$$

(4.9.5)

The frequency distribution in matrix, $\mathbb{C}_3^{\mathrm{T}}$ the transpose of $\mathbb{C}_3$ presents two important benefit characteristics of cost saving and reduction in pain and suffering. The cost-saving component generates a matrix, $\mathbb{P}_3^{\mathrm{T}}$ whose entry P_{ij} represents a unit value savings due to reduction in needed care. This is represented in a matrix representation in eqn.(4.9.6) . The values in eqn. (4.9.6) are estimates per person for one year. Care should be taken for their interpretation and use when the framework is applied in actual situation. For example, P_{81} = \$780 is the estimate monetary value of a year's benefit of a fruitful job change for a young female person who has passed through a treatment regime of mental illness. The value, $P_{12,4} =$ \$1800, is an estimate of realized saving due to an avoidance of funeral cost because one old, mentally ill male is estimated to be saved by a treatment regime. The matrix, $\mathbb{P}_3$, is thus a matrix of prices.

$$\mathbb{P}_3^{\mathrm{T}} = \begin{bmatrix} P_{81} & P_{12,1} & P_{13,1} & P_{14,1} & P_{19,5} \\ P_{82} & P_{12,2} & P_{13.2} & P_{14,2} & P_{19,6} \\ P_{83} & P_{12.3} & P_{13,3} & P_{14,3} & P_{19,7} \\ P_{84} & P_{12,4} & P_{13,4} & P_{14,4} & P_{19,8} \\ 0 & P_{12,5} & P_{13,5} & P_{14,5} & 0 \\ 0 & P_{12,6} & P_{13,6} & P_{14,5} & 0 \\ 0 & P_{12,7} & P_{13,7} & P_{14,7} & 0 \end{bmatrix} = \begin{bmatrix} 780 & 1200 & 2000 & 250 & 450 \\ 1040 & 1200 & 2000 & 250 & 200 \\ 3120 & 1800 & 2000 & 400 & 300 \\ 3900 & 1800 & 2000 & 400 & 250 \\ 0 & 2500 & 2000 & 700 & 0 \\ 0 & 2500 & 2000 & 700 & 0 \\ 0 & 2500 & 2000 & 700 & 0 \end{bmatrix}$$

$$= \left[P_{rs}\right]$$

(4.9.6)

Given the time matrix, T_2 , of eqn. (4.9.5) we compute a new time matrix, $\hat{T}_2$, for aggregate cohort time savings for a weak by combining it with matrix, $\mathbb{C}_2$ from eqn. (4.9.2). The individual entries are computed as $t_{ij} = T_{ij} b_{ij}$; thus $t_{31} = T_{31} b_{31} = (2)(1) = 2$ or $t_{18,10} = T_{18,10} b_{18,10} = (40)(3) = 120$.

$$\hat{T}_2 = \begin{bmatrix} t_{31} & t_{32} & t_{33} & t_{34} \\ t_{41} & t_{42} & t_{43} & t_{44} \\ t_{51} & t_{52} & t_{53} & t_{54} \\ t_{16,8} & t_{16,9} & t_{16,10} & t_{16,11} \\ t_{17,8} & t_{17,9} & t_{17,10} & t_{17,11} \\ t_{18,8} & t_{18,9} & t_{18,10} & t_{18,11} \\ t_{20,8} & t_{20,9} & t_{20,10} & t_{20,11} \\ t_{21,8} & t_{21,9} & t_{21,10} & t_{21,11} \\ t_{22,1} & t_{22,2} & t_{22,3} & t_{22,4} \\ t_{22,5} & t_{22,6} & t_{22,7} & t_{22,8} \end{bmatrix} = \begin{bmatrix} 2 & 4 & 75 & 76 \\ 15 & 40 & 960 & 1200 \\ 25 & 24 & 240 & 320 \\ 0 & 200 & 400 & 600 \\ 0 & 120 & 600 & 400 \\ 0 & 80 & 120 & 120 \\ 120 & 160 & 200 & 200 \\ 120 & 160 & 80 & 80 \\ 60 & 100 & 900 & 1530 \\ 120 & 120 & 40 & 0 \end{bmatrix} = \left[t_{ij}\right]$$

(4.9.7)

Corresponding to matrix, $\hat{T}_2$ we must find the corresponding price matrix, $\mathbb{P}_2$. The prices here are the average hourly wage rates which have to be obtained. These prices may be market imputed (indirect method of revealed preference) or elicited through an instrument (direct method of contingent valuation) or may be obtained by a fuzzy-decision construct. The following price matrix, $\mathbb{P}_2$, is presented for the purpose of this example.

$$\mathbb{P}_2 = \begin{bmatrix} w_{31} & w_{32} & w_{33} & w_{34} \\ w_{41} & w_{42} & w_{43} & w_{44} \\ w_{51} & w_{52} & w_{53} & w_{54} \\ w_{16,8} & w_{16,9} & w_{16,10} & w_{16,11} \\ w_{17,8} & w_{17,9} & w_{17,10} & w_{17,11} \\ w_{18,8} & w_{18,9} & w_{18,10} & w_{18,11} \\ w_{20,8} & w_{20,9} & w_{20,10} & w_{20,11} \\ w_{21,8} & w_{21,9} & w_{21,10} & w_{21,8} \\ w_{22,1} & w_{22,2} & w_{22,3} & w_{22,4} \\ w_{22,5} & w_{22,6} & w_{22,7} & w_{22,8} \end{bmatrix} = \begin{bmatrix} 1.80 & 1.80 & 2.25 & 2.25 \\ 1.80 & 1.80 & 2.25 & 2.25 \\ 1.35 & 12.0 & 12.0 & 1.80 \\ 12.0 & 12.0 & 12.0 & 12.0 \\ 20.0 & 20.0 & 20.0 & 12.0 \\ 12.0 & 10.0 & 10.0 & 20.0 \\ 10.0 & 10.0 & 10.0 & 10.0 \\ 10.0 & 10.0 & 10.0 & 10.0 \\ 1.80 & 1.80 & 2.25 & 2.25 \\ 10.0 & 10.0 & 10.0 & 10.0 \end{bmatrix}$$

$$= \left[w_{ij} \right]$$

(4.9.8)

The first three horizontal price vectors are obtained by combining the minimum wage rate, \$4.50, as representing minimum average productivity with estimated productivity of corresponding cohorts that a recovered mentally ill person belongs. The corresponding productivity vectors are (0.4, 0.4, 0.5, 0.5), (0.4, 0.4, 0.5, 0.5) and (0.3, 0.3, 0.5, 0.4). For example, $w_{31} = (0.4)(4.50) = 1.80$ and $w_{32} = (0.3)(4.50) = 1.35$.

The monetary value of benefits from any treatment regime for mental illness is of two components of (1) savings obtained through reduction in property damages, cost of social services and cost of other medical services besides cost of treating mental illness; and (2) increased income resulting from productivity increases consequent on improved mental state of a treatment. The data required to compute these values without the corresponding intensity assessment are provided in eqns. (4.9.3) and (4.9.6 - 4.9.8). eqn. (4.9.7) presents the distribution of weekly total time savings which must be projected for the whole working year. The number of paid weeks is taken, on the average, to be 35 weeks in a year for this example. The monetary total value without the distributional weights may be computed as (see Eq. (4.5.1):

$$B_{ST}^{M} = 35\sum_{i}\left(\hat{T}_2\mathbb{P}_2^{\mathrm{T}}\right) + \sum_{r}\mathbb{C}_3^{\mathrm{T}}\mathbb{P}_3 = 35\left(\sum_{i}\sum_{j} t_{ij}w_{ij}\right) = \sum_{r}\sum_{s} P_{rs}b_{rs}$$

(4.9.9)

4.9.2 On Fuzzy Computing and Distributional Weights of Benefit Intensities and Aggregation

We now seek α_{ij} for the benefit intensities of each i - j cohort such that the aggregate benefit value may be computed as in eqn. 4.7.12 where

$$B_{ST}^{M} = \sum\sum P_{ij}\alpha_{ij}^{*}b_{ij}$$

(4.9.10)

The implied reason for the distributional weights is that all social decisions involve weighting of benefits and costs where such weights may differ over benefits and costs, as well as between the same monetary values of cost and benefit. The social evaluation of such weights (benefit intensities) may be obtained by means of fuzzy information elicitation from relevant classes, set of experts and decision-making units. The elicited information is then processed to provide the optimal class weights through the logic of fuzzy aggregation [2, 11, 27, 32, 62, 74, 75, 926].

In line of this, let us consider the intensity of benefit of one dollar value over social classes. Furthermore, let us consider the social assessment of intensities of different benefit characteristics viewed in terms of dollar value. We then combine these two assessments of distributional weights of classes and benefit characteristics to develop a matrix of cohort benefit intensity weights. For the purpose of our example, let us consider two expert evaluators for the relative intensities of the benefit characteristics and another two expert evaluators for the relative intensities of one dollar's benefit over the social classes. The experts for evaluating the intensities over the benefit characteristics may or may not be the same as those evaluating the benefit intensities over classes. From the expert evaluators we elicit fuzzy information regarding intensities. Additionally, we ascertain fuzzy information on the degree of expertness of the evaluators in providing true evaluations of the intensities of benefit characteristics and the degree to which social classes are affected. The degrees of expertness of the evaluators are the confidence weights that the society attaches to the acceptance of the experts' valuations.

To compute the relative intensity valuations of the benefit characteristics we combine set, $\mathbb{E}$ of eqn. (4.7.2) with set $\mathbb{H}$ that specifies the degree of expertness to obtain a Cartesian product set, $\mathbb{G}_l$ as :

$$\mathbb{G}_l = \mathbb{E}_l \otimes \mathbb{H}_l = \left\{ \left(x,\mu(x)\right) \mid x \in \mathbb{E}_l,\ \mu(\cdot) \in \mathbb{H}_l \text{ and } l \in \mathbb{L} \right\}$$

where $\mathbb{E}$ is the set of elicited values of benefit intensities from the experts with index set $\mathbb{L}$, and $\otimes$ represents Cartesian product. Regarding the valuation of differential class benefit intensities, we specify a set, $\mathbb{F}$ as defining the elicited values from the experts. By constructing a Cartesian product set with $\mathbb{F}$ and $\mathbb{H}$ we obtain a fuzzy set $\mathbb{Z}$ in general and $\mathbb{Z}_q$ for the qth expert as:

$$\mathbb{Z}_q = \mathbb{F}_q \otimes \mathbb{H}_q = \left\{\left(y,\mu(y)\right) \mid y \in \mathbb{F}_q,\ \mu(\cdot) \in \mathbb{H}_q \text{ and } q \in \mathbb{L}\right\}$$

(4.9.12)

The following values are provided in Tables (4.9.3) and (4.9.4) to correspond to eqn. (4.9.11) and eqn. (4.9.12) where $l = 1,2$ and q=1,2 and the l th evaluator is not necessarily the same as the qth evaluator, where x_j, y_i are the elicited benefit intensity values of classes and benefit characteristics respectively and $\mu(x)$ and $\mu(y)$ are their corresponding degrees of social acceptance as measured by the degree of expertness of the evaluators.

Table 4.9.3
Elicited fuzzy Distributional Weights Over Social Classes: An Example

$x \mid \mu(x)$		G_1	G_2	G_3	G_4	G_5	G_6	G_7	G_8	G_9	G_{10}	G_{11}
$\mathbb{G}_1$	$\mathbb{E}_1$	4	3	2	2.0	1.0	2.0	2.0	1.0	1.5	1.0	1.0
	$\mathbb{H}_1$	0.9	0.6	0.7	0.8	0.6	0.7	0.8	0.8	1.0	0.9	1.0
$\mathbb{G}_2$	$\mathbb{E}_2$	3.0	3.0	2.0	1.0	2.0	3.0	3.0	1.5	1.5	1.0	1.0
	$\mathbb{H}_2$	0.7	0.7	0.6	0.6	1.0	0.9	0.8	1.0	0.9	0.9	1.0

Given the values in Table 4.9.3 and 4.9.4 we construct a set of membership values by means of an algebraic product composition for each of the evaluators as:

$$\mathbb{R}_k = \mathbb{E}_l \odot \mathbb{F}_q = \left\{ z = x \bullet y \mid x \in \mathbb{E}_l \ , y \in \mathbb{F}_q \ , k = l \odot q \text{ and } l, q \in \mathbb{L} \right\}$$
(4.9.13)

where k is a cohort indexing. In a two by two (2 × 2) problem, the k indexes from 1 to 4 where $\mathbb{R}_k$ is a fuzzy relational matrix of product evaluations distributed over the cohorts by accounting for degrees of intensities over social groups and benefit characteristics. Corresponding to $\mathbb{R}_k$, we construct the needed

Table 4.9.4 Elicited Fuzzy Distributional Weights over Benefit Characteristics: Example

$y \backslash \mu(y)$		b_1	b_2	b_3	b_4	b_5	b_6	b_7	b_8	b_9	b_{10}	b_{11}	b_{12}	b_{13}	b_{14}	b_{15}	b_{16}	b_{17}	b_{18}	b_{19}	b_{20}	b_{21}	b_{22}
$\mathbb{Z}_1$	$\mathbb{F}_1$	3.0	20	1.0	20	1.0	1.0	0.9	1.0	1.0	1.0	1.5	3.0	25	3.0	20	1.0	0.8	1.0	1.0	1.5	1.0	20
	$\mathbb{H}_1$	0.8	0.9	0.7	0.8	0.9	1.0	1.0	0.8	0.9	0.8	0.7	0.6	0.8	0.9	0.9	1.0	1.0	1.0	0.9	0.8	0.9	1.0
$\mathbb{Z}_2$	$\mathbb{F}_2$	25	20	20	1.5	1.0	1.5	1.5	1.0	1.0	1.0	1.5	40	3.0	3.0	20	1.0	1.0	1.0	1.0	1.5	1.0	20
	$\mathbb{H}_2$	0.8	0.7	0.9	0.7	0.8	0.9	1.0	1.0	0.9	0.6	0.6	0.7	1.0	1.0	0.8	0.9	0.9	1.0	1.0	0.9	1.0	1.0

corresponding membership matrix, $\mu_{\mathbb{R}_k}(x,y) = \mu_{\mathbb{R}_k}(z)$ by a fuzzy product composition as:

$$\mu_{\mathbb{R}_k}(x,y) = \left\{ \mu_{\mathbb{R}_k}(z) = \mu_{\mathbb{E}_l}(x) \bullet \mu_{\mathbb{F}_q}(y) \mid \mu_{\mathbb{E}_L}, \mu_{\mathbb{F}_q} \in \mathbb{H} \text{ and } \mu_{\mathbb{R}_k} \in [0,1] \right\}$$
(4.9.14)

The size of k and hence the number of matrices to be constructed for the decision problem increases as the number of evaluators increases. For reasons of computational simplicity in the case of this example we assume that there are two evaluators for the classes and benefit characteristics and that the fuzzy product set is evaluationally non-interactive between the two different evaluators. In other words, in $l \odot q$, $k = l$ if $l = q$ and k is undetermined if $l \neq q$. By using eqns. (4.9.15) and (4.9.14) with the data in Tables (4.9.3) and (4.9.4) we can construct four matrices $\mathbb{R}_k$ and $\mu_{\mathbb{R}_k}(\cdot)$, $k = 1,2$ that define the fuzzy matrices of evaluation and their corresponding degrees of social confidence for the two evaluators. Thus

$$\mathbb{R}_1 = \begin{bmatrix}
12.0 & 9.0 & 6.0 & 6.0 & 3.0 & 6.0 & 6.0 & 3.0 & 4.5 & 3.0 & 3.0 \\
8.0 & 6.0 & 4.0 & 4.0 & 2.0 & 4.0 & 4.0 & 2.0 & 3.0 & 2.0 & 2.0 \\
4.0 & 3.0 & 2.0 & 2.0 & 1.0 & 2.0 & 2.0 & 1.0 & 1.5 & 1.0 & 1.0 \\
8.0 & 6.0 & 4.0 & 4.0 & 2.0 & 4.0 & 4.0 & 2.0 & 3.0 & 2.0 & 2.0 \\
4.0 & 3.0 & 2.0 & 2.0 & 1.0 & 2.0 & 2.0 & 1.0 & 1.5 & 1.0 & 1.0 \\
4.0 & 3.0 & 2.0 & 2.0 & 1.0 & 2.0 & 2.0 & 1.0 & 1.5 & 1.0 & 1.0 \\
3.6 & 2.7 & 1.8 & 1.8 & 0.9 & 1.8 & 1.8 & 0.9 & 1.35 & 0.9 & 0.9 \\
4.0 & 3.0 & 2.0 & 2.0 & 1.0 & 2.0 & 2.0 & 1.0 & 1.5 & 1.0 & 1.0 \\
4.0 & 3.0 & 2.0 & 2.0 & 1.0 & 2.0 & 2.0 & 1.0 & 1.5 & 1.0 & 1.0 \\
4.0 & 3.0 & 2.0 & 2.0 & 1.0 & 2.0 & 2.0 & 1.0 & 1.5 & 1.0 & 1.0 \\
6.0 & 4.5 & 3.0 & 3.0 & 1.5 & 3.0 & 3.0 & 1.5 & 2.25 & 1.5 & 1.5 \\
12.0 & 9.0 & 6.0 & 6.0 & 3.0 & 6.0 & 6.0 & 3.0 & 4.5 & 3.0 & 3.0 \\
10.0 & 7.5 & 5.0 & 5.0 & 2.5 & 5.0 & 5.0 & 2.5 & 3.75 & 2.5 & 2.5 \\
12.0 & 9.0 & 6.0 & 6.0 & 3.0 & 6.0 & 6.0 & 3.0 & 4.5 & 3.0 & 3.0 \\
8.0 & 6.0 & 4.0 & 4.0 & 2{,}0 & 4.0 & 4.0 & 2.0 & 3.0 & 2.0 & 2.0 \\
4.0 & 3.0 & 2.0 & 2.0 & 1.0 & 2.0 & 2.0 & 1.0 & 1.5 & 1.0 & 1.0 \\
3.2 & 2.4 & 1.6 & 1.6 & 0.8 & 1.6 & 1.6 & 0.8 & 1.2 & 0.8 & 0.8 \\
4.0 & 3.0 & 2.0 & 2.0 & 1.0 & 2.0 & 2.0 & 1.0 & 1.5 & 1.0 & 1.0 \\
4.0 & 3.0 & 2.0 & 2.0 & 1.0 & 2.0 & 2.0 & 1.0 & 1.5 & 1.0 & 1.0 \\
6.0 & 4.5 & 3.0 & 3.0 & 1.5 & 3.0 & 3.0 & 1.5 & 2.25 & 1.5 & 1.5 \\
4.0 & 3.0 & 2.0 & 2.0 & 1.0 & 2.0 & 2.0 & 1.0 & 1.5 & 1.0 & 1.0 \\
8.0 & 6.0 & 4.0 & 4.0 & 2.0 & 4.0 & 4.0 & 2.0 & 3.0 & 2.0 & 2.0
\end{bmatrix} = \left[R_{ij}\right]_1$$

(4.9.15)

$$\mu_{\mathbb{R}_1}(x,y) = \begin{bmatrix}
0.72 & 0.48 & 0.56 & 0.64 & 0.48 & 0.56 & 0.64 & 0.64 & 0.80 & 0.72 & 0.80 \\
0.81 & 0.54 & 0.63 & 0.72 & 0.54 & 0.63 & 0.72 & 0.72 & 0.90 & 0.81 & 0.90 \\
0.63 & 0.42 & 0.49 & 0.56 & 0.42 & 0.49 & 0.56 & 0.56 & 0.70 & 0.63 & 0.70 \\
0.72 & 0.48 & 0.56 & 0.64 & 0.48 & 0.56 & 0.64 & 0.64 & 0.80 & 0.72 & 0.80 \\
0.81 & 0.54 & 0.63 & 0.72 & 0.54 & 0.63 & 0.72 & 0.72 & 0.90 & 0.81 & 0.90 \\
0.90 & 0.60 & 0.70 & 0.80 & 0.60 & 0.70 & 0.80 & 0.80 & 1.00 & 0.90 & 1.00 \\
0.90 & 0.60 & 0.70 & 0.80 & 0.60 & 0.70 & 0.80 & 0.80 & 1.00 & 0.90 & 1.00 \\
0.72 & 0.48 & 0.56 & 0.64 & 0.48 & 0.56 & 0.64 & 0.64 & 0.80 & 0.72 & 0.80 \\
0.81 & 0.54 & 0.63 & 0.72 & 0.54 & 0.63 & 0.72 & 0.72 & 0.90 & 0.81 & 0.90 \\
0.72 & 0.48 & 0.56 & 0.64 & 0.48 & 0.56 & 0.64 & 0.64 & 0.80 & 0.72 & 0.80 \\
0.63 & 0.42 & 0.49 & 0.56 & 0.42 & 0.49 & 0.56 & 0.56 & 0.70 & 0.63 & 0.70 \\
0.54 & 0.36 & 0.42 & 0.48 & 0.36 & 0.42 & 0.48 & 0.48 & 0.60 & 0.54 & 0.60 \\
0.72 & 0.48 & 0.56 & 0.64 & 0.48 & 0.56 & 0.64 & 0.64 & 0.80 & 0.72 & 0.80 \\
0.81 & 0.54 & 0.63 & 0.72 & 0.54 & 0.63 & 0.72 & 0.72 & 0.90 & 0.81 & 0.90 \\
0.81 & 0.54 & 0.63 & 0.72 & 0.54 & 0.63 & 0.72 & 0.72 & 0.90 & 0.81 & 0.90 \\
0.90 & 0.60 & 0.70 & 0.80 & 0.60 & 0.70 & 0.80 & 0.80 & 1.00 & 0.90 & 1.00 \\
0.90 & 0.60 & 0.70 & 0.80 & 0.60 & 0.70 & 0.80 & 0.80 & 1.00 & 0.90 & 1.00 \\
0.90 & 0.60 & 0.70 & 0.80 & 0.60 & 0.70 & 0.80 & 0.80 & 1.00 & 0.90 & 1.00 \\
0.81 & 0.54 & 0.63 & 0.72 & 0.54 & 0.63 & 0.72 & 0.72 & 0.90 & 0.81 & 0.90 \\
0.72 & 0.48 & 0.56 & 0.64 & 0.48 & 0.56 & 0.64 & 0.64 & 0.80 & 0.72 & 0.80 \\
0.81 & 0.54 & 0.63 & 0.72 & 0.54 & 0.63 & 0.72 & 0.72 & 0.90 & 0.81 & 0.90 \\
0.90 & 0.60 & 0.70 & 0.80 & 0.60 & 0.70 & 0.80 & 0.80 & 1.00 & 0.90 & 1.00
\end{bmatrix}$$

(4.9.16)

$$\mathbb{R}_2 = \begin{bmatrix}
7.50 & 7.5 & 5.0 & 2.5 & 5.0 & 7.5 & 7.5 & 3.75 & 3.75 & 2.5 & 2.5 \\
6.0 & 6.0 & 4.0 & 2.0 & 4.0 & 6.0 & 6.0 & 3.0 & 3.0 & 2.0 & 2.0 \\
4.5 & 4.5 & 3.0 & 1.5 & 3.0 & 4.5 & 4.5 & 2.25 & 2.25 & 1.5 & 1.5 \\
3.0 & 3.0 & 2.0 & 1.0 & 2.0 & 3.0 & 3.0 & 1.5 & 1.5 & 1.0 & 1.0 \\
4.5 & 4.5 & 3.0 & 1.5 & 3.0 & 4.5 & 4.5 & 2.25 & 2.25 & 1.5 & 1.5 \\
4.5 & 4.5 & 3.0 & 1.5 & 3.0 & 4.5 & 4.5 & 2.25 & 2.25 & 1.5 & 1.5 \\
3.0 & 3.0 & 2.0 & 1.0 & 2.0 & 3.0 & 3.0 & 1.5 & 1.5 & 1.0 & 1.0 \\
3.0 & 3.0 & 2.0 & 1.0 & 2.0 & 3.0 & 3.0 & 1.5 & 1.5 & 1.0 & 1.0 \\
3.0 & 3.0 & 2.0 & 1.0 & 2.0 & 3.0 & 3.0 & 1.5 & 1.5 & 1.0 & 1.0 \\
4.5 & 4.5 & 3.0 & 1.5 & 3.0 & 4.5 & 4.5 & 2.25 & 2.25 & 1.5 & 1.5 \\
12.0 & 12.0 & 8.0 & 4.0 & 8.0 & 12.0 & 12.0 & 6.0 & 6.0 & 4.0 & 4.0 \\
9.0 & 9.0 & 6.0 & 3.0 & 6.0 & 9.0 & 9.0 & 4.5 & 4.5 & 3.0 & 3.0 \\
9.0 & 9.0 & 6.0 & 3.0 & 6.0 & 9.0 & 9.0 & 4.5 & 4.5 & 3.0 & 3.0 \\
9.0 & 9.0 & 6.0 & 3.0 & 6.0 & 9.0 & 9.0 & 4.5 & 4.5 & 3.0 & 3.0 \\
6.0 & 6.0 & 4.0 & 2.0 & 4.0 & 6.0 & 6.0 & 3.0 & 3.0 & 2.0 & 2.0 \\
3.0 & 3.0 & 2.0 & 1.0 & 2.0 & 3.0 & 3.0 & 1.5 & 1.5 & 1.0 & 1.0 \\
4.5 & 4.5 & 3.0 & 1.5 & 3.0 & 4.5 & 4.5 & 2.25 & 2.25 & 1.5 & 1.5 \\
3.0 & 3.0 & 2.0 & 1.0 & 2.0 & 3.0 & 3.0 & 1.5 & 1.5 & 1.0 & 1.0 \\
4.5 & 4.5 & 3.0 & 1.5 & 3.0 & 4.5 & 4.5 & 2.25 & 2.25 & 1.5 & 1.5 \\
3.0 & 3.0 & 2.0 & 1.0 & 2.0 & 3.0 & 3.0 & 1.5 & 1.5 & 1.0 & 1.0 \\
3.0 & 3.0 & 2.0 & 1.0 & 2.0 & 3.0 & 3.0 & 1.5 & 1.5 & 1.0 & 1.0 \\
4.5 & 4.5 & 3.0 & 1.5 & 3.0 & 4.5 & 4.5 & 2.25 & 2.25 & 1.5 & 1.5 \\
3.0 & 3.0 & 2.0 & 1.0 & 2.0 & 3.0 & 3.0 & 1.5 & 1.5 & 1.0 & 1.0 \\
6.0 & 6.0 & 4.0 & 2.0 & 4.0 & 6.0 & 6.0 & 3.0 & 3.0 & 2.0 & 2.0
\end{bmatrix} = \left[R_{ij}\right]_2$$

(4.9.17)

$$\mu_{\mathbb{R}_2}(x,y) = \begin{bmatrix}
0.56 & 0.56 & 0.48 & 0.48 & 0{,}80 & 0.72 & 0.64 & 0.80 & 0.72 & 0.72 & 0.80 \\
0.49 & 0.49 & 0.42 & 0.42 & 0.70 & 0.63 & 0.56 & 0.70 & 0.63 & 0.63 & 0.70 \\
0.63 & 0.63 & 0.54 & 0.54 & 0.90 & 0.81 & 0.72 & 0.90 & 0.81 & 0.81 & 0.90 \\
0.49 & 0.49 & 0.42 & 0.42 & 0.70 & 0.63 & 0.56 & 0.70 & 0.63 & 0.63 & 0.70 \\
0.56 & 0.56 & 0.48 & 0.48 & 0.80 & 0.72 & 0.64 & 0.80 & 0.72 & 0.72 & 0.80 \\
0.63 & 0.63 & 0.54 & 0.54 & 0.90 & 0.81 & 0.72 & 0.90 & 0.81 & 0.81 & 0.90 \\
0.70 & 0.70 & 0.60 & 0.60 & 1.00 & 0.90 & 0.80 & 1.00 & 0.90 & 0.90 & 1.00 \\
0.70 & 0.70 & 0.60 & 0.60 & 1.00 & 0.90 & 0.80 & 1.00 & 0.90 & 0.90 & 1.00 \\
0.63 & 0.63 & 0.54 & 0.54 & 0.90 & 0.81 & 0.72 & 0.90 & 0.81 & 0.81 & 0.90 \\
0.42 & 0.42 & 0.36 & 0.36 & 0.60 & 0.54 & 0.48 & 0.60 & 0.54 & 0.54 & 0.60 \\
0.42 & 0.42 & 0.36 & 0.36 & 0.60 & 0.54 & 0.48 & 0.60 & 0.54 & 0.54 & 0.60 \\
0.49 & 0.49 & 0.42 & 0.42 & 0.70 & 0.63 & 0.56 & 0.70 & 0.63 & 0.63 & 0.70 \\
0.70 & 0.70 & 0.60 & 0.60 & 1.00 & 0.90 & 0.80 & 1.00 & 0.90 & 0.90 & 1.00 \\
0.70 & 0.70 & 0.60 & 0.60 & 1.00 & 0.90 & 0.80 & 1.00 & 0.90 & 0.90 & 1.00 \\
0.56 & 0.56 & 0.48 & 0.48 & 0.80 & 0.72 & 0.64 & 0.80 & 0.72 & 0.72 & 0.80 \\
0.63 & 0.63 & 0.54 & 0.54 & 0.90 & 0.81 & 0.72 & 0.90 & 0.81 & 0.81 & 0.90 \\
0.63 & 0.63 & 0.54 & 0.54 & 0.90 & 0.81 & 0.72 & 0.90 & 0.81 & 0.81 & 0.90 \\
0.70 & 0.70 & 0.60 & 0.60 & 1.00 & 0.90 & 0.80 & 1.00 & 0.90 & 0.90 & 1.00 \\
0.70 & 0.70 & 0.60 & 0.60 & 1.00 & 0.90 & 0.80 & 1.00 & 0.90 & 0.90 & 1.00 \\
0.63 & 0.63 & 0.54 & 0.54 & 0.90 & 0.81 & 0.72 & 0.90 & 0.81 & 0.81 & 0.90 \\
0.70 & 0.70 & 0.60 & 0.60 & 1.00 & 0.90 & 0.80 & 1.00 & 0.90 & 0.90 & 1.00 \\
0.70 & 0.70 & 0.60 & 0.60 & 1.00 & 0.90 & 0.80 & 1.00 & 0.90 & 0.90 & 1.00
\end{bmatrix}$$

(4.9.18)

The next step in the computational process is to combine the values in $\mathbb{R}_k$ and $\mu_{\mathbb{R}_k}(\cdot)$ to develop the optimal values of benefit intensities which will constitute the distributional weights. We utilize Theorems 4.7.3(A) and (B) in terms of the following algorithmic process so as to arrive at the final matrix of intensity weights (see also Chapter 3, section 3.5.4.2 and Chapters 5 and 6 of this volume.) The theorems require that we select values whose evaluators have the highest degree of expertness. To proceed we specify the matrices, $\mu_{\mathbb{R}_k}(\cdot)$ and $\mathbb{R}_k$ in terms of their column vectors, ν_{kj} and w_{kj} respectively. Thus

$$\mu_{\mathbb{R}_k}(\cdot) = \left[\nu_{k1}, \nu_{k2}, \cdots, \nu_{kj}, \cdots, \nu_{k,11}\right]$$

(4.9.19)

$$\mathbb{R}_k = \left[w_{k1}, w_{k2}, \cdots, w_{kj}, \cdots, w_{kj}\right]$$

(4.9.20)

There are as many pairs of matrices of eqns. (4.9.19) and (4.9.20) as there are independent evaluators. For the example under consideration we have two pairs of matrices of the form

$$\left.\begin{aligned} \mathbb{R}_1 &= \left[w_{11}, w_{12}, \cdots, w_{1j}, \cdots, w_{1,11}\right] \\ \mu_{\mathbb{R}_1}(\bullet) &= \left[\nu_{11}, \nu_{12}, \cdots, \nu_{1j}, \cdots, \nu_{1,11}\right] \end{aligned}\right\} \quad (4.9.21)$$

$$\left.\begin{aligned} \mathbb{R}_2 &= \left[w_{21}, w_{22}, \cdots, w_{2j}, w_{2,11}\right] \\ \mu_{\mathbb{R}_2}(\bullet) &= \left[\nu_{21}, \nu_{22}, \cdots \nu_{2j}, \cdots, \nu_{2,11}\right] \end{aligned}\right\} \quad (4.9.22)$$

For each class and all evaluators we have a matrix $\mu_{\mathbb{R}_j}(\bullet)$ and $\mathbb{R}_j$ and for any fixed j. For example, the number one class with k evaluators and a given number of benefit characteristics has a matrix whose column vectors are:

$$\mu_{\mathbb{R}_{k1}}(\bullet) = \left[\nu_{11}, \nu_{21}, \nu_{31}, \cdots, \nu_{k1}\right] \quad (4.9.23a)$$

and

$$\mathbb{R}_{k1} = \left[w_{11}, w_{21}, w_{31}, \cdots, w_{k1}\right] \quad (4.9.23b)$$

Generally,

$$\mathbb{R}_{kj} = \left[w_{1j}, w_{2j}, w_{3j}, \cdots, w_{kj}, \cdots\right] \quad (4.9.24a)$$

$$\mu_{\mathbb{R}_{kj}}(\bullet) = \left[\nu_{1j}, \nu_{2j}, \nu_{3j}, \cdots, \nu_{kj} \cdots\right] \quad (4.9.24b)$$

Define a fuzzy produce space, $\mathbb{R}_j$, and the corresponding membership functions as

$$\mathbb{R}_j = \bigotimes_{k \in \mathbb{L}} w_{kj}$$

$$\mu_{\mathbb{R}_j}(\bullet) = \bigotimes_{k \in \mathbb{L}} \nu_{kj}$$

where $\mathbb{L}$ is the index set of evaluators and j is fixed in $\mathbb{N}$, the index set of social classes. From eqns. (4.9.23a &b) we construct a maximizing matrix by means of a first projection, $\left(P^1_{roj}\right)$ of a rank relation, R_P on $\mathbb{R}_j$ as

$$\nu_{\bullet j} = P^{1}_{roj}\left[R_P \mid \mathbb{R}\right] = \sup_{x \in \mathbb{R}_j} \bigotimes_{k \in \mathbb{L}} \nu_{kj}$$

(4.9.26)

and corresponding to it we ascertain a vector $W_{\bullet j}$ of distributional weights where $\nu_{\bullet j}$ and $W_{\bullet j}$ are column vectors of new matrices, $\mu_{\mathbb{R}}(\bullet)$ and $\mathbb{R}$, respectively.

By application of the set of computational algorithms of eqns. (4.9.23) - (4.9.26) to the values provided in Eqs. (4.9.16) and (4.9.18) we can construct the matrix of optimal intensity weights, $\mathbb{R}$ as distributed over the cohorts with corresponding social degree of acceptance, $\mu_{\mathbb{R}}(\bullet)$ as in eqns (4.9.27) and (4.9.28).

The final aggregation algorithm requires that we partition the matrices, $\mathbb{R}$ and $\mu_{\mathbb{R}}(\bullet)$ into four pair-wise sub-matrices that correspond to sub-matrices $\mathbb{C}_1$, (eqn. (4.9.1)), $\hat{\mathbb{T}}_2$, (eqn. (4.9.7)), $\mathbb{C}_3^{T}$, (eqn. (4.9.3) and $\mathbb{C}_4$, (eqn. (4.9.4)). The relevant partition and sub-matrices are: A_1, A_2, A_3 and A_4, respectively, with corresponding matrices of the forms shown in eqns. (4.9.29 – 4.9.36).

$$\mathbb{R} = \begin{bmatrix}
12.0 & 7.5 & 6.0 & 6.0 & 5.0 & 7.0 & 7.5 & 3.75 & 4.5 & 2.5 & 3.0 \\
8.0 & 6.0 & 4.0 & 4.0 & 4.0 & 4.0 & 4.0 & 2.0 & 3.0 & 2.0 & 2.0 \\
4.5 & 4.5 & 3.0 & 2.0 & 3.0 & 4.5 & 4.5 & 2.25 & 2.5 & 1.5 & 1.5 \\
8.6 & 3.0 & 4.0 & 4.0 & 2.0 & 3.0 & 4.0 & 1.5 & 3.0 & 2.0 & 2.0 \\
4.0 & 4.5 & 2.0 & 2.0 & 3.0 & 4.5 & 2.0 & 2.25 & 1.5 & 1.0 & 1.0 \\
4.0 & 4.5 & 2.0 & 2.0 & 3.0 & 4.5 & 2.0 & 2.25 & 1.5 & 1.0 & 1.0 \\
3.6 & 3.0 & 1.8 & 1.8 & 2.0 & 3.0 & 3.0 & 1.5 & 1.35 & 1.0 & 1.0 \\
4.0 & 3.0 & 2.0 & 2.0 & 2.0 & 3.0 & 3.0 & 1.5 & 1.5 & 1.0 & 1.0 \\
4.0 & 3.0 & 2.0 & 2.0 & 2.0 & 3.0 & 2.0 & 1.5 & 1.5 & 1.0 & 1.0 \\
4.0 & 4.5 & 2.0 & 2.0 & 3.0 & 4.5 & 2.0 & 1.0 & 1.5 & 1.0 & 1.0 \\
6.0 & 4.5 & 3.0 & 3.0 & 8.0 & 2.0 & 3.0 & 6.0 & 2.25 & 1.5 & 1.5 \\
12.0 & 9.0 & 6.0 & 6.0 & 6.0 & 9.0 & 9.0 & 4.5 & 4.5 & 3.0 & 3.0 \\
10.0 & 9.0 & 6.0 & 5.0 & 6.0 & 9.0 & 9.0 & 4.5 & 4.5 & 3.0 & 3.0 \\
12.0 & 9.0 & 6.0 & 6.0 & 6.0 & 9.0 & 9.0 & 4.5 & 4.5 & 3.0 & 3.0 \\
8.0 & 6.0 & 4.0 & 4.0 & 4.0 & 6.0 & 4.0 & 3.0 & 3.0 & 2.0 & 2.0 \\
4.0 & 3.0 & 2.0 & 2.0 & 2.0 & 3.0 & 2.0 & 1.5 & 1.5 & 1.0 & 1.0 \\
3.2 & 4.5 & 1.6 & 1.6 & 3.0 & 4.5 & 1.6 & 2.25 & 1.2 & 0.8 & 0.8 \\
4.0 & 3.0 & 2.0 & 2.0 & 2.0 & 3.0 & 3.0 & 1.5 & 1.5 & 1.0 & 1.0 \\
4.0 & 3.0 & 2.0 & 2.0 & 2.0 & 3.0 & 3.0 & 1.5 & 1.5 & 1.0 & 1.0 \\
6.0 & 4.5 & 3.0 & 3.0 & 3.0 & 4.5 & 4.5 & 2.25 & 2.5 & 1.5 & 1.5 \\
4.0 & 3.0 & 2.0 & 2.0 & 2.0 & 3.0 & 3.0 & 1.5 & 1.5 & 1.0 & 1.0 \\
8.0 & 6.0 & 4.0 & 4.0 & 4.0 & 6.0 & 4.0 & 2.0 & 3.0 & 2.0 & 2.0
\end{bmatrix}$$

(4.9.27)

$$\mu_{\mathbb{R}}(\bullet) = \begin{bmatrix} 0.72 & 0.56 & 0.56 & 0.64 & 0.80 & 0.72 & 0.64 & 0.80 & 0.80 & 0.72 & 0.80 \\ 0.81 & 0.54 & 0.63 & 0.72 & 0.70 & 0.63 & 0.72 & 0.72 & 0.90 & 0.81 & 0.90 \\ 0.63 & 0.63 & 0.54 & 0.56 & 0.90 & 0{,}81 & 0.72 & 0.90 & 0.81 & 0.81 & 0.90 \\ 0.72 & 0.49 & 0.56 & 0.64 & 0.70 & 0.63 & 0.64 & 0.70 & 0.81 & 0.72 & 0.80 \\ 0.81 & 0.56 & 0.63 & 0.72 & 0.80 & 0.72 & 0.72 & 0.80 & 0.90 & 0.81 & 0.90 \\ 0.90 & 0.63 & 0.70 & 0.80 & 0.90 & 0.81 & 0.80 & 0.90 & 1.00 & 0.90 & 1.00 \\ 0.90 & 0.70 & 0.70 & 0.80 & 1.00 & 0.90 & 0.80 & 1.00 & 1.00 & 0.90 & 1.00 \\ 0.72 & 0.70 & 0.60 & 0.64 & 1.00 & 0.90 & 0.80 & 1.00 & 0.90 & 0.81 & 1.00 \\ 0.81 & 0.63 & 0.63 & 0.72 & 0.90 & 0.81 & 0.72 & 0.90 & 0.90 & 0.81 & 0.90 \\ 0.72 & 0.48 & 0.56 & 0.64 & 0.60 & 0.56 & 0.64 & 0.64 & 0.80 & 0.72 & 0.80 \\ 0.63 & 0.42 & 0.49 & 0.56 & 0.60 & 0.54 & 0.56 & 0.60 & 0.70 & 0.63 & 0.70 \\ 0.54 & 0.49 & 0.42 & 0.48 & 0.70 & 0.63 & 0.56 & 0.70 & 0.63 & 0.63 & 0.70 \\ 0.72 & 0.70 & 0.60 & 0.64 & 1.00 & 0.90 & 0.80 & 1.00 & 0.90 & 0.90 & 1.00 \\ 0.81 & 0.70 & 0.63 & 0.72 & 1.00 & 0.90 & 0.80 & 1.00 & 0.90 & 0.90 & 1.00 \\ 0.81 & 0.56 & 0.63 & 0.72 & 0.80 & 0.72 & 0.72 & 0.81 & 0.90 & 0.81 & 0.90 \\ 0.90 & 0.63 & 0.70 & 0.80 & 0.90 & 0.81 & 0.80 & 0.90 & 1.00 & 0.90 & 1.00 \\ 0.90 & 0.63 & 0{,}70 & 0.80 & 0.90 & 0.81 & 0.80 & 0.90 & 1.00 & 0.90 & 1.00 \\ 0.90 & 0.70 & 0.70 & 0.80 & 1.00 & 0.90 & 0.80 & 1.00 & 1.00 & 0.90 & 1.00 \\ 0.81 & 0.70 & 0.63 & 0.72 & 1.00 & 0.90 & 0.80 & 1.00 & 0.90 & 0.90 & 1.00 \\ 0.72 & 0.63 & 0.56 & 0.64 & 0.90 & 0.81 & 0.72 & 0.90 & 0.81` & 0.81 & 0.90 \\ 0.81 & 0.70 & 0.63 & 0.72 & 1.00 & 0.9 & 0.80 & 1.00 & 0.90 & 0.90 & 1.00 \\ 0.90 & 0.70 & 0.70 & 0.80 & 1.00 & 0.90 & 0.80 & 1.00 & 1.00 & 0.90 & 1.00 \end{bmatrix}$$

(4.9.28)

$$A_1 = \begin{bmatrix} \alpha_{11} & \alpha_{12} & \alpha_{13} & \alpha_{14} \\ \alpha_{21} & \alpha_{22} & \alpha_{23} & \alpha_{24} \\ \alpha_{91} & \alpha_{92} & \alpha_{93} & \alpha_{94} \\ \alpha_{10,1} & \alpha_{10,2} & \alpha_{10,3} & \alpha_{10,4} \\ \alpha_{11,1} & \alpha_{11,1} & \alpha_{11,1} & \alpha_{11,4} \end{bmatrix} = \begin{bmatrix} 12 & 7.5 & 6.0 & 6.0 \\ 8.0 & 6.0 & 4.0 & 4.0 \\ 4.0 & 3.0 & 2.0 & 2.0 \\ 4.0 & 4.5 & 2.0 & 2.0 \\ 6.0 & 4.5 & 3.0 & 3.0 \end{bmatrix} = \left[\alpha_{lk}\right]$$

(4.9.29)

$$\mu_{A_1}(\bullet) = \begin{bmatrix} \beta_{11} & \beta_{12} & \beta_{13} & \beta_{14} \\ \beta_{21} & \beta_{22} & \beta_{23} & \beta_{24} \\ \beta_{91} & \beta_{92} & \beta_{93} & \beta_{94} \\ \beta_{10,1} & \beta_{10,2} & \beta_{10,3} & \beta_{10,4} \\ \beta_{11,1} & \beta_{11,2} & \beta_{11,3} & \beta_{11.4} \end{bmatrix} = \begin{bmatrix} 0.72 & 0.52 & 0.64 & 0.80 \\ 0.81 & 0.54 & 0.63 & 0.72 \\ 0.81 & 0.63 & 0.63 & 0.72 \\ 0.72 & 0.48 & 0.56 & 0.64 \\ 0.63 & 0.42 & 0.49 & 0.56 \end{bmatrix} = \left[\beta_{lk}\right]$$

(4.9.30)

$$A_2 = \begin{bmatrix} \alpha_{31} & \alpha_{32} & \alpha_{33} & \alpha_{34} \\ \alpha_{41} & \alpha_{42} & \alpha_{43} & \alpha_{44} \\ \alpha_{51} & \alpha_{52} & \alpha_{53} & \alpha_{54} \\ \alpha_{16,8} & \alpha_{16,9} & \alpha_{16,10} & \alpha_{16,11} \\ \alpha_{17,8} & \alpha_{17,9} & \alpha_{17,10} & \alpha_{17,11} \\ \alpha_{18.8} & \alpha_{18,9} & \alpha_{18,10} & \alpha_{18,11} \\ \alpha_{20,8} & \alpha_{20,9} & \alpha_{20,10} & \alpha_{20,11} \\ \alpha_{21,8} & \alpha_{21,9} & \alpha_{21,10} & \alpha_{21,11} \\ \alpha_{22,1} & \alpha_{22,2} & \alpha_{22,3} & \alpha_{22.4} \\ \alpha_{22,5} & \alpha_{22,6} & \alpha_{22,7} & \alpha_{22,8} \end{bmatrix} = \begin{bmatrix} 4.50 & 4.50 & 3.00 & 2.00 \\ 8.00 & 3.00 & 4.00 & 4.00 \\ 4.00 & 4.50 & 2.00 & 2.00 \\ 1.50 & 1.50 & 1.00 & 1.00 \\ 2.25 & 1.20 & 0.80 & 0.80 \\ 1.50 & 1.50 & 1.00 & 1.00 \\ 2.25 & 2.50 & 1.50 & 1.50 \\ 1.50 & 1.50 & 1.00 & 1.00 \\ 8.00 & 6.00 & 4.00 & 4.00 \\ 4.00 & 6.00 & 4.00 & 2.00 \end{bmatrix} = \left[w_{ij}\right]$$

(4.9.31)

$$\mu_{A_2}(\bullet) = \begin{bmatrix} \beta_{31} & \beta_{32} & \beta_{33} & \beta_{34} \\ \beta_{41} & \beta_{42} & \beta_{43} & \beta_{44} \\ \beta_{51} & \beta_{52} & \beta_{53} & \beta_{54} \\ \beta_{16,8} & \beta_{16,9} & \beta_{16,10} & \beta_{16,11} \\ \beta_{17,8} & \beta_{17,9} & \beta_{17,10} & \beta_{17,11} \\ \beta_{18,8} & \beta_{18,9} & \beta_{18,10} & \beta_{18,11} \\ \beta_{20,8} & \beta_{20,9} & \beta_{20.10} & \beta_{20,11} \\ \beta_{21,8} & \beta_{21,9} & \beta_{21,10} & \beta_{21,11} \\ \beta_{22,1} & \beta_{22,2} & \beta_{22,3} & \beta_{22,4} \\ \beta_{22,5} & \beta_{22,6} & \beta_{22,7} & \beta_{22,8} \end{bmatrix} = \begin{bmatrix} 0.63 & 0.63 & 0.54 & 0.56 \\ 0.72 & 0.49 & 0.56 & 0.64 \\ 0.81 & 0.56 & 0.63 & 0.72 \\ 0.90 & 1.00 & 0.90 & 1.00 \\ 0.90 & 1.00 & 0.90 & 1.00 \\ 1.00 & 1.00 & 0.90 & 1.00 \\ 0.90 & 0.81 & 0.80 & 0.90 \\ 1.00 & 0.90 & 0.90 & 1.00 \\ 0.90 & 0.70 & 0.70 & 0.80 \\ 1.00 & 0.90 & 0.80 & 1.00 \end{bmatrix} = \left[\beta_{ij}\right]$$

(4.9.32)

$$A_3 = \begin{bmatrix} \alpha_{81} & \alpha_{12,1} & \alpha_{13,1} & \alpha_{14.1} & \alpha_{19,5} \\ \alpha_{82} & \alpha_{12,2} & \alpha_{13,2} & \alpha_{14,2} & \alpha_{19,6} \\ \alpha_{83} & \alpha_{12,3} & \alpha_{13,3} & \alpha_{14,3} & \alpha_{19,7} \\ \alpha_{84} & \alpha_{12,4} & \alpha_{13,4} & \alpha_{14,4} & \alpha_{19,8} \\ 0 & \alpha_{12,5} & \alpha_{13,5} & \alpha_{14,5} & 0 \\ 0 & \alpha_{12,6} & \alpha_{13,6} & \alpha_{14,6} & 0 \\ 0 & \alpha_{12,7} & \alpha_{13,7} & \alpha_{14,7} & 0 \end{bmatrix} = \begin{bmatrix} 4.0 & 12.0 & 10.0 & 12.0 & 2.0 \\ 3.0 & 9.0 & 9.0 & 9.0 & 3.0 \\ 2.0 & 6.0 & 6.0 & 6.0 & 3.0 \\ 2.0 & 6.0 & 5.0 & 6.0 & 1.5 \\ 0 & 6.0 & 6.0 & 6.0 & 0 \\ 0 & 9.0 & 9.0 & 9.0 & 0 \\ 0 & 9.0 & 9.0 & 9.0 & 0 \end{bmatrix} = \left[\alpha_{rs}\right]$$

(4.9.33)

$$\mu_{A_3}(\cdot) = \begin{bmatrix} \beta_{81} & \beta_{12,1} & \beta_{13,1} & \beta_{14,1} & \beta \\ \beta_{82} & \beta_{12,2} & \beta_{13,2} & \beta_{14,2} & \beta \\ \beta_{83} & \beta_{12,3} & \beta_{13,3} & \beta_{14,3} & \beta \\ \beta_{84} & \beta_{12,4} & \beta_{13,4} & \beta_{14,4} & \beta \\ 0 & \beta_{12,5} & \beta_{13,5} & \beta_{14,5} & 0 \\ 0 & \beta_{12,6} & \beta_{13,6} & \beta_{14,6} & 0 \\ 0 & \beta_{12,7} & \beta_{13,7} & \beta_{14,7} & 0 \end{bmatrix} = \begin{bmatrix} 0.72 & 0.54 & 0.72 & 0.81 & 1.00 \\ 0.70 & 0.49 & 0.70 & 0.70 & 0.90 \\ 0.60 & 0.49 & 0.60 & 0.63 & 0.80 \\ 0.64 & 0.48 & 0.64 & 0.72 & 1.00 \\ 0.00 & 0.70 & 1.00 & 1.00 & 0.00 \\ 0.00 & 0.63 & 0.90 & 0.90 & 0.00 \\ 0.00 & 0.56 & 0.80 & 0.90 & 0.00 \end{bmatrix} = \left[\beta_{rs}\right]$$

(4.9.34)

$$A_4 = \begin{bmatrix} \alpha_{61} & \alpha_{62} & \alpha_{63} & \alpha_{64} & 0 & 0 & 0 \\ \alpha_{71} & \alpha_{72} & \alpha_{73} & \alpha_{74} & 0 & 0 & 0 \\ \alpha_{15,1} & \alpha_{15,2} & \alpha_{15,3} & \alpha_{15,4} & \alpha_{15,5} & \alpha_{15,6} & \alpha_{15,7} \end{bmatrix} = \begin{bmatrix} 4.0 & 4.5 & 2.0 & 2.0 & 0.0 & 0.0 & 0.0 \\ 3.6 & 3.0 & 1.8 & 1.8 & 0.0 & 0.0 & 0.0 \\ 8.0 & 6.0 & 4.0 & 4.0 & 4.0 & 6.0 & 4.0 \end{bmatrix} = \left[\alpha_{qh}\right]$$

(4.9.35)

$$\mu_{A_4} = \begin{bmatrix} \beta_{61} & \beta_{62} & \beta_{63} & \beta_{64} & 0 & 0 & 0 \\ \beta_{71} & \beta_{72} & \beta_{73} & \beta_{74} & 0 & 0 & 0 \\ \beta_{15,1} & \beta_{15,2} & \beta_{15,3} & \beta_{15,4} & \beta_{15,5} & \beta_{15,6} & \beta_{15,7} \end{bmatrix}$$

$$= \begin{bmatrix} 0.09 & 0.60 & 0.70 & 0.80 & 0 & 0 & 0 \\ 0.90 & 0.70 & 0.70 & 0.80 & 0 & 0 & 0 \\ 0.81 & 0.56 & 0.63 & 0.72 & 0.80 & 0.72 & 0.72 \end{bmatrix} = \left[\beta_{qh}\right]$$

(4.9.36)

We may now present the final benefit information set by combining the relevant matrices. The total monetary benefit value with distributional weights may be computed by combining Eq. (4.9.1) with those of eqns. (4.9.31) - (4.9.34) to obtain an adjusted value, $\hat{\boldsymbol{B}}_{ST}^{M}$ as:

$$\hat{\boldsymbol{B}}_{ST}^{M} = 35\left(\sum_{i}\sum_{j}\alpha_{ij}\beta_{ij}t_{ij}w_{ij}\right) + \sum_{r}\sum_{s}\alpha_{rs}\beta_{rs}p_{rs}b_{rs}$$

$$= 35(101, 913,68)+1106052.0 = 3566978.80+1106052.0 = \$4673030.80$$

Without the distributional weights, the total monetary benefit value is obtained as:

$\boldsymbol{B}_{ST}^{M} = 35(679806.20) = 367300 = 2376717 + 367300.0 = \$2744017.00$

The one year total monetary benefit for treating mental illness of our hypothetical community by accounting for benefit intensities and the degree of expertness of the evaluators is \$4673030.80. If benefit intensities are not accounted for the monetary value of the benefits is 2744017.00 and the difference $\Delta\boldsymbol{B}_{ST} = 1929013.80$. It may be noted that the monetary benefit value is of two components of the part accruing to the treated person and the part accruing to the society.

In addition to this monetary benefit information, we also have important benefit information subsets which are displayed in eqns. (4.9.1) and (4.9.4) with their corresponding matrices of intensity weights and the degrees of social acceptance of these weights as measured by the degree of expertness of the evaluators. The information subsets are such that $\mathbb{C}_1$, [eqn. 4.9.1)], A_1 [eqn. (4.9.29)] and $\mu_{A_1}(\bullet)$, [Eq. (4.9.3)] are corresponding subsets while $\mathbb{C}_4$ [eqn. (4.9.4), A_4 [eqn. (4.9.35) and $\mu_{A_4}(\bullet)$, [eqn. (4.9.36) constitute another corresponding sets of non-monetary information. Together they constitute a complete information set on benefit of a treatment regime of mental illness. This computation for benefit must be repeated for all relevant decision periods to obtain a time series of benefit data whose present value equivalence may be constructed conditional on a given social discount rate and process. The required information set may also be forecasted from the initial benefit imputation and other relevant hard and soft data.

5 Essential Mathematics for Fuzzy Decision Processes

There are certain mathematical structures that we have used in integrating cost-benefit analysis and the theory of fuzzy decision into a unified theory of decision making. The fuzzy decision theory requires the use of logic and the mathematics of fuzzy processes. The mathematical orientation of the fuzzy processes is currently known as *Fuzzy mathematics* with its own logic of reasoning and inference. It provides a new mathematical viewpoint that constitutes a well structured body of concepts, techniques, algorithms, and inferential processes that allow crisp conclusions to be derived about fuzzy phenomena. To assist in an easy reference to the body of knowledge relevant to the work done here we shall present the essentials of fuzzy mathematics in this Chapter.

5.1 The Basis of the Theory of Fuzzy Mathematics

Consider a space, Ω, of measurable values in some scientific sense. We may take this space to be $\mathbb{R}^n = \Omega$

Definition 5.1.1

The space Ω is said to be fuzzy if either some or all the concepts and measurements defining the structure of the space admit of either shades of meaning or degrees of accuracy or subjectivity or combinations of them.

Definition 5.1.2

A number $\alpha \in [0,1]$ is said to be a fuzzy numerical value defined over space, Ω if for each element in the space it establishes a rank order of either shades of meaning of concepts or shades of exactness of measurement or degrees of accuracy of statements.

Definition 5.1.3

A function that assigns numerical value, $\alpha \in [0,1]$ to express either degrees or shades of belonging of elements around a value $x \in \Omega$ is referred to as a *membership function* which may also be viewed as a fuzzy distribution function.

Definition 5.1.4

If α is a fuzzy number associated with $x \in \Omega$, then $\exists$ a set $\mathbb{A}$ that is a fuzzy set covering of x with a membership function

$$\mu_{\mathbb{A}} : \mathbb{A} \rightarrow [0,1] \text{ such that } \exists\, \hat{x} \in \mathbb{A} \subseteq \Omega \text{ with } \mu_{\mathbb{A}}(\hat{x}) = \alpha .$$

Definition 5.1.5

A fuzzy variable, x is any quantity which may take on any of the values in a specified domain, $\mathbb{A} \subset \Omega$, with an associated distribution of degrees of belonging over $\mathbb{A}$ where the variables always has an associated degree of belonging, accuracy or shades of meaning.

Note:

Every fuzzy variable is characterized by a pair of values of actual value and a corresponding membership value of belonging to the fuzzy covering. The fuzzy covering provides a distribution of degrees of belonging to a defined set. One may think of the structure of fuzzy variable in terms the structure of random variable where there is an actual value an associated probability that belongs to the random covering which provides a distribution of degrees of uncertainty of outcome.

Definition 5.1.6

A space, Ω is said to be fuzzy measurable if there exist a membership function $\mu : \Omega \rightarrow [0,1]$ such that a fuzzy measured value $\alpha \in [0,1]$ can be attached to each element in the space

Definition 5.1.7

A fuzzy subset $\mathbb{A} \in \Omega$ is a collection of pairs of values of the form

$$\mathbb{A} = \{(\mathrm{x}, \mu_{\mathbb{A}}(\mathrm{x})) \mid \mathrm{x} \in \Omega \text{ and } \mu_{\mathbb{A}}(\mathrm{x}) \in [0,1]\}$$

Proposition 5.1.1

Every fuzzy set or subset is a fuzzy covering of a defined fuzzy variable and every fuzzy covering is a fuzzy set or fuzzy subset.

Definition 5.1.8

If $x \in \Omega$ then x is said to belong to a subset of $\mathbb{A} \subset \Omega$, surely if $\mu_{\mathbb{A}}(x) = 1$. The element, x, is said not to belong to $\mathbb{A}$ surely if $\mu_{\mathbb{A}}(x) = 0$ where $\mu_{\mathbb{A}}(x) \in [0,1]$.

Example 5.1.1

An example of a membership function may take the form such as

$$\mu_{\mathbb{A}}(x) = \frac{x}{1+x}, \ x \in \mathbb{A} \subset \mathbb{R}$$

where $\mathbb{R}$ is the set of all real numbers.

Definition 5.1.9

A set

$\mathbb{M} = \left\{ \mu_{\mathbb{A}_i}(x) \mid x \in \mathbb{A}_i \text{ and } 0 \le \mu_{\mathbb{A}_i}(x) \le \mu_{\Omega}(x) \le 1, i \in \mathbb{I} \right\}$ is called membership set where $\mathbb{A}$ is a fuzzy subset in Ω.

Example 5.1.2

An example of a fuzzy subset, $\mathbb{A}$ of Ω may take the form like

$$\mathbb{A} = \left\{ (x, \mu_{\mathbb{A}}(x)) \mid \mu_{\mathbb{A}}(x) = x\left[k(1+x)\right]^{-1}, \ k \ge 1, \ x \in \Omega \right\}$$

The geometry of Example 5.1.2 is shown in Figure 5.1.1

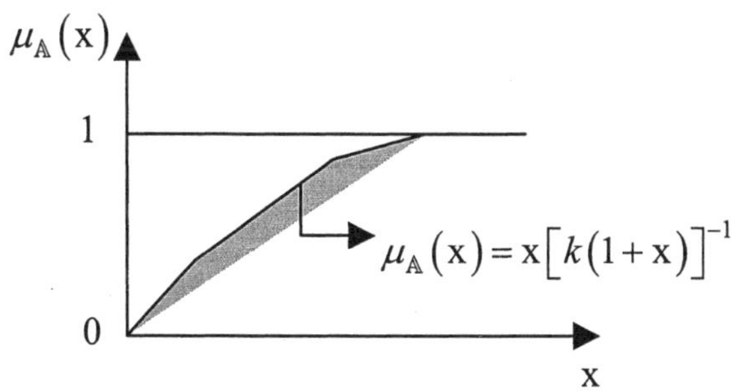

Figure 5.1.1 A Geometry of a Fuzzy Set

A simple numerical character of fuzzy set may be given as an example. Let a linguistic variable be "love" that is expressed by John to four females of Adjoa, Jennie, Akua and Theresa. The degree of love is expressed by a fuzzy number. Here the concept of "love" is considered as a fuzzy set that may be expressed as

$$\mathbb{A} = \{(\text{Adjoa}, 0.9), (\text{Jennie}, 0.65), (\text{Akua}, 0.70), (\text{Theresa}, 0.7)\}$$

where the numbers express degree of love attached to each female. $\mathbb{A}$ is a fuzzy set with a membership distribution regarding the concept of love.

The concept of fuzzy set is a direct extension of the classical concept of ordinary set and subset. If we consider an ordinary set, A relative to a parent set Ω, then for all elements $x \in \Omega$, one and only one of these statements is true in terms of classical logical validity. That is, for all x, $\mu_A(x) = 1$ or $\mu_A(x) = 0$, $x \in \Omega$. By the method of Boolean algebra all the important set operations may be obtained by using the fuzzy membership function. Let us consider ordinary sets A and B with membership functions

$$\mu_A(x) = \begin{cases} 1 & \text{if } x \in A \\ 0 & \text{if } x \notin A \end{cases}$$

(5.1.1)

$$\mu_B(x) = \begin{cases} 1, & \text{if } x \in B \\ 0, & \text{if } x \notin B \end{cases}$$

(5.1.2)

It may immediately be established that

$$\mu_{A \cap B}(x) = \begin{cases} 1, & \text{if } x \in A \text{ and } x \in B \\ 0, & \text{if } x \in A \text{ or } x \in B \text{ but not both} \end{cases}$$

(5.1.3)

Proposition 5.1.2

$$A \cap B = \{(x, \mu_{A \cap B}(x)) \mid \mu_{A \cap B}(x) = \mu_A(x) \bullet \mu_B(x), \ x \in \Omega\}$$

(5.1.4)

where $(\bullet)$ *is a Boolean product operator.*

The statement (5.1.2) may be demonstrated by using Figure 5.1.2

A (columns), B (rows)

$(\bullet)$	0	1
0	0	0
1	0	1

Figure 5.1.2 Boolean Product Operator for Set Intersection

Similarly the union of ordinary sets A and B written as $A \cup B$ may be defined by first specifying the membership function of the union as

$$\mu_{A \cup B}(\mathrm{x}) = \begin{cases} 1 \ , & \text{if } \mathrm{x} \in A \text{ or } \mathrm{x} \in B \\ 0 \ , & \text{if } \mathrm{x} \notin A \text{ and } \mathrm{x} \notin B \end{cases}$$

(5.1.5)

The union of the two ordinary sets may thus be specified in terms of Boolean sum operator, $(\dot{+})$, on the membership functions as

$$A \cup B = \{(\mathrm{x}, \mu_{A \cup B}(\mathrm{x})) \mid \mu_{A \cup B}(\mathrm{x}) = \mu_A(\mathrm{x}) \dot{+} \mu_B(\mathrm{x}) \ , \ \mathrm{x} \in \Omega\}$$

(5.1.6)

The statement of eqn. (5.1.6) is computationally represented by Figure 5.1.3.

B (columns), A (rows)

$\dot{+}$	0	1
0	0	1
1	1	1

Figure 5.1.3 Boolean Sum Operator for Set Union

It must be remembered that the operations of the Boolean product and sum are equivalent to

$$\mu_{A \cap B}(x) = \mu_A(x) \bullet \mu_B(x) = \min\left[\mu_A(x), \mu_B(x)\right]$$

(5.1.7)

$$\mu_{A \cup B}(x) = \mu_A(x) \dot{+} \mu_B(x) = \max\left[\mu_A(x), \mu_B(x)\right]$$

(5.1.8)

The membership function of the complementary set A' of A is written as

$$\mu_{A'}(x) = 1 - \mu_A(x)$$

(5.1.9)

We shall represent minimum (min) and maximum (max) operator by $\wedge$ and $\vee$ for the Boolean product and sum respectively.

5.2 Simple Operations on Fuzzy Sets

Consider fuzzy sets $\mathbb{A}$ and $\mathbb{B}$ relative to the parent set Ω and $\mathbb{M}$ its associated membership set.

Definition 5.2.1

A fuzzy set $\mathbb{B}$, is said to contain another fuzzy set, $\mathbb{A}$, if for all $x \in \Omega$

$$\mu_{\mathbb{A}}(x) \le \mu_{\mathbb{B}}(x)$$

(5.2.1)

Definition 5.2.2

A fuzzy set $\mathbb{A}$ is said to be equal to another fuzzy set $\mathbb{B}$ if for all $x \in \Omega$ $(\mathbb{A} = \mathbb{B})$ with

$$\mu_{\mathbb{A}}(x) = \mu_{\mathbb{B}}(x)$$

(5.2.2)

Definition 5.2.3

A fuzzy set $\mathbb{A}$ is said to be the complement of another fuzzy set, $\mathbb{B}$, if for all $x \in \Omega$ we have

$$\mu_{\mathbb{A}}(x) = 1 - \mu_{\mathbb{B}}(x)$$

(5.2.3)

$$\mu_{\mathbb{B}}(x) = 1 - \mu_{\mathbb{A}}(x)$$

(5.2.4)

Definition 5.2.4

The union of two fuzzy sets $\mathbb{A}$ and $\mathbb{B}$ written as $(\mathbb{A} \cup \mathbb{B})$ is the smallest fuzzy set that contains both $\mathbb{A}$ and $\mathbb{B}$ and equipped with a membership function of the form

$$\mu_{\mathbb{A}\cup\mathbb{B}}(x) = \mu_{\mathbb{A}} \vee \mu_{\mathbb{B}} \;,\; x \in \Omega$$

(5.2.5a)

Alternatively

$$\mathbb{A} \cup \mathbb{B} = \left\{ \left(x \,,\, \mu_{\mathbb{A}\cup\mathbb{B}}(x) \right) \mid \mu_{\mathbb{A}\cup\mathbb{B}}(x) = \mu_{\mathbb{A}}(x) \vee \mu_{\mathbb{B}}(x) \;,\; x \in \Omega \right\}$$

(5.2.5b)

Definition 5.2.5

The intersection of two sets $\mathbb{A}$ and $\mathbb{B}$ written as $\mathbb{A} \cap \mathbb{B}$ is the largest new fuzzy set formed such that it is contained by both $\mathbb{A}$ and $\mathbb{B}$ and has a membership function of the form

$$\mu_{\mathbb{A}\cap\mathbb{B}}(x) = \mu_{\mathbb{A}}(x) \wedge \mu_{\mathbb{B}}(x) \;,\; x \in \Omega$$

(5.2.6a)

Alternatively we can write

$$\mathbb{A} \cap \mathbb{B} = \left\{ \left(x, \mu_{\mathbb{A}\cap\mathbb{B}}(x) \right) \mid \mu_{\mathbb{A}\cap\mathbb{B}}(x) = \mu_{\mathbb{A}}(x) \wedge \mu_{\mathbb{B}}(x) \;,\; x \in \Omega \right\}$$

(5.2.6b)

Definition 5.2.6

In general if $\mathbb{A}_i$ is a member of a family of fuzzy sets, F, indexed by $i \in \mathbb{I}$ where $\mathbb{I}$ is an index set of $\mathbb{A}$ then the union and intersection of the members may be written respectively as

$$\bigcup_{i\in\mathbb{I}} \mathbb{A}_i = \left\{ \left(x, \mu_{\bigcup_{i\in\mathbb{I}} \mathbb{A}_i}(x) \right) \mid \mu_{\bigcup_{i\in\mathbb{I}} \mathbb{A}_i}(x) = \bigvee_{i\in\mathbb{I}} \left[\mu_{\mathbb{A}_i}(x) \right] \,,\, x \in \Omega \text{ and } \mathbb{A}_i \subset \mathsf{F} \right\}$$

(5.2.7)

$$\bigcap_{i\in\mathbb{I}} \mathbb{A}_i = \left\{ \left(x, \mu_{\bigcap_{i\in\mathbb{I}} \mathbb{A}_i}(x) \right) \middle| \mu_{\bigcap_{i\mathbb{I}} \mathbb{A}_i}(x) = \bigwedge_{i\in\mathbb{I}} \left[\mu_{\mathbb{A}_i}(x) \right], x \in \Omega, \mathbb{A}_i \subset \mathsf{F} \right\}$$

(5.2.8)

From the above definitions one may easily check the following set identities where $(')$ represents complementation on fuzzy sets $\mathbb{A}, \mathbb{B}$ and $\mathbb{C}$.

1. $\left.\begin{array}{l} \mathbb{A} \cap \mathbb{B} = \mathbb{B} \cap \mathbb{A} \\ \mathbb{A} \cup \mathbb{B} = \mathbb{B} \cup \mathbb{A} \end{array}\right\}$ Commutative Law

2. $\left.\begin{array}{l} (\mathbb{A} \cap \mathbb{B}) \cap \mathbb{C} = \mathbb{A} \cap (\mathbb{B} \cap \mathbb{C}) \\ (\mathbb{A} \cup \mathbb{B}) \cup \mathbb{C} = \mathbb{A} \cup (\mathbb{B} \cup \mathbb{C}) \end{array}\right\}$ Associative Law

3. $\left.\begin{array}{l} \mathbb{A} \cup \mathbb{A} = \mathbb{A} \\ \mathbb{A} \cap \mathbb{A} = \mathbb{A} \end{array}\right\}$ Idempotent Law

4. $\left.\begin{array}{l} \mathbb{A} \cup (\mathbb{A} \cap \mathbb{B}) = \mathbb{A} \\ \mathbb{A} \cap (\mathbb{A} \cup \mathbb{B}) = \mathbb{A} \end{array}\right\}$ Absorptive Law

5. $\left.\begin{array}{l} \mathbb{A} \cup (\mathbb{B} \cap \mathbb{C}) = (\mathbb{A} \cup \mathbb{B}) \cap (\mathbb{A} \cup \mathbb{C}) \\ \mathbb{A} \cap (\mathbb{B} \cup \mathbb{C}) = (\mathbb{A} \cap \mathbb{B}) \cup (\mathbb{A} \cap \mathbb{C}) \end{array}\right\}$ Distributive Law

6 $\left.\begin{array}{l} (\mathbb{A} \cap \mathbb{B})' = \mathbb{A}' \cup \mathbb{B}' \\ (\mathbb{A} \cup \mathbb{B})' = \mathbb{A}' \cap \mathbb{B}' \end{array}\right\}$ De Morgan Law

7. $\mathbb{A} \cup \varnothing = \mathbb{A}$, $\mathbb{A} \cap \varnothing = \varnothing$, $\mathbb{A} \cap \Omega = \mathbb{A}$

where $\varnothing$ is the null set and Ω is the universal set.

Similarly, certain identities become reachable by using the min. and max. operators $(\wedge, \vee)$, on the corresponding membership functions to arrive at the following theorems.

Theorem 5.2.1

Let $\mu_{\mathbb{A}}(\mathrm{x}), \mu_{\mathbb{B}}(\mathrm{x})$ and $\mu_{\mathbb{C}}(\mathrm{x})$ *be membership functions that define fuzzy sets* $\mathbb{A}$, $\mathbb{B}$ and $\mathbb{C}$ *relative to universal or parent set* Ω *respectively then*

1.
$$\left.\begin{aligned} \mu_{\mathbb{A}}(\mathrm{x}) \vee \mu_{\mathbb{B}}(\mathrm{x}) = \mu_{\mathbb{B}}(\mathrm{x}) \vee \mu_{\mathbb{A}}(\mathrm{x}) \\ \mu_{\mathbb{A}}(\mathrm{x}) \wedge \mu_{\mathbb{B}}(\mathrm{x}) = \mu_{\mathbb{B}}(\mathrm{x}) \wedge \mu_{\mathbb{A}}(\mathrm{x}) \end{aligned}\right\}$$

2.
$$\left.\begin{aligned} \left[\mu_{\mathbb{A}}(\mathrm{x}) \vee \mu_{\mathbb{B}}(\mathrm{x})\right] \vee \mu_{\mathbb{C}}(\mathrm{x}) = \mu_{\mathbb{A}}(\mathrm{x}) \vee \left[\mu_{\mathbb{B}}(\mathrm{x}) \vee \mu_{\mathbb{C}}(\mathrm{x})\right] \\ \left[\mu_{\mathbb{A}}(\mathrm{x}) \wedge \mu_{\mathbb{B}}(\mathrm{x})\right] \wedge \mu_{\mathbb{C}}(\mathrm{x}) = \mu_{\mathbb{A}}(\mathrm{x}) \wedge \left[\mu_{\mathbb{B}}(\mathrm{x}) \wedge \mu_{\mathbb{C}}(\mathrm{x})\right] \end{aligned}\right\}$$

3.
$$\left.\begin{aligned} \mu_{\mathbb{A}}(\mathrm{x}) \wedge \mu_{\mathbb{A}}(\mathrm{x}) = \mu_{\mathbb{A}}(\mathrm{x}) \\ \mu_{\mathbb{A}}(\mathrm{x}) \vee \mu_{\mathbb{A}}(\mathrm{x}) = \mu_{\mathbb{A}}(\mathrm{x}) \end{aligned}\right\}$$

4.
$$\mu_{\mathbb{A}}(\mathrm{x}) \vee \left[\mu_{\mathbb{A}}(\mathrm{x}) \wedge \mu_{\mathbb{B}}(\mathrm{x})\right] = \mu_{\mathbb{A}}(\mathrm{x}) \wedge \left[\mu_{\mathbb{A}}(\mathrm{x}) \vee \mu_{\mathbb{B}}(\mathrm{x})\right] = \mu_{\mathbb{A}}(\mathrm{x})$$

5.
$$\left.\begin{aligned} &\mu_{\mathbb{A}}(\mathrm{x}) \vee \left[\mu_{\mathbb{B}}(\mathrm{x}) \wedge \mu_{\mathbb{C}}(\mathrm{x})\right] \\ &\qquad = \left[\mu_{\mathbb{A}}(\mathrm{x}) \vee \mu_{\mathbb{B}}(\mathrm{x})\right] \wedge \left[\mu_{\mathbb{A}}(\mathrm{x}) \vee \mu_{\mathbb{C}}(\mathrm{x})\right] \\ &\mu_{\mathbb{A}}(\mathrm{x}) \wedge \left[\mu_{\mathbb{B}}(\mathrm{x}) \vee \mu_{\mathbb{C}}(\mathrm{x})\right] \\ &\qquad = \left[\mu_{\mathbb{A}}(\mathrm{x}) \wedge \mu_{\mathbb{B}}(\mathrm{x})\right] \vee \left[\mu_{\mathbb{A}}(\mathrm{x}) \wedge \mu_{\mathbb{C}}(\mathrm{x})\right] \end{aligned}\right\}$$

6.
$$\left.\begin{aligned} \mu_{\mathbb{A}}(\mathrm{x}) \vee 0 = \mu_{\mathbb{A}}(\mathrm{x}) \\ \mu_{\mathbb{A}}(\mathrm{x}) \wedge 0 = 0 \end{aligned}\right\}$$

7.
$$\left.\begin{aligned} \mu_{\mathbb{A}}(\mathrm{x}) \vee 1 = 1 \\ \mu_{\mathbb{A}}(\mathrm{x}) \wedge 1 = \mu_{\mathbb{A}}(\mathrm{x}) \end{aligned}\right\}$$

And generally when we consider a family of fuzzy sets, F *with members* $\mathbb{B}_i$ such that $i \in \mathbb{I}$ and $\mathbb{A}$ *is another fuzzy set then*

8. $$\mu_{\mathbb{A}}(\mathrm{x}) \wedge \left[\bigvee_{i \in \mathbb{I}} \mu_{\mathbb{B}_i}(\mathrm{x})\right] = \bigvee_{i \in \mathbb{I}} \left[\mu_{\mathbb{A}}(\mathrm{x}) \wedge \mu_{\mathbb{B}_i}(\mathrm{x})\right]$$

9. $$\mu_{\mathbb{A}}(\mathrm{x}) \vee \left[\bigwedge_{i \in \mathbb{I}} \mu_{\mathbb{B}_i}(\mathrm{x})\right] = \bigwedge_{i \in \mathbb{I}} \left[\mu_{\mathbb{A}}(\mathrm{x}) \vee \mu_{\mathbb{B}_i}(\mathrm{x})\right]$$

Definition 5.2.7

A fuzzy set, $\mathbb{A}$ is said to be empty if $\mu_{\mathbb{A}}(\mathrm{x}) = 0$ and said to be non-empty if $0 < \mu_{\mathbb{A}}(\mathrm{x}) \leq 1$ for all $\mathrm{x} \in \Omega$.

5.2.1 Visual Representation of Fuzzy-Set Operations

To have an intuitive and visual appreciation of the logic of fuzzy set operations consider fuzzy sets, $\mathbb{A}$ and $\mathbb{B}$ as specified as

$$\mathbb{A} = \left\{\left[\mathrm{x}, \mu_{\mathbb{A}}(\mathrm{x})\right] \mid \mu_{\mathbb{A}}(\mathrm{x}) = \mathrm{e}^{-k\mathrm{x}}, \ k > 0, \ \mathrm{x} \in \Omega\right\}$$

(5.2.1.1)

$$\mathbb{B} = \left\{\left[\mathrm{x}, \ \mu_{\mathbb{B}}(\mathrm{x})\right] \mid \mu_{\mathbb{B}}(\mathrm{x}) = 1 - \mathrm{e}^{-k(\mathrm{x}-\alpha)^2}, \ k > 0, \ \mathrm{x} > \alpha, \ \mathrm{x} \in \Omega\right\}$$

(5.2.1.2)

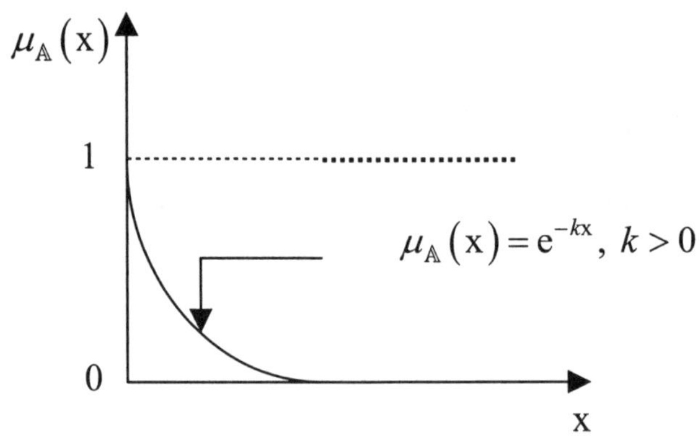

Figure 5.2.2 a The Geometry of Fuzzy Set $\mathbb{A}$

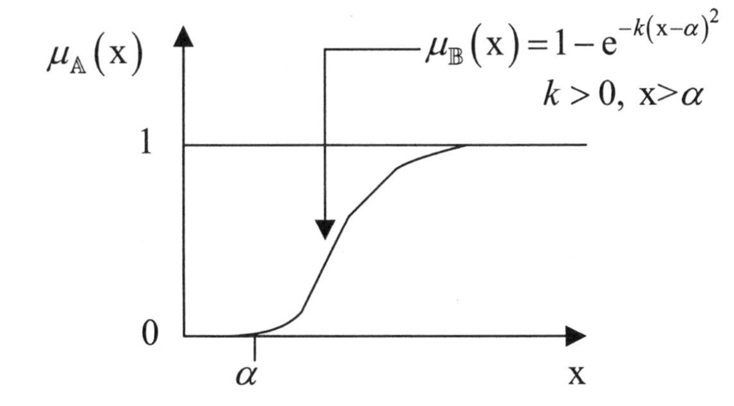

Figure 5.2.2b The Geometry of Fuzzy Set $\mathbb{B}$

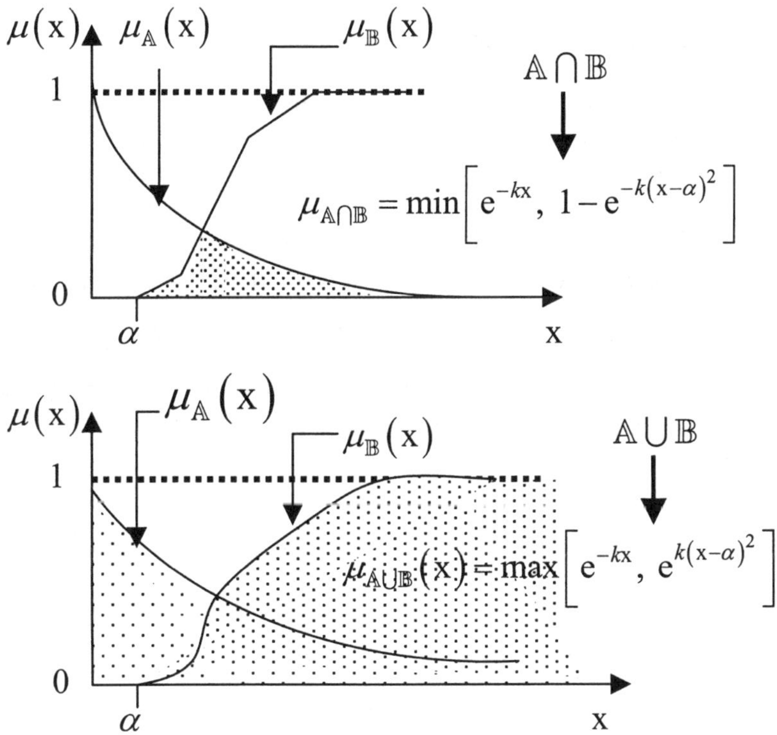

Figure 5.2.3 The Geometry of $\mathbb{A} \cap \mathbb{B}$ and $\mathbb{A} \cup \mathbb{B}$

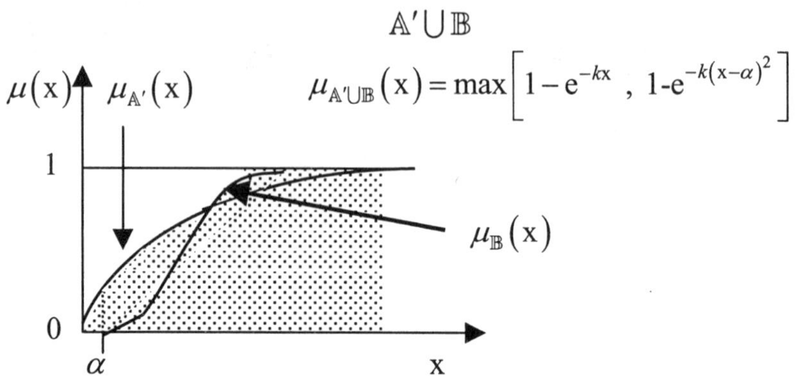

Figure 5.2.4 The Geometry of $\mathbb{A}' \cup \mathbb{B}$

5.2.2 Distance Functions on Fuzzy Sets

In some analysis of decision problems, we may sometimes be confronted with a question as to how far are we away from the selected target or a known ideal. This has some notion of distance concept and measurement. We, therefore, introduce a notion and its measurement over fuzzy sets with this question. To deal with this distance question let us consider two families, F_1 and F_2 of fuzzy sets $\mathbb{A}$ and $\mathbb{B}$ with corresponding membership functions $\mu_{\mathbb{A}}(x)$ and $\mu_{\mathbb{B}}(x)$. Let the members of the fuzzy sets $\mathbb{A}$ and $\mathbb{B}$ in the family be indexed by an index set $\mathbb{I}$ such that $x_i, i \in \mathbb{I}$ are generic elements of the two sets.

Definition 5.2.2.1

The Hamming distance, $d(\mathbb{A},\mathbb{B})$ of two fuzzy sets $\mathbb{A}$ abd $\mathbb{B}$ with membership functions $\mu_{\mathbb{A}}(\bullet)$, $\mu_{\mathbb{B}}(\bullet) \in [0,1]$ respectively is a quantity whose value is defined as

$$d(\mathbb{A},\mathbb{B}) = \sum_{i \in \mathbb{I}} \left| \mu_{\mathbb{A}}(x_i) \mu_{\mathbb{B}}(x_i) \right|$$

(5.2.2.1)

Definition 5.2.2.2

The relative Hamming distance, $\delta(\mathbb{A},\mathbb{B})$ of two fuzzy sets $\mathbb{A}$ and $\mathbb{B}$ with membership functions $\mu_{\mathbb{A}}(\bullet)$ and $\mu_{\mathbb{B}}(\bullet) \in [0,1]$ and generic element, x, is a quantity whose value is defined as

$$\delta(\mathbb{A},\mathbb{B}) = \frac{1}{n}\left[\delta(\mathbb{A},\mathbb{B})\right] = \frac{1}{n}\sum_{i\in\mathbb{I}}\left|\mu_{\mathbb{A}}(x_i) - \mu_{\mathbb{B}}(x_i)\right|$$

(5.2.2.2)

Definition 5.2.2.3

The Euclidean distance (or quadratic distance), $e(\mathbb{A},\mathbb{B})$ of two fuzzy sets $\mathbb{A}$ and $\mathbb{B}$ with membership functions $\mu_{\mathbb{A}}(\bullet)$ and $\mu_{\mathbb{B}}(\bullet) \in [0,1]$ and generic element, x, is a quantity whose value is defined as

$$e(\mathbb{A}\mathbb{B}) = \left\{\sum_{i\in\mathbb{I}}\left[\mu_{\mathbb{A}}(x_i) - \mu_{\mathbb{B}}(x_i)\right]^2\right\}^{\frac{1}{2}}$$

(5.2.2.3)

or

$$\left[e(\mathbb{A}\mathbb{B})\right]^2 = \sum_{i\in\mathbb{I}}\left[\mu_{\mathbb{A}}(x) - \mu_{\mathbb{B}}(x)\right]^2$$

(5.2.2.4)

Definition 5.2.2.4

The relative Euclidean distance, $\mathcal{E}(\mathbb{A},\mathbb{B})$, of two fuzzy sets $\mathbb{A}$, and $\mathbb{B}$ with membership functions $\mu_{\mathbb{A}}(\bullet)$, $\mu_{\mathbb{B}}(\bullet) \in [0,1]$ respectively, $\#\mathbb{I} = n$, and a generic element, x, is a quantity whose value is defined as

$$\mathcal{E}(\mathbb{A},\mathbb{B}) = n^{-\frac{1}{2}}\left[e(\mathbb{A},\mathbb{B})\right] = \left\{\frac{1}{n}\sum_{i\in\mathbb{I}}\left[\mu_{\mathbb{A}}(x) - \mu_{\mathbb{B}}(x)\right]^2\right\}^{\frac{1}{2}}$$

(5.2.2.5)

or

$$\left[\mathcal{E}(\mathbb{AB})\right]^2 = \frac{1}{n}\left[e(\mathbb{A},\mathbb{B})\right] = \frac{1}{n}\sum_{i\in\mathbb{I}}\left[\mu_{\mathbb{A}}(x_i)\mu_{\mathbb{B}}(x_i)\right]^2$$

(5.2.2.6)

Observation 5.2.2.1

All the distance functions that have been defined may be shown to satisfy the minimum mathematical necessities of

$$\left.\begin{array}{ll} \text{a)} & d(A,B) \geq 0 \\ \text{b)} & d(A,B) = d(B,A) \\ \text{c)} & d(A,C) \leq d(A,B) * d(B,C) \\ \text{d)} & d(A,A) = 0 \end{array}\right\}$$

(5.2.2.7)

where A, B and C are ordinary sets and $(*)$ is an operator associated with the notion of distance, $d(\bullet)$.

Observation 5.2.2.2

Notice from the definitions (5.2.3.1) and (5.2.3.3) that

$$\left|\mu_{\mathbb{A}}(x_i) - \mu_{\mathbb{B}}(x_i)\right| = \max\left[\mu_{\mathbb{A}}(x_i), \mu_{\mathbb{B}}(x_i)\right] - \min\left[\mu_{\mathbb{A}}(x_i), \mu_{\mathbb{B}}(x_i)\right]$$

(5.2.3.8)

where $\mu_{\mathbb{A}}(\bullet), \mu_{\mathbb{B}}(\bullet) \in [0,1]$. Furthermore, if $\#\mathbb{I} = n$ then

$$\left.\begin{array}{ll} \text{a)} & 0 \leq d(\mathbb{A},\mathbb{B}) \leq n \\ \text{b)} & 0 \leq \delta(\mathbb{A},\mathbb{B}) \leq 1 \\ \text{c)} & 0 \leq e(\mathbb{A},\mathbb{B}) \leq n^{\frac{1}{2}} \\ \text{d)} & 0 \leq \mathcal{E}(\mathbb{A},\mathbb{B}) \leq 1 \end{array}\right\}$$

(5.2.2.9)

It also follows from the Boolean property of $a^2 = a,\ a \in [0,1]$ that

$$\left.\begin{aligned} e^2(\mathbb{AB}) &= \mathrm{d}(\mathbb{AB}) \\ \mathcal{E}(\mathbb{AB}) &= \delta(\mathbb{AB}) \end{aligned}\right\}$$

(5.2.2.10)

Observation 5.2.2.3

In the case where the reference set is denumerable and the series converges then we can write

$$\mathrm{d}(\mathbb{AB}) = \sum_{i=1}^{\infty} \left| \mu_{\mathbb{A}}(\mathrm{x}_i) - \mu_{\mathbb{B}}(\mathrm{x}_i) \right|$$

(5.2.2.11)

and if the reference set is the set of the real numbers then

$$\mathrm{d}(\mathbb{A},\mathbb{B}) = \int_{-\infty}^{\infty} \left| \mu_{\mathbb{A}}(\mathrm{x}) - \mu_{\mathbb{B}}(\mathrm{x}) \right| \mathrm{dx}$$

(5.2.2.12)

Similarly, the Euclidean distance may be written as

$$e(\mathbb{A},\mathbb{B}) = \left\{ \sum_{i=1}^{\infty} \left[\mu_{\mathbb{A}}(\mathrm{x}_i) - \mu_{\mathbb{B}}(\mathrm{x}_i) \right]^2 \right\}^{\frac{1}{2}}$$

(5.2.2.13)

or

$$e(\mathbb{A},\mathbb{B}) - \left\{ \int_{-\infty}^{\infty} \left[\mu_{\mathbb{A}}(\mathrm{x}_i) \quad \mu_{\mathbb{B}}(\mathrm{x}_i) \right]^2 \mathrm{dx} \right\}^{\frac{1}{2}}$$

(5.2.2.14)

The geometry of the distance functions may be shown as in Figure 5.2.2.1

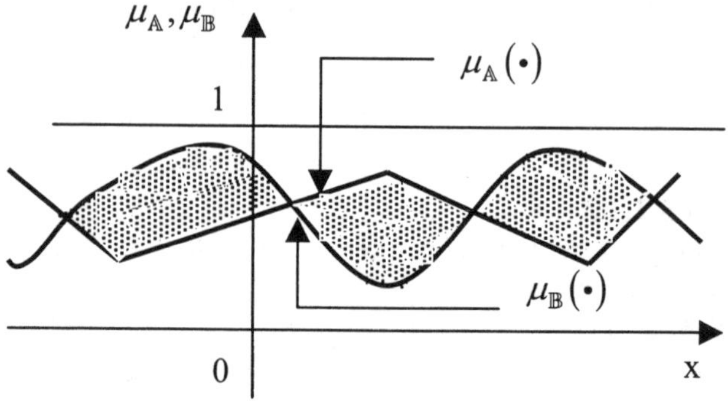

Figure 5.2.2.1 Geometry of the Distance Function

It may also be observed that if the reference set, Ω, is a subset of the set of real numbers $\mathbb{R}$, and that the reference set is bounded between α and β then equations (5.2.2.12) and (5.2.2.14) are convergent and hence both the Hamming distance and Euclidean distance are finite and may be represented respectively as

$$\delta(\mathbb{AB}) = \mathrm{d}(\mathbb{AB})(\beta - \alpha)^{-1}, \; \alpha < \beta \tag{5.2.2.15}$$

$$\mathcal{E}(\mathbb{AB}) = e(\mathbb{AB})[\beta - \alpha]^{-\frac{1}{2}}, \; \alpha < \beta \tag{5.2.2.16}$$

Observation 5.2.2.4

There are important advantages of the notion and measurement of distance in the analysis of decisions either in ordinal or cardinal sense. In some social decision problems we may not be able to compare alternatives in the choice set given the social target. Nonetheless, it might be possible to answer a question as to how far is the result away from the target given each alternative where the target is known. In this respect, we can compute the distance between the result and the target for each applicable alternative and then induce a ranking order. The decision problem may be to select that alternative which has the smallest or the largest distance value depending on the nature of the problem.

5.2.3 Basic Indexes of Fuzziness in Sets

For each fuzzy problem or a logical representation of fuzzy problem, one may want to know the degree of fuzziness contained in the environment of decision. Such knowledge allows us not only to rank problems in terms of their fuzziness but to relate the degree of fuzziness to measures of risk. In this respect an index of fuzziness is useful and required for effective analysis and choice.

Definition 5.2.3.1

A set $\hat{\mathsf{A}}$ is said to be the ordinary subset nearest to a fuzzy set $\mathbb{A}$ if it has the smallest norm (either Hamming or Euclidean) in that

$$\mathsf{A} = \left\{ \left(x , \mu_{\hat{\mathsf{A}}}(x) \right) \mid \ d\left(\mathbb{A}, \hat{\mathsf{A}} \right) \to \min \right\}$$

(5.2.3.1)

with an α − level set specified as

$$\mu_{\hat{\mathsf{A}}}(x) = \begin{cases} 0 \ , \ \text{if} \ \mu_{\mathbb{A}}(x) < \alpha \\ 1 \ , \ \text{if} \ \mu_{\mathbb{A}}(x) \geq \alpha \end{cases} , \quad \alpha \in [0,1]$$

(5.2.3.2)

Definition 5.2.3.2

A linear index of fuzziness of a fuzzy set $\mathbb{A}$ is a number $\nu(\mathbb{A})$ whose value is defined as

$$\nu(\mathbb{A}) = 2n^{-1}\left[d\left(\mathbb{A}, \hat{\mathsf{A}} \right) \right]$$

(5.2.3.3)

where $\hat{\mathsf{A}}$ is the ordinary subset nearest to fuzzy set, $\mathbb{A}$. The quadratic index of fuzziness is a number, $\eta(\mathbb{A})$ defined as

$$\eta(\mathbb{A}) = 2n^{-\frac{1}{2}}\left[e\left(\mathbb{A}, \hat{\mathbf{A}} \right) \right]$$

(5.2.3.4)

Observation 5.2.3.1

It may be verified with 2 in equations (5.2.3.3) and (5.2.3.4) that

$$0 \leq \delta\left(\mathbb{A}, \hat{\mathbf{A}}\right) \leq \frac{1}{2} \text{ and } 0 \leq \mathcal{E}\left(\mathbb{A}, \hat{\mathbf{A}}\right) \leq \frac{1}{2}$$

(5.2.3.5)

and hence we have

$$0 \leq \nu\left(\mathbb{A}\right) \leq 1 \text{ and } 0 \leq \eta\left(\mathbb{A}\right) \leq 1$$

(5.2.3.6)

Furthermore, the following properties of the nearest ordinary subset hold

$$\left.\begin{aligned} \widehat{\mathsf{A} \cap \mathsf{B}} &= \hat{\mathsf{A}} \cap \hat{\mathsf{B}} \\ \widehat{\mathsf{A} \cup \mathsf{B}} &= \hat{\mathsf{A}} \cup \hat{\mathsf{B}} \end{aligned}\right\}$$

(5.2.3.7)

Similarly,

$$\left|\mu_{\mathbb{A}}\left(x_i\right)\mu_{\mathsf{A}}\left(x_i\right)\right| = \mu_{\mathbb{A}\cap\mathbb{A}'}\left(x_i\right) \quad \text{, for all } x \in \Omega$$

(5.2.3.8)

It may finally be shown that

$$\nu\left(\mathbb{A}\right) = 2n^{-1}\left[\sum_{i\in\mathbb{I}} \mu_{\mathbb{A}\cap\mathbb{A}'}\left(x_i\right)\right]$$

(5.2.3.9)

and that

$$\nu\left(\mathbb{A}'\right) = \nu\left(\mathbb{A}\right)$$

(5.2.3.10)

In reasoning through these statements one must keep in mind the definitions of complements of fuzzy set and ordinary subset nearest to a fuzzy set. The concept of an ordinary set nearest to a given fuzzy set has a number of useful applications in decision analysis where cut offs become necessary in defining the relevant set for analysis. This may involve defining the set of most qualified experts among all experts for a specific element of analysis or the set of best controllers from all possible controllers for a specific task.

Definition 5.2.3.3

Let $x_i \in \mathbb{A}$, $i \in \mathbb{I}$, then an individual proportionality of fuzziness, $\pi_{\mathbb{A}}(\bullet)$ in the fuzzy set, $\mathbb{A}$, is

$$\pi_{\mathbb{A}}(x_i) = \frac{\mu_{\mathbb{A}}(x_i)}{\sum_{i \in \mathbb{I}} \mu_{\mathbb{A}}(x_i)}$$

(5.2.3.11)

and $\Pi_{\pi_{\mathbb{A}}}(x)$ as distribution function for proportionality of fuzziness.

Definition 5.2.3.4

The entropy index of fuzziness of a fuzzy set, $\mathbb{A}$, is defined by a function of the form

$$\mathcal{H}\left(\pi_{\mathbb{A}}(x_i) \mid i \in \mathbb{I}\right) = -[\log \#\mathbb{I}]^{-1} \left[\sum_{i \in \mathbb{I}} \pi_{\mathbb{A}}(x_i) \log \pi_{\mathbb{A}}(x_i)\right]$$

$$\mathcal{H}(\bullet) = \left[(\log \#\mathbb{I}) \sum_{i \in \mathbb{I}} \mu_{\mathbb{A}}(x_i)\right]^{-1} \left\{\left[\left(\sum_{i \in \mathbb{I}} \mu_{\mathbb{A}}(x_i)\right)\left(\log \sum_{i \in \mathbb{I}} \mu_{\mathbb{A}}(x_i)\right)\right]\right\}$$
$$-\left\{\left[\log \#\mathbb{I} \sum_{i \in \mathbb{I}} \mu_{\mathbb{A}}(x_i)\right]^{-1} \left[\sum_{i \in \mathbb{I}} \mu_{\mathbb{A}}(x) \log \mu_{\mathbb{A}}(x)\right]\right\}$$

(5.2.3.12)

Observation 5.2.3.2

The value $\mathcal{H}\left(\pi_{\mathbb{A}}(x_i) \mid i \in \mathbb{I}\right)$ is sometimes referred to as *fuzzy entropy function*. It may be viewed as a measure of fuzzy information contained in the fuzzy set. The fuzzy entropy function may be linked by construction to an overall risk function due to fuzziness in the variables contained in the fuzzy set. The resulting risk function may then be examined in terms of the effects of fuzziness on varying degree of risk.

Example 5.2.3.1

Consider a fuzzy set, $\mathbb{A} = \left(x, \mu_{\mathbb{A}}(x)\right)$

x	$\mu_{\mathbb{A}}(x)$	$\pi_{\mathbb{A}}(x)$	$\log\pi_{\mathbb{A}}(x)$	$\pi_{\mathbb{A}}(x)\log\pi_{\mathbb{A}}(x)$
x_1	0.70	$\frac{0.70}{3.70}=0.19$	-1.67	−0.32
x_2	0.90	$\frac{0.90}{3.70}=0.02$	−3.72	−0.07
x_3	0.00	$\frac{0.00}{3.70}=0.00$	0.00	0.00
x_4	0.60	$\frac{0.60}{3.70}=0.16$	−1.82	−0.29
x_5	0.50	$\frac{0.50}{3.70}=0.14$	−2.00	−0.03
x_6	1.00	$\frac{1.00}{3.70}=0.27$	−1.31	−0.35
TOTAL	3.70			−1.06

$$\mathcal{H}\left(\pi_{\mathbb{A}}(x)\,|\,i=1,\cdots,6\right)=-\left[\frac{\sum_{i=1}^{6}\pi_{\mathbb{A}}(x)\log\pi_{\mathbb{A}}(x)}{\log 6}\right]=0.87$$

One can compute from the values of the fuzzy set $\mathbb{A}$, its complement, the linear index of fuzziness, the quadratic index.

5.3 The α-Level Sets and the Decomposition Theorem

Some decision problems in fuzzy systems using fuzzy sets require a specification of the relevant portions of the fuzzy sets. The relevant set may be viewed as the ordinary subset nearest to the fuzzy set. The relevant set for decision analysis may be obtained from the fuzzy set through the decomposition of the fuzzy set and the use of α-level cuts.

Definition 5.3.1

A general set A is said to be crisp if its membership function, $\mu_{\mathsf{A}}(x_i),\ i\in\mathbb{I}$ with generic element $x\in\mathsf{A}$ is such that $\mu_{\mathsf{A}}(x_i)\in\{0,1\}$ for all $i\in\mathbb{I}$. A crisp set is also referred to as an ordinary set. The set is said to be fuzzy if $\mu_{\mathsf{A}}(x)\in[0,1]$ in which case A is represented by the notation, $\mathbb{A}$.

Definition 5.3.2

An α -level set of a given fuzzy set, $\mathbb{A}$, is an ordinary set denoted as $\mathbb{A}_\alpha$ whose members are of the form

$$\mathbb{A}_\alpha = \{(x, \mu_{\mathbb{A}}(x)) \mid \mu_{\mathbb{A}}(x) \geq \alpha, \alpha \in [0,1], x \in \Omega\}$$

where Ω is a reference set of $\mathbb{A}$.

Note:

The concept of α -level cut involves either upper cut referred to as upper-level cut or lower cut referred to as lower-level cut. The definition above is an upper-level cut. The lower-level and middle-level cuts may be written respectively
as

$$\mathbb{A}_\alpha = \{(x, \mu_{\mathbb{A}}) \mid \mu_{\mathbb{A}}(X) \leq \alpha, x \in \Omega, \alpha \in [0,1]\}$$

and

$$\mathbb{A}_{(\alpha_1,\alpha_2)}(x) = \{(x, \mu_{\mathbb{A}}(x)) \mid \mu_{\mathbb{A}}(x) \in [\alpha_1, \alpha_2], \alpha_1 < \alpha_2 \in [0,1] \text{ and } x \in \Omega\}$$

The concept of α -level set may be illustrated graphically as in Figures 5.3.1 and 5.3.2 and with fuzzy set in example 5.3.1

Example 5.3.1

Consider a discrete fuzzy set $\mathbb{A}$ specified as

$$\mathbb{A} = \begin{Bmatrix} (x_1,0.6),(x_2,0.8),(x_3,0.2),(x_4,0.5),(x_5,0.8) \\ (x_6,0.4),(x_7,0.7),(x_8,0.9),(x_9,0.3),(x_{10},0.1) \end{Bmatrix}$$

From the fuzzy set $\mathbb{A}$ we define an α -level set of upper cut also referred to as an *increasing-fixed-level* cut

$$\mathbb{A}_\alpha = \{(x, \mu_{\mathbb{A}}(x)) \mid \mu_{\mathbb{A}}(x) \geq 0.6\}$$

$$= \{x_1, x_2, x_5, x_7, x_8\}$$

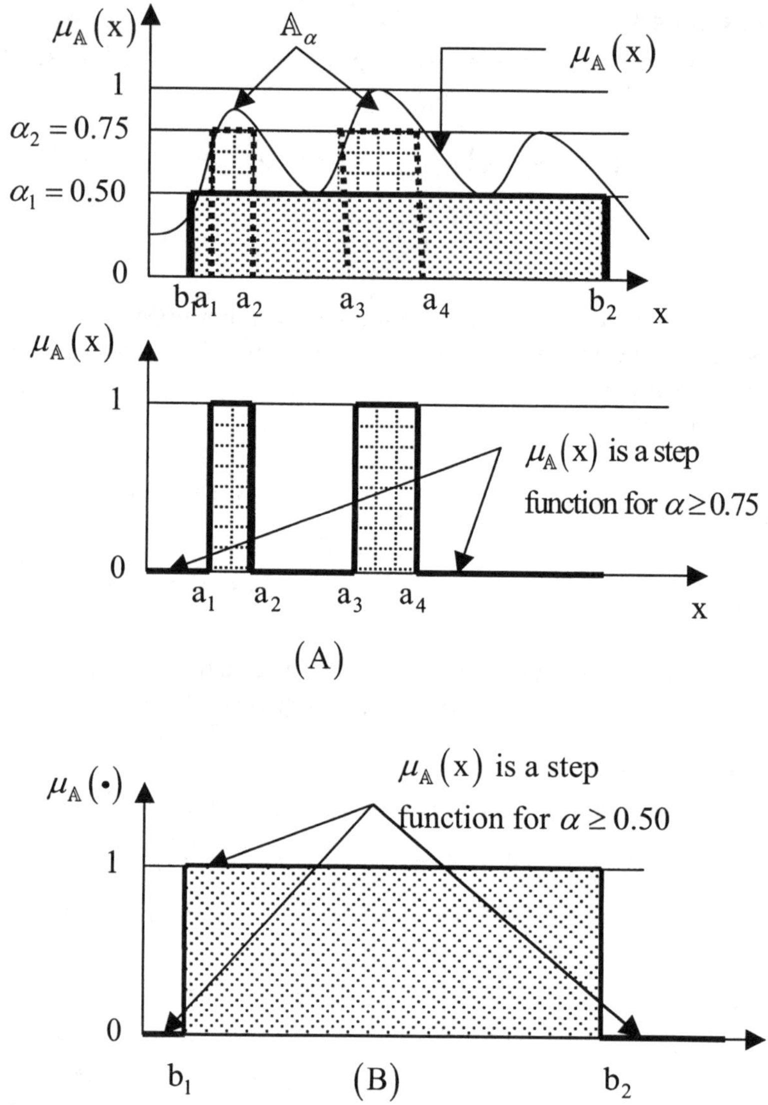

Figure 5.3.1 An Example of a Geometry of an α -Level Set In Panel (A) $\alpha \geq 0.75$ and in Panel (B) $\alpha \geq 0.50$

where the membership function may be written as

$$\mu_{\mathbb{A}_\alpha}(x)\begin{cases} =1\,, \text{ if } \mu_{\mathbb{A}}(x) \geq 0.6 \\ =0\,, \text{ if } \mu_{\mathbb{A}}(x) < 0.6 \end{cases}$$

The α -level set of lower cut also referred to as *decreasing-fix-level cut* may be defined as

$$\mathbb{A}_{\bar{\alpha}} = \{(\mathrm{x}, \mu_{\mathbb{A}}(\mathrm{x})) \mid \mu_{\mathbb{A}}(\mathrm{x}) \le 0.4\}$$

$$\mathbb{A}_{\bar{\alpha}} = \{\mathrm{x}_3, \mathrm{x}_6, \mathrm{x}_9, \mathrm{x}_{10}\}$$

where the membership function is defined as

$$\mu_{\mathbb{A}_{\bar{\alpha}}}(\mathrm{x}) \begin{cases} =1 \text{ , if } \mu_{\mathbb{A}}(\mathrm{x}) \le 0.4 \\ =0 \text{ , if } \mu_{\mathbb{A}}(\mathrm{x}) > 0.4 \end{cases}$$

Similarly, we may define the middle-level cut also referred to as *double-level-cut* as

$$\mathbb{A}\alpha_1\alpha_2 = \{(\mathrm{x}, \mu_{\mathbb{A}}(\mathrm{x})) \mid \alpha_1 \le \mu_{\mathbb{A}}(\mathrm{x}) \le \alpha_2\}$$

$$= \{(\mathrm{x}, \mu_{\mathbb{A}}(\mathrm{x})) \mid 0.4 \le \mu_{\mathbb{A}}(\mathrm{x}) \le 0.8\}$$

$$= \{\mathrm{x}_1, \mathrm{x}_2, \mathrm{x}_4, \mathrm{x}_5, \mathrm{x}_6, \mathrm{x}_7\}$$

where membership function of $\mathbb{A}_{\alpha_1\alpha_2}$ may be written as

$$\mu_{\mathbb{A}}(\mathrm{x}) \begin{cases} =1 \text{ , if } \mu_{\mathbb{A}}(\mathrm{x}) \in [\alpha_1, \alpha_2] \\ =0 \text{ , if } \mu_{\mathbb{A}}(\mathrm{x}) \notin [\alpha_1, \alpha_2] \end{cases}, \alpha_1 < \alpha_2.$$

Observation 5.3.1

A fuzzy set, $\mathbb{A}$, is said to have $\alpha -$cuts if an α -level set $\mathbb{A}_\alpha$ is definable over the fuzzy set $\mathbb{A}$ and $\mu_{\mathbb{A}_\alpha}(\mathrm{x}) \ge \alpha$. The $\alpha = \mu_{\mathbb{A}}(\mathrm{x})$ is said to be its lower-support if it is an upper cut. It is said to be its upper-support if it is a lower cut in which case $\mu_{\mathbb{A}_\alpha}(\mathrm{x}) \le \alpha$.

Lemma 5.3.1

If $\alpha_i \ge \alpha_j$, $\forall \alpha_i\ \alpha_j \in [0,1]$ then $\mathbb{A}_{\alpha_i} \subseteq \mathbb{A}_{\alpha_j}$, *for upper-level cuts, and for lower-level cuts* $\mathbb{A}_{\alpha_j} \subseteq \mathbb{A}_{\alpha_i}$.

Proof

Consider the membership functions of $\mathbb{A}_{\alpha_i}$ and $\mathbb{A}_{\alpha_j}$, with $\mu_{\mathbb{A}_{\alpha_i}}$ and $\mu_{\mathbb{A}_{\alpha_j}}$ as mappings from the parent fuzzy set $\mathbb{A}$ respectively. For each element $x \in \Omega$ and for upper-level cut it is the case that

$$\mu_{\mathbb{A}}(x) \geq \alpha_i \text{ if } x \in \mathbb{A}_{\alpha_i}$$

Similarly

$$\mu_{\mathbb{A}}(x) \geq \alpha_j \text{ , if } x \in \mathbb{A}_{\alpha_j}$$

Since $\mu_{\mathbb{A}}(x) = \alpha_i \geq \mu_{\mathbb{A}}(x) = \alpha_j$ it must be the case that if $x \in \mathbb{A}_{\alpha_j}$ and $x \notin \mathbb{A}_{\alpha_i}$ then $\mu_{\mathbb{A}}(x) \in [\alpha_j, \alpha_i)$

Thus for all $x \in \mathbb{A}_{\alpha_i}$, $x \in \mathbb{A}_{\alpha_j}$

However $\exists\, x \in \mathbb{A}_{\alpha_j} \ni x \notin \mathbb{A}$ since for $\mu_{\mathbb{A}_{\alpha_i}}(x) < \alpha_i, \mu_{\mathbb{A}_i}(x) = 0$. The value $\mu_{\mathbb{A}_{\alpha_i}}(x) > 0$ violates the definition of upper cut and hence $\mathbb{A}_{\alpha_i} \subseteq \mathbb{A}_{\alpha_j}$. A similar logical reasoning leads us to the conclusion that for a lower-level cut

$$\mu_{\mathbb{A}}(x) \leq \alpha_i \text{ if } x \in \mathbb{A}_{\alpha_i} \text{ and } \mu_{\mathbb{A}}(x) \leq \alpha_j \text{ if } x \in \mathbb{A}_{\alpha_j}$$

Since $\mu_{\mathbb{A}}(x) = \alpha_j \leq \mu_{\mathbb{A}}(x) = \alpha_i$ it follows that $\mu_{\mathbb{A}}(x) \in (\alpha_j, \alpha_i]$, if $x \in \mathbb{A}_{\alpha_j}$ and $x \notin \mathbb{A}_{\alpha_i}$. Thus $x \in \mathbb{A}_{\alpha_j}$ it is the case that $x \in \mathbb{A}_{\alpha_i}$. However $\exists\, x \in \mathbb{A}_{\alpha_i} \ni x \notin \mathbb{A}_{\alpha_j}$. Since for all membership values such that $\mu_{\mathbb{A}_{\alpha_j}}(x) > \alpha_j$ it is the case that $\mu_{\mathbb{A}_j}(x) = 0\ \forall x \in \Omega$. The value $\mu_{\mathbb{A}_{\alpha_j}}(x) > 0$ violates the condition of lower cut. It thus follows that $\mathbb{A}_{\alpha_j} \subseteq \mathbb{A}_{\alpha_i}$.

Theorem 5.3.1

If $\mathbb{A}_{\alpha_i}$, $\alpha_i \in [0,1]$ are α *-cuts of a fuzzy set* $\mathbb{A}$ *then the union of all the crisp sets of all the* α *-cuts is equal to the fuzzy set. That is*

$$\mathbb{A} = \bigcup_{\alpha_i} \left[\alpha_i \bullet \mathbb{A}_{\alpha_i} \right].$$

Proof

For general proof we must specify two different membership functions for upper-cut and lower-cut conditions

$$\mu_{\mathbb{A}\alpha_i}(\mathrm{x}) \begin{cases} =1 & \text{if } \mu_{\mathbb{A}}(\mathrm{x}) \geq \alpha_i \\ =0 & \text{if } \mu_{\mathbb{A}}(\mathrm{x}) < \alpha_i \end{cases}, \text{ for upper-level cut}$$

Similarly
$$\mu_{\mathbb{A}_{\alpha_i}}(\mathrm{x}) \begin{cases} =1 & \text{if } \mu_{\mathbb{A}}(\mathrm{x}) \leq \alpha_i \\ =0 & \text{if } \mu_{\mathbb{A}}(\mathrm{x}) > \alpha_i \end{cases}, \text{ for lower-level cut}$$

By the method of Boolean operation on fuzzy sets we can write

$$\max_{\alpha_i} \mu_{\left[\alpha_i \bullet \mathbb{A}_{\alpha_i}\right]}(\mathrm{x}) = \bigvee_{\alpha_i} \left[\alpha_i \bullet \mathbb{A}_{\alpha_i} \right]$$

$$= \bigvee_{\alpha_i \leq \mu_{\mathbb{A}}(\mathrm{x})} [\alpha_i] \text{ for upper-level cut}$$

$$= \bigvee_{\mu_{\mathbb{A}}(\mathrm{x}) \geq \alpha_i} [\alpha_i] \text{ for lower-level cut}$$

$$= \mu_{\mathbb{A}}(\mathrm{x}) \Rightarrow \mathbb{A} = \bigvee_{\alpha_i} \left[\alpha_i \bullet \mathbb{A}_{\alpha_i} \right]$$

Observation 5.3.2

The α -level cut of fuzzy set may be used to decompose the fuzzy set into its α -level sets. As such the theorem is often referred to as decomposition theorem on fuzzy sets. Furthermore this theorem will be found to be very useful in proving some important theorem in fuzzy mathematical programming and optimization.

Corollary 5.3.1

If $\mu_{\mathbb{A}_{\alpha_i}}(\mathrm{x})$, $\alpha_i \in [0,1]$ *are the membership functions of an ordered sequence of* $\alpha -$*level sets of a fuzzy sets* $\mathbb{A}$ $\big[$that is, either $(\alpha_1 > \alpha_2 > \cdots > \alpha_i > \cdots)$ *or* $\alpha_1 < \alpha_2 < \cdots < \alpha_i \cdots\big]$ *with a membership function* $\mu_{\mathbb{A}}(\mathrm{x})$ *then*

$$\mu_{\mathbb{A}}(\mathrm{x}) = \bigvee_{\alpha_i}\left[\alpha_i \bullet \mu_{\mathbb{A}_{\alpha_i}}(\mathrm{x})\right]$$

Proof

Define $$\mu_{\mathbb{A}_{\alpha_i}} \begin{cases} =1 \ , \ \text{if } \mu_{\mathbb{A}}(\mathrm{x}) \geq \alpha_i \\ =0 \ , \ \text{if } \mu_{\mathbb{A}}(\mathrm{x}) < \alpha_i \end{cases}$$

Thus for each α_i, it is the case that

$$\alpha_i \bullet \mu_{\mathbb{A}_{\alpha_i}}(\mathrm{x}) \begin{cases} =\alpha_i \ , \ \text{if } \mu_{\mathbb{A}}(\mathrm{x}) \geq \alpha_i \\ =0 \ , \ \text{if } \mu_{\mathbb{A}}(\mathrm{x}) < \alpha_i \end{cases}$$

and hence

$$\bigvee_{\alpha_i \in [0,1]}\left[\alpha_i \bullet \mu_{\mathbb{A}_{\alpha_i}}(\mathrm{x})\right] = \bigvee_{\alpha_i \leq \mu_{\mathbb{A}}(\mathrm{x})}[\alpha_i] = \mu_{\mathbb{A}}(\mathrm{x})$$

5.4 Fuzzy Graphs and Fuzzy Relations

The notion of fuzzy graph is important in understanding geometric relations of problems and their solutions in theories and applications of fuzzy sets. This is also true in other areas of scientific modeling, explanation and visual representation of theoretical constructs. In fuzzy modeling, fuzzy graphs and relations play fundamental role in the applications of fuzzy mathematics to diverse problems. Other important and fundamental concepts and notions in fuzzy mathematics are fuzzy correspondence and fuzzy relations. In other to introduce these concepts and notions, consider two fuzzy sets, $\mathbb{A}_1$ and $\mathbb{A}_2$ with generic elements $x \in \mathbb{A}_1$, $y \in \mathbb{A}_2$ and $(x,y) \in \mathbb{A} \otimes \mathbb{A}$

Definition 5.4.1

A fuzzy set, $\mathbb{G}$ such that for all $(x,y) \in \mathbb{A}_1 \otimes \mathbb{A}_2$ and $\mu_{\mathbb{G}}(x,y) \in \mathsf{M}$ where M is a membership set of $\mathbb{A}_1 \otimes \mathbb{A}_2$ is said to be a fuzzy graph of $\mathbb{A}_1 \otimes \mathbb{A}_2$. The $\mu_{\mathbb{G}}(x,y)$ is called the value of the graph given the pair (x,y).

Definition 5.4.2

A fuzzy matrix H is an (mxn) set of numbers, $\alpha_{ij} \in [0,1]$ which are arranged in m-rows and n-columns to produce a rectangle of the form

$$\mathsf{H} = \begin{pmatrix} \alpha_{11} & \alpha_{12} & \cdots & \alpha_{1n} \\ \alpha_{21} & \alpha_{22} & \cdots & \alpha_{2n} \\ \vdots & & \alpha_{ij} & \vdots \\ \alpha_{m1} & \alpha_{m2} & \cdots & \alpha_{mn} \end{pmatrix}$$

(5.4.1)

The numbers $\alpha_{ij} \in [0,1]$ are relational values defined by joint membership functions.

Observation 5.4.1

The fuzzy matrices belong to a class of *fractional matrices.* Any fuzzy graph, $\mathbb{G} \subseteq \underset{i \in \mathbb{I}}{\otimes} \mathbb{A}_i$ may be represented by a fuzzy matrix, $\mathsf{H}_{\mathbb{G}}$ that may be defined as

$$\mathsf{H}_{\mathbb{G}} = \begin{pmatrix} \alpha_{11} & \cdots & \alpha_{1n} \\ \vdots & \alpha_{ij} & \vdots \\ \alpha_{m1} & \cdots & \alpha_{mn} \end{pmatrix}$$

(5.4.2)

and whose (i,j)th entries are $\mu_{\mathbb{G}}(\mathrm{x}_i,\mathrm{x}_j)=\alpha_{ij}\in[0,1]$.

Example 5.4.1

An example of a $(3\mathrm{x}3)$ fuzzy matrix for a fuzzy graph is

	y_1	y_2	y_3
x_1	1.0	0.8	0.6
x_2	0.2	1.0	0.4
x_3	0.0	1.0	0.7

$$\Rightarrow \mathsf{H}_{\mathbb{G}}=\begin{pmatrix}1.0 & 0.8 & 0.6\\ 0.2 & 1.0 & 0.4\\ 0.0 & 1.0 & 0.7\end{pmatrix}$$

where $\mu_{\mathbb{G}}(\mathrm{x}_1,\mathrm{y}_3)=0.6,\ \mu_{\mathbb{G}}(\mathrm{x}_2,\mathrm{y}_2)=1$ and others.

Definition 5.4.3

A binary fuzzy relation, $\mathbb{R}$, between two sets A and B with generic elements $\mathrm{x}\in\mathsf{A}$ and $\mathrm{y}\in\mathsf{B}$ is a fuzzy subset of $\mathsf{A}\otimes\mathsf{B}$ with a membership function $\mu_{\mathbb{R}}(\mathrm{x},\mathrm{y})\in\mathsf{M}$, a membership set.

Observation 5.4.2

The value of the joint membership function, $\mu_{\mathbb{R}}(\mathrm{x},\mathrm{y})$ of a fuzzy relation, $\mathbb{R}$, is the *intensity* of the relation as it is related to a particular concept. The values may have many interpretations such as degree of association, conformity, conflict and others depending on the problem at hand. If $\mu_{\mathbb{R}}(\mathrm{x},\mathrm{y})=0\Rightarrow$ no relation. A perfect relation implies that $\mu_{\mathbb{R}}(\mathrm{x},\mathrm{y})=1$. The intensity of the relation increases from 0 to a maximum of one. Furthermore, any fuzzy relation, $\mathbb{R}$ defined on a Cartesian product $\underset{i\in\mathbb{I}}{\otimes}\mathbb{A}_i$ can be represented in a fuzzy relational matrix, $\mathsf{H}_{\mathbb{R}}$ whose $(i-j)$th entries are given by the values of the joint membership function $\mu_{\mathbb{R}}(\mathrm{x}_i,\mathrm{x}_j\mid i,j\in\mathbb{I})$ and $\mathbb{R}\subseteq\underset{i\in\mathbb{I}}{\otimes}\mathbb{A}_i$. The fuzzy relational matrix may be written as

$$H_{\mathbb{R}} = \begin{pmatrix} \beta_{11} & \cdots & \beta_{1n} \\ \vdots & \beta_{ij} & \vdots \\ \beta_{m1} & \cdots & \beta_{mn} \end{pmatrix}$$

(5.4.3)

Example 5.4.2

An example of fuzzy relation is given below

$\mathbb{R}(x,y)$

	y_1	y_2	y_3
x_1	0.3	0.5	0.4
x_2	0.4	0.0	0.0
x_3	0.2	1	0.1
x_4	0	0.9	0.8

where $\mu_{\mathbb{R}}(x_3,y_3) = 0.1$, $\mu_{\mathbb{R}}(x_4,y_2) = 0.9$ and hence the matrix, $H_{\mathbb{R}}$ is

$$H_{\mathbb{R}} = \begin{pmatrix} 0.3 & 0.5 & 0.4 \\ 0.4 & 0.0 & 0.0 \\ 0.2 & 1.0 & 0.1 \\ 0.0 & 0.9 & 0.8 \end{pmatrix}$$

5.4.1 Joint and Marginal Membership Functions

Consider sets A and B with a product set $A \otimes B$. Let $\mathbb{R}$ be a fuzzy relation on $A \otimes B$.

Definition 5.4.1.1

A membership function, $\mu_{\mathbb{R}}(x,y)$, established over $A \otimes B$ is said to be joint if for all $x \in A$ and $y \in B$, $\mu_{\mathbb{R}}(x,y) \in M$ is defined.

Definition 5.4.1.2A

Let A and B be finite sets of the form $\mathsf{A} = \{x_i \mid i \in \mathbb{I}\}$ and $\mathsf{B} = \{y_j \mid j \in \mathbb{J}\}$ where $\mathbb{I}$ and $\mathbb{J}$ are the corresponding index sets along with a fuzzy relation $\mathbb{R}$ defined on $\mathsf{A} \otimes \mathsf{B}$ with a joint membership function $\mu_{\mathbb{R}}(x,y)$. The *upward* and *downward* projection of $\mathbb{R}$ on any of the sets, written as $[\mathbb{R} \swarrow\!\!\nearrow]$ is a new fuzzy set formed with a membership function that disregards all the membership values in the joint except those that have strongest and lowest relation in the joint. For an upward projection we shall write $[\mathbb{R} \nearrow]$ and for a downward projection write $[\mathbb{R} \swarrow]$

Definition 5.4.1.2B

If $\mu_{\mathbb{R}}(x,y)$, $x \in \mathsf{A}$ and $y \in \mathsf{B}$ is a fuzzy relation in $\mathsf{A} \otimes \mathsf{B}$ then the upward x-projection and downward x-projection of $\mathbb{R}$ are the x-marginal membership functions written respectively as

$$\mu_{\mathbb{R}^x_{\nearrow}}(x|y) = \bigvee_{y \in \mathsf{B}} \mu_{\mathbb{R}}(x,y)$$

(5.4.1.1a)

$$\mu_{\mathbb{R}^x_{\swarrow}}(x|y) = \bigwedge_{y \in \mathsf{B}} \mu_{\mathbb{R}}(x,y)$$

(5.4.1.1b)

The y-projections of $\mathbb{R}$ are the y-marginal membership functions written respectively as

$$\mu_{\mathbb{R}^y_{\nearrow}}(y|x) = \bigvee_{x \in \mathsf{A}} \mu_{\mathbb{R}}(x,y)$$

(5.4.1.2a)

$$\mu_{\mathbb{R}^y_{\swarrow}}(y|x) = \bigwedge_{x \in \mathsf{A}} \mu_{\mathbb{R}}(x,y)$$

(5.4.1.2b)

The global upward projection and downward or upward and downward x-y projection on $\mathbb{R}$ is computationally defined as

$$h\left(\mathbb{R}^{xy}_{\nearrow}\right)=\bigvee_{y\in B}\bigvee_{x\in A}\mu_{\mathbb{R}}(x,y)=\bigvee_{x\in A}\bigvee_{y\in B}\mu_{\mathbb{R}}(x,y)$$

$$h\left(\mathbb{R}^{xy}_{\swarrow}\right)=\bigwedge_{y\in B}\bigwedge_{x\in A}\mu_{\mathbb{R}}(x,y)=\bigwedge_{x\in A}\bigwedge_{y\in B}\mu_{\mathbb{R}}(x,y)$$

Example 5.4.1.1

Consider a fuzzy x-y relational matrix of the form

$\mathbb{R}(x,y)$ → $\left[\mathbb{R}^{x}_{\nearrow}\right]$ → $\mu_{\mathbb{R}^{x}_{\nearrow}}(x|y)=\bigvee_{y}\mu_{\mathbb{R}}(x,y)$

	y_1	y_2	y_3	y_4	$\mu_{\mathbb{R}^{x}_{\nearrow}}(x\vert y)$
x_1	0.1	0.2	0.9	0.3	0.9
x_2	0.6	0.8	0.0	0.1	0.8
x_3	0.0	0.9	0.3	0.5	0.9
x_4	0.9	0.1	0.8	0.0	0.9
x_5	0.5	0.7	0.0	0.6	0.7

$\left[\mathbb{R}^{y}_{\nearrow}\right]$

$$\mu_{\mathbb{R}^{y}_{\nearrow}}(y|x)=\bigvee_{x}\mu_{\mathbb{R}}(x,y)=\quad[0.9\ \ 0.9\ \ 0.9\ \ 0.6]$$

The upward global projection is $h(\mathbb{R})=0.9$. From the same data of Example 5.4.1.1 we can compute downward x- and y-projections and the corresponding downward global projection respectively as

$\mathbb{R}(x,y)$ → $\left[\mathbb{R}^{x}_{\swarrow}\right]$ → $\mu_{\mathbb{R}^{x}_{\swarrow}}(x|y)=\bigwedge_{y}\mu_{\mathbb{R}}(x,y)$

	y_1	y_2	y_3	y_4	$\mu_{\mathbb{R}^{x}_{\swarrow}}(x\vert y)$
x_1	0.1	0.2	0.9	0.3	0.1
x_2	0.6	0.8	0.0	0.1	0.1
x_3	0.0	0.9	0.3	0.5	0.0
x_4	0.9	0.1	0.8	0.0	0.0
x_5	0.5	0.7	0.0	0.6	0.0

$\left[\mathbb{R}^{y}_{\swarrow}\right]$

$$\mu_{\mathbb{R}^{y}_{\swarrow}}(y|x)=\bigwedge_{x}\mu_{\mathbb{R}}(x,y)=\quad[0.1\ \ 0.1\ \ 0.0\ \ 0.0]$$

The downward global projection is $h(\mathbb{R}_{\swarrow}) = 0.0$.

In applied problems we encounter some analytical demands where links of relations or conclusions of one relation is required on the basis of a given relation. The analytical procedure requires combinations and compositions of relations in the fuzzy inference. Among such compositions are union and intersection of relations.

Definition 5.4.1.3

A union of two fuzzy relations $\mathbb{R}_1$ and $\mathbb{R}_2$ is a new fuzzy relation formed, and represented as $\mathbb{R}_1 \cup \mathbb{R}_2$ with a membership function defined as

$$\mu_{\mathbb{R}_1 \cup \mathbb{R}_2}(x,y) = \mu_{\mathbb{R}_1}(x,y) \vee \mu_{\mathbb{R}_2}(x,y)$$

(5.4.1.3a)

Alternatively we may write

$$\mathbb{R}_1 \cup \mathbb{R}_2 = \{((x,y), \mu_{\mathbb{R}_1 \cup \mathbb{R}_2}(x,y)) \mid \mu_{\mathbb{R}_1 \cup \mathbb{R}_2}(x,y) = \mu_{\mathbb{R}_1}(x,y) \vee \mu_{\mathbb{R}_2}(x,y)\}$$

(5.4.1.3b)

Definition 5.4.1.4

An intersection of two fuzzy relations $\mathbb{R}_1$ and $\mathbb{R}_2$ written as $\mathbb{R}_1 \cap \mathbb{R}_2$ is a new fuzzy set formed with a membership function defined as

$$\mu_{\mathbb{R}_1 \cap \mathbb{R}_2}(x,y) = \mu_{\mathbb{R}_1}(x,y) \wedge \mu_{\mathbb{R}_2}(x,y)$$

(5.4.1.4a)

Alternatively we may write

$$\mathbb{R}_1 \cap \mathbb{R}_2 = \{((x,y), \mu_{\mathbb{R}_1 \cap \mathbb{R}_2}(x,y)) \mid \mu_{\mathbb{R}_1 \cap \mathbb{R}_2}(x,y) = \mu_{\mathbb{R}_1}(x,y) \wedge \mu_{\mathbb{R}_2}(x,y)\}$$

(5.4.1.4b)

Observation 5.4.1.1

Let a family of fuzzy relations R be specified as $\mathsf{R} = \{\mathbb{R}_i \mid i \in \mathbb{I}\}$ where $\mathbb{I}$ is an index set. The union and intersection of the family of the relations may be written respectively as new fuzzy sets of the form

$$\bigcup_{i\in\mathbb{I}} \mathbb{R}_i = \left\{ \left((x,y), \mu_{\bigcup_{i\in\mathbb{I}} \mathbb{R}_i}(x,y) \right) \mid \mu_{\bigcup_{i\in\mathbb{I}} \mathbb{R}_i}(x,y) = \bigvee_{i\in\mathbb{I}} \mu_{\mathbb{R}_i}(x,y) \right\}$$

(5.4.1.5)

$$\bigcap_{i\in\mathbb{I}} \mathbb{R}_i = \left\{ \left((x,y), \mu_{\bigcap_{i\in\mathbb{I}} \mathbb{R}_i}(x,y) \right) \mid \mu_{\bigcap_{i\in\mathbb{I}} \mathbb{R}_i}(x,y) = \bigwedge_{i\in\mathbb{I}} \mu_{\mathbb{R}_i}(x,y) \right\}$$

(5.4.1.6)

Definition 5.4.1.5

A complement of a fuzzy relation $\mathbb{R}$, denoted as $\mathbb{R}'$ is a fuzzy set formed with a membership function defined as

$$\mathbb{R}' = \{(x,y), \mu_{\mathbb{R}'}(x,y) \mid \mu_{\mathbb{R}'}(x,y) = 1 - \mu_{\mathbb{R}}(x,y)\}$$

(5.4.1.7)

$\forall x \in \mathsf{A}$, $\forall y \in \mathsf{B}$ and $(x,y) \in \mathsf{A} \otimes \mathsf{B}$.

Definition 5.4.1.6

A fuzzy relation $\mathbb{R}_1$ is said to be contained in a fuzzy relation, $\mathbb{R}_2$, if for all $x \in \mathsf{A}$ and $y \in \mathsf{B}$ it is the case that $\mu_{\mathbb{R}_1}(x,y) \le \mu_{\mathbb{R}_2}(x,y)$, $\forall (x, y) \in \mathsf{A} \otimes \mathsf{B}$. We write $\mathbb{R}_1 \subset \mathbb{R}_2$ and hence $\mathbb{R}_2$ is said to be an envelope or enclosure of $\mathbb{R}_1$.

5.4.2 Decomposition and Fuzzy Relations

The concept of decomposition of relation into degrees of belonging is very important in understanding a number of scientific problem solving. A typical case is decomposing a huge and difficult problem into subproblems into sequential forms. Such decomposition takes place via the method of fix-level sets with a defined membership values. It is also useful in cluster analysis to form matrix of crisp cohort and cohort analysis.

Definition 5.4.2.1

The nearest ordinary set of α -level of a relation $\mathbb{R}$ on $\mathsf{A} \otimes \mathsf{B}$ is an ordinary set or relation $\mathbb{R}_\alpha$ defined by its α -level set $\forall \alpha \in [0,1]$ in the form

$$\mathbb{R}_\alpha = \{(x,y) \mid \mu_{\mathbb{R}}(x,y) \geq \alpha,\ x \in \mathsf{A},\ y \in \mathsf{B} \text{ and } (x,y) \in \mathsf{A} \otimes \mathsf{B}\}$$

(5.4.2.1)

Observation 5.4.2.1

It is interesting to observe that one can specify the union and intersections of fuzzy relations in terms of the corresponding relational matrices. One should also refer to the development of the theory of computable cost-benefit identification matrices presented in Chapter 2 of this volume. Let $\mathbb{R}_1$ and $\mathbb{R}_2$ be two fuzzy binary relation defined on a Cartesian product set $\mathsf{A} \otimes \mathsf{B}$ with corresponding relational matrices, $\mathsf{H}_{\mathbb{R}_1}$ and $\mathsf{H}_{\mathbb{R}_2}$ specified as

$$\mathsf{H}_{\mathbb{R}_1} = \begin{pmatrix} \beta_{11} & \cdots & \beta_{1n} \\ \vdots & \beta_{ij} & \vdots \\ \beta_{m1} & \cdots & \beta_{mn} \end{pmatrix}, \quad \mu_{\mathbb{R}_1}(x_i, y_j) = \beta_{ij}$$

$$\mathsf{H}_{\mathbb{R}_2} = \begin{pmatrix} \alpha_{11} & \cdots & \alpha_{1n} \\ \vdots & \alpha_{ij} & \vdots \\ \alpha_{m1} & \cdots & \alpha_{mn} \end{pmatrix}, \quad \mu_{\mathbb{R}_2}(x_i, y_j) = \alpha_{ij}$$

Then the relational matrix of the union of the fuzzy relations $\mathbb{R}_1$ and $\mathbb{R}_2$ may be computed as

$$\mathsf{H}_{\mathbb{R}_1 \cup \mathbb{R}_2} = \mathsf{H}_{\mathbb{R}_1} \oplus \mathsf{H}_{\mathbb{R}_2}$$

where $\oplus$ is a Boolean sum operator. The matrix of intersection of the fuzzy relations $\mathbb{R}_1$ and $\mathbb{R}_2$ may be computed as

$$\mathsf{H}_{\mathbb{R}_1 \cap \mathbb{R}_2} = \mathsf{H}_{\mathbb{R}_1} \odot \mathsf{H}_{\mathbb{R}_2}$$

where $\odot$ is a Boolean product operator. In the case of a family of relations $\mathsf{R} = \{\mathbb{R}_i \mid i \in \mathbb{I}\}$ we can write the union and intersection respectively as

$$\mathsf{H}_{\bigcup_{l\in\mathbb{I}} \mathbb{R}_l} = \left\{ \bigoplus_{l\in\mathbb{I}} \mathsf{H}_{\mathbb{R}_l} \mid \mu_{\bigcup_{l\in\mathbb{I}} \mathbb{R}_l}(\mathrm{x}_i, \mathrm{y}_j) = \beta_{ij} \right\}$$

$$\mathsf{H}_{\bigcap_{l\in\mathbb{I}} \mathbb{R}_l} = \left\{ \bigodot_{l\in\mathbb{I}} \mathsf{H}_{\mathbb{R}_l} \mid \mu_{\bigcap_{l\in\mathbb{I}} \mathbb{R}_l}(\mathrm{x}_i, \mathrm{y}_j) = \hat{\alpha}_{ij} \right\}$$

For more extensive discussions on fuzzy relations and decompositions see [1149] [1152] [1153a] [1165] [1229].

Example 5.4.2.2

Let $\mathbb{R}_1$ and $\mathbb{R}_2$ be two fuzzy binary relations defined on a Cartesian product set $\mathsf{A} \otimes \mathsf{B}$ with corresponding relational matrices of the form

$$\mathsf{H}_{\mathbb{R}_1} = \begin{pmatrix} 0.3 & 0.2 & 0.9 & 0.0 \\ 0.8 & 0.9 & 0.0 & 0.2 \\ 0.5 & 0.0 & 0.4 & 0.0 \end{pmatrix}$$

$$\mathsf{H}_{\mathbb{R}_2} = \begin{pmatrix} 0.3 & 0.0 & 0.7 & 0.0 \\ 0.1 & 0.8 & 0.9 & 0.9 \\ 0.6 & 0.7 & 0.3 & 0.2 \end{pmatrix}$$

From the data given by $\mathsf{H}_{\mathbb{R}}$ and $\mathsf{H}_{\mathbb{R}}$ we can compute the union, intersection, and complements $\mathbb{R}'$ of the relations as

$$\mathsf{H}_{\mathbb{R}_1 \cup \mathbb{R}_2} = \begin{pmatrix} 0.3 & 0.2 & 0.9 & 0.0 \\ 0.8 & 0.9 & 0.0 & 0.2 \\ 0.5 & 0.0 & 0.4 & 0.0 \end{pmatrix} \oplus \begin{pmatrix} 0.3 & 0.0 & 0.7 & 0.0 \\ 0.1 & 0.8 & 0.9 & 0.9 \\ 0.6 & 0.7 & 0.3 & 0.2 \end{pmatrix}$$

$$= \begin{pmatrix} 0.3 & 0.2 & 0.9 & 0.0 \\ 0.8 & 0.9 & 0,9 & 0.9 \\ 0.6 & 0.7 & 0.4 & 0.2 \end{pmatrix}$$

$$\mathsf{H}_{\mathbb{R}_1 \cap \mathbb{R}_2} = \begin{pmatrix} 0.3 & 0.2 & 0.9 & 0.0 \\ 0.8 & 0.9 & 0.0 & 0.2 \\ 0.5 & 0.0 & 0.4 & 0.0 \end{pmatrix} \odot \begin{pmatrix} 0.3 & 0.0 & 0.7 & 0.0 \\ 0.1 & 0.8 & 0.9 & 0.9 \\ 0.6 & 0.7 & 0.3 & 0.2 \end{pmatrix}$$

$$= \begin{pmatrix} 0.3 & 0.0 & 0.7 & 0.0 \\ 0.1 & 0.8 & 0.0 & 0.2 \\ 0.5 & 0.0 & 0.3 & 0.0 \end{pmatrix}$$

$$\mathsf{H}_{\mathbb{R}_1'} = \begin{pmatrix} 0.7 & 0.8 & 0.1 & 0.1 \\ 0.2 & 0.1 & 1.0 & 0.8 \\ 0.5 & 1.0 & 0.6 & 1.0 \end{pmatrix} , \quad \mathsf{H}_{\mathbb{R}_2'} = \begin{pmatrix} 0.7 & 1.0 & 0.3 & 1.0 \\ 0.9 & 0.2 & 0.1 & 0.1 \\ 0.4 & 0.3 & 0.7 & 0.8 \end{pmatrix}$$

Observation 5.4.2.2

Notice that the ordinary set of fuzzy relation may be written in terms of its membership function as

$$\mu_{\mathbb{R}_\alpha}(\mathrm{x,y}) \begin{cases} =1 & \text{if } \mu_{\mathbb{R}}(\mathrm{x,y}) \geq \alpha \\ =0 & \text{if } \mu_{\mathbb{R}}(\mathrm{x,y}) < \alpha \end{cases}$$

and by fix-level cut we immediately notice that if $\alpha_1 \geq \alpha_2$ then $\mathbb{R}_{\alpha_1} \subseteq \mathbb{R}_{\alpha_2}$. The meaning of containment, intersection and union of fuzzy relation and its scientific relevance will be dictated by the problem of social interest.

Theorem 5.4.2.1

Any fuzzy relation, $\mathbb{R}$ is decomposable into its α -level cuts (fix-level sets) such that

$$\mathbb{R} = \bigvee_{\alpha} \alpha \bullet \mathbb{R}_\alpha , \ \forall \alpha \in [0,1]$$

where

$$\mu_{\mathbb{R}_\alpha}(x,y)\begin{cases}=1 & \text{if } \mu_{\mathbb{R}}(x,y)\geq\alpha\\ =0 & \text{if } \mu_{\mathbb{R}}(x,y)<\alpha\end{cases}$$

Proof

By the properties of fix-level sets we can write

$$\mu_{\bigcup_\alpha \alpha\bullet\mathbb{R}_\alpha}(x,y)=\bigvee_\alpha \alpha\bullet\mu_{\mathbb{R}_\alpha}(x,y)$$

$$=\bigvee_{\alpha\leq\mu_{\mathbb{R}}(x,y)}\alpha$$

$$=\mu_{\mathbb{R}}(x,y)$$

Observation 5.4.2.3

The following properties may be established for fuzzy relations. Consider a product set, A^2

1.

$\forall(x,y)\in A^2$ if $\mu_{\mathbb{R}}(x,y)=\alpha\in[0,1]\Rightarrow\mu_{\mathbb{R}}(y,x)=\alpha$

then $\mathbb{R}$ is *symmetric.*

2. $\forall(x,x)\in A^2$ if $\mu_{\mathbb{R}}(x,x)=1$ then $\mathbb{R}$ is said to be *reflexive.*

3. $\forall x,y,z\in A^2$ if $\mu_{\mathbb{R}}(x,z)\geq\bigvee_y\left[\mu_{\mathbb{R}}(x,y)\wedge\mu_{\mathbb{R}}(y,z)\right]$

then $\mathbb{R}$ is *transitive.*

Definition 5.4.2.2

A fuzzy relation $\mathbb{R}$ defined on A^2 is said to be a *similitude* or *fuzzy equivalence relation* if for all $x,y,z\in A^2$ it is symmetric, reflexive and transitive. That is

1. $\mu_{\mathbb{R}}(x,y)=\alpha=\mu_{\mathbb{R}}(y,x)$ **(symmetric)**
2. $\mu_{\mathbb{R}}(x,x)=1$ **(reflexive)**
3. $\mu_{\mathbb{R}}(x,z)\geq\bigvee_y\left[\mu_{\mathbb{R}}(x,y)\wedge\mu_{\mathbb{R}}(y,z)\right]$ **(transitive)**

Definition 5.4.2.3

A fuzzy binary relation defined on a product set A^2 is said to be a *dissimilitude* relation if $(x,y,z) \in \mathsf{A}^2$ then

1. $\mu_{\mathbb{R}'}(x,y) = \mu_{\mathbb{R}'}(y,x)$ **(symmetric)**
2. $\mu_{\mathbb{R}'}(x,x) = 0$ **(anti-reflective)**
3. $\mu_{\mathbb{R}'}(x,y) \le \bigwedge_{y}\left[\mu_{\mathbb{R}'}(x,y) \vee \mu_{\mathbb{R}'}(y,z)\right]$ **(min-max transitive)**

where $\mathbb{R}'$ is the complement of $\mathbb{R}$ and $\mu_{\mathbb{R}'}(x,y) = 1 - \mu_{\mathbb{R}}(x,y)$.

Definition 5.4.2.4

A fuzzy relation, $\mathbb{R}$ defined on A^2 is said to be *resemblance relation or fuzzy equivalence relations* if for $(x,y,z) \in \mathsf{A}^2$, $\mathbb{R}$, is reflexive and symmetric and min-max transitive. That is:

1. $\mu_{\mathbb{R}}(x,x) = 1$ **(reflexive)**
2. $\mu_{\mathbb{R}}(x,y) = \mu_{\mathbb{R}}(y,x)$ **(symmetric)**
3. $\mu_{\mathbb{R}}(x,z) \ge \bigvee_{y \in \mathbb{A}}\left[\mu_{\mathbb{R}}(x,y) \wedge \mu_{\mathbb{R}}(y,z)\right]$ **(max-min transitivity)**

Definition 5.4.2.5

A fuzzy relation, $\mathbb{R}$, defined on A^2 is said to be dissemblance if for all $(x,y,z) \in \mathsf{A}^2, \mathbb{R}$ is anti-reflexive and symmetric. That is

1. $\mu_{\mathbb{R}}(x,x) = 0$ **(anti-reflexive)**
2. $\mu_{\mathbb{R}}(x,y) = \mu_{\mathbb{R}}(y,x)$ **(symmetric)**
3. $\mu_{\mathbb{R}}(x,y) \le \bigwedge_{y \in \mathbb{A}}\left[\mu_{\mathbb{R}}(x,y) \vee \mu_{\mathbb{R}}(x,y)\right]$ **(min-max transitivity)**

Care must be exercised in distinction and use of fuzzy equivalence relation and the dissemblance relation in axiomatic decision constructs. In economic, engineering, financial and environmental analysis the fuzzy equivalence relation provide a logical process to induce and study equivalence classes in terms of iso-benefit, iso-cost, iso-risk, iso-control technologies, iso-pollutant production technologies iso-contingency, and many others. These equivalence classes can then be analyzed and ranked by fuzzy order relation.

Definition 5.4.2.6

A fuzzy relation, $\mathbb{R}$ defined on $\mathbb{A}^2$ is said to be fuzzy order relation if given $(x,y,z) \in \mathbb{A}$, it is reflexive, anti-symmetric and min-max transitive. Thus

1. $\mu_{\mathbb{R}}(x,x) = 1$ **(reflexive)**
2. $\mu_{\mathbb{R}}(x,y) \neq \mu_{\mathbb{R}}(y,x) \Rightarrow \mu_{\mathbb{R}}(x,y) = \mu_{\mathbb{R}}(y,x) = 0$ **(anti-symmetry)**
3. $\mu_{\mathbb{R}}(x,z) \geq \bigvee_{y \in \mathbb{A}} \left[\mu_{\mathbb{R}}(x,y) \wedge \mu_{\mathbb{R}}(y,z)\right]$ **(min-max transitivity)**

To use the fuzzy ranking to order the equivalence classes the x and y must be interpreted as equivalence classes in the same domain of analysis. In this way the equivalence classes can be ordered in the relevant spaces.

5.4.3 Some other Useful Fuzzy Set Operations

There are other useful operations on fuzzy sets that provide analytical strength in many scientific investigations into problems of both human and non-human systems. Some of these operations may be useful in risk and contingency analysis (see Chapter 2 of this volume).

Definition 5.4.3.1

The product of two fuzzy sets $\mathbb{A}$ and $\mathbb{B}$ is a new set written as $\mathbb{A} \bullet \mathbb{B}$ with a membership function defined as:

$$\mu_{\mathbb{A} \bullet \mathbb{B}}(x) = \mu_{\mathbb{A}}(x) \bullet \mu_{\mathbb{B}}(x)$$

(5.4.3.1)

where $\mathbb{A}$ and $\mathbb{B}$ has common generic element with membership functions $\mu_{\mathbb{A}}(x)$ and $\mu_{\mathbb{B}}(x)$ respectively.

Definition 5.4.3.2

The m-power fuzzy set of a fuzzy set $\mathbb{A}$ with a membership function $\mu_{\mathbb{A}}(x)$ is a new fuzzy set, $\mathbb{A}^{\alpha}$ formed with a membership function that may be defined as

$$\mu_{\mathbb{A}^m}(x) = \left[\mu_{\mathbb{A}}(x)\right]^m, \ m \in \mathsf{R}^+$$

(5.4.3.2)

where R^+ is the positive reals

Definition 5.4.3.3

If $\mathbb{A}$ is fuzzy set with membership function, $\mu_{\mathbb{A}}(x)$ then its concentration, $\left[\mathrm{CON}(\mathbb{A})\right]$ and dilation $\left[\mathrm{DIL}(\mathbb{A})\right]$ are new fuzzy sets formed whose membership functions may be computed respectively as:

$$\mu_{\mathrm{CON}(\mathbb{A})}(x) = \left[\mu_{\mathbb{A}}(x)\right]^2$$

(5.4.3.3)

$$\mu_{\mathrm{DIL}(\mathbb{A})}(x) = \left[\mu_{\mathbb{A}}(x)\right]^{\frac{1}{2}}$$

(5.4.3.4)

Definition 5.4.3.4

The contrast intensification (INT) of a fuzzy set, $\mathbb{A}$, with a membership function $\mu_{\mathbb{A}}(x)$ is a new fuzzy set formed written as $\mathrm{INT}(\mathbb{A})$ whose membership function may be specified as

$$\mu_{\mathrm{INT}(\mathbb{A})}(x) \begin{cases} = \mu_{\mathrm{CON}(\mathbb{A})}(x), & \forall x \text{ such that } \mu_{\mathbb{A}}(x) < 0.5 \\ = \mu_{\mathrm{DIL}(\mathbb{A})}(x), & \forall x \text{ such that } \mu_{\mathbb{A}}(x) \geq 0.5 \end{cases}$$

(5.4.3.5)

Note:

The measure of concentration, dilation and intensification are used as modifiers of meaning in hedging processes, fuzzy outcomes or generally of fuzzy sets in decision or non-decision processes. For more discussions and areas of usage see also [695] [735], [809], [810][1152].

Definition 5.4.3.5

The blurring of a fuzzy set $\mathbb{A}$, with a membership function, $\mu_{\mathbb{A}}(x)$ is a new fuzzy set formed written as, $\mathrm{BLR}(\mathbb{A})$, whose membership function may be defined as

$$\mu_{\mathrm{BLR}(\mathbb{A})}(x)\begin{cases} = \mu_{\mathrm{DIL}(\mathbb{A})}(x), & \text{such that } \forall\ x\ \mu_{\mathbb{A}}(x) < 0.5 \\ = \mu_{\mathrm{CON}(\mathbb{A})}(x), & \text{such that } \forall\ x\ \mu_{\mathbb{A}}(x) \geq 0.5 \end{cases}$$

(5.4.3.6)

Definition 5.4.3.6

A convex combination of two fuzzy sets $\mathbb{A}$ and $\mathbb{B}$ with membership functions $\mu_{\mathbb{A}}(x)$ and $\mu_{\mathbb{B}}(x)$ respectively is a new fuzzy set formed and written as $\mathbb{C} = \alpha\mathbb{A} + (1-\alpha)\mathbb{B}$ and with a membership function defined as

$$\mu_{\mathbb{C}}(x) = \alpha\mu_{\mathbb{A}}(x) + (1-\alpha)\mu_{\mathbb{B}}(x), \quad 0 \leq \alpha \leq 1$$

(5.4.3.7)

Definition 5.4.3.7

The algebraic sum of two fuzzy sets $\mathbb{A}$ and $\mathbb{B}$ with membership functions $\mu_{\mathbb{A}}(x)$ and $\mu_{\mathbb{B}}(x)$ respectively is a new fuzzy set formed and written as $\mathbb{A} \oplus \mathbb{B}$ and whose membership function is defined as

$$\mu_{\mathbb{A}\oplus\mathbb{B}}(x) = \mu_{\mathbb{A}}(x) + \mu_{\mathbb{B}}(x) - \mu_{\mathbb{A}}(x) \bullet \mu_{\mathbb{B}}(x)$$

(5.4.3.8)

5.5 Categories and Graphs of some Common Membership Functions

The following membership functions and graphs are presented here for quick reference particularly for the advantage of the users of this book whose are not familiar with the on going work on fuzzy phenomenon. For extensive presentation the reader is referred to [809] [810][1151][1152][1153b]. Consider reference sets of non-negative reals, R^+ and integers, N^+. Three classes of fuzzy sets with corresponding membership functions may be examined: a) "x is small", b) "x is large" and c) "x is closer to a" where $x \in \mathsf{R}^+$ or $x \in \mathsf{N}^+$ and "a" is a given positive constant. The system can also be constructed for the negative reals and integers. Let $\mu_{\bullet}(x)$ be a generic membership function defined over either R^+ or N^+, and then the following membership functions may be specified. We shall first deal with membership functions that express the concept of "x is small". We can also speak of "medium small", "medium big" and others in terms of negative and positive values.

5.5.1 Membership Functions for "x is small"

Let us examine a class of membership functions that relate to statement of "x is small". The statement may be extended to "x is medium small" or "x is very small" and many relevant linguistic statements. These statements are considered as linguistic variables.

5.5.1.1 *Step Membership Function for "x is Small"*

$$\mu(x)=\begin{cases}1 , & \leq x \leq a\\ 0 , & x>a\end{cases} \qquad \text{Step Membership Function}$$

(5.5.1.1)

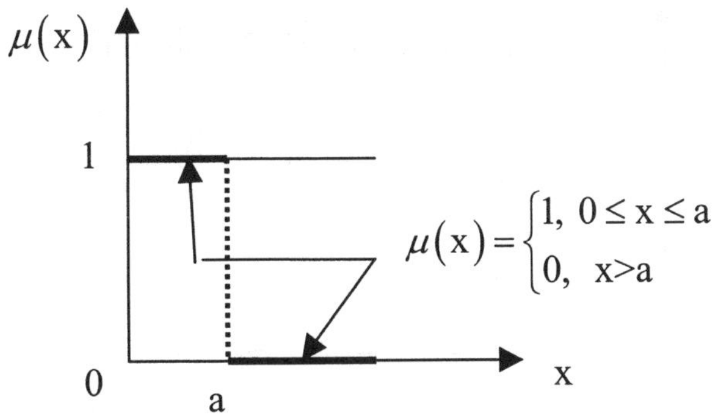

Figure 5.5.1.1 Graph of Step Membership Function for "x ix small"

5.5.1.2 *Exponential Membership Functions for "x is Small"*

$$\mu(x) = e^{-kx} \quad , \; k > 0$$

(5.5.1.2.1)

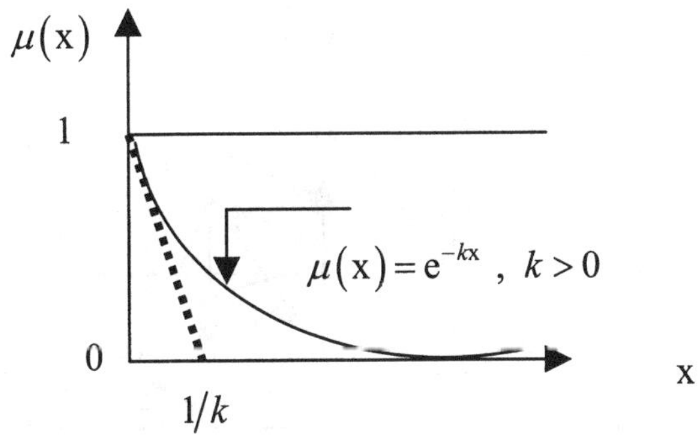

Figure 5.5.1.2.1 A Graph of an Exponential Membership Function for "x is Small"

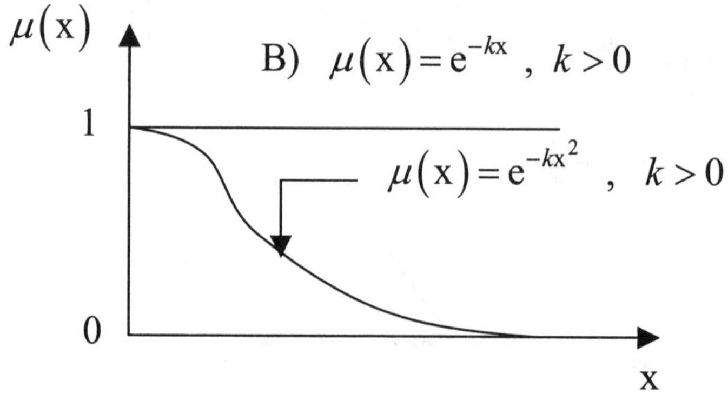

Figure 5.5.1.2.2 A Graph of an Exponential Membership function for "x is small" $\mu(\bullet)$

5.5.1.3 The Z and Ramp Membership Functions for "x Is Small"

$$\mu(x)\begin{cases} =1 & , x\in[0,a_1] \\ =\dfrac{a_2-x}{a_2-a_1} & , x\in[a_1,a_2] \\ =0 & , x\in(a_2,\infty] \end{cases}$$

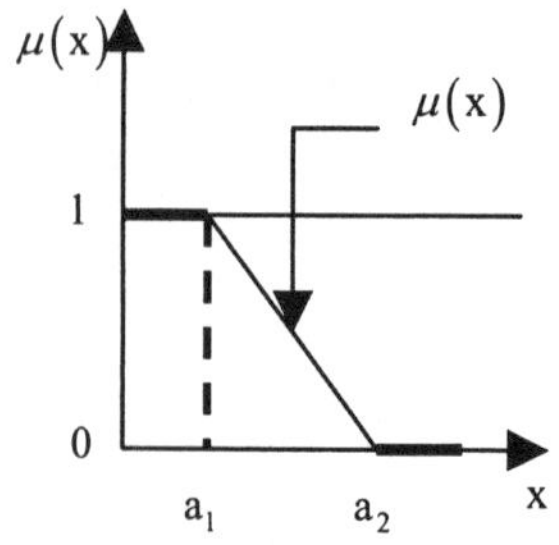

Figure 5.5.1.3.1 Geometry of the Z-Membership Function

$$\mu(x)\begin{cases} =1-ax^k, & x\in\left[0\,,\,a^{-\left(\frac{1}{k}\right)}\right] \\ =0 & , x\in\left[a^{\left(\frac{1}{k}\right)},\,\infty\right] \end{cases}$$

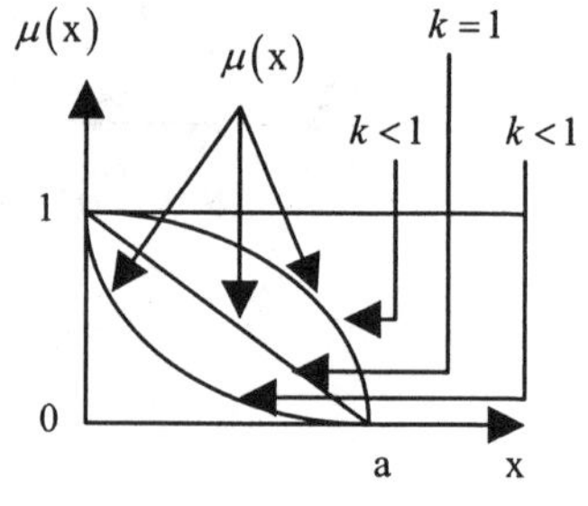

Figure 5.5.1.3.2 Geometry of Ramp Membership Function

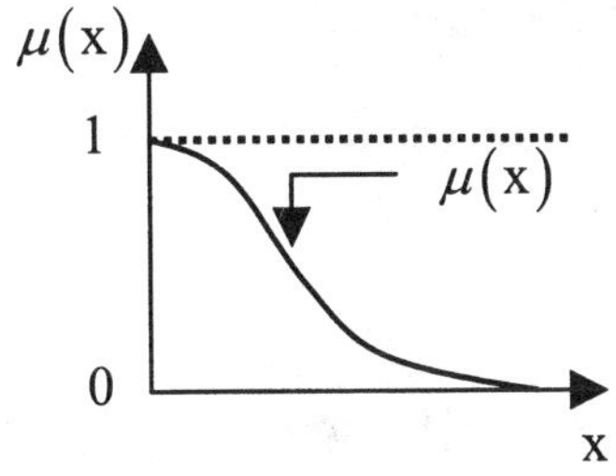

Figure 5.5.1.3.3 An Inverted S-membership Function

equation. to Figure 5.5.1.3.3

$$\mu(x) = \frac{1}{1 + kx^2} \quad , k > 1$$

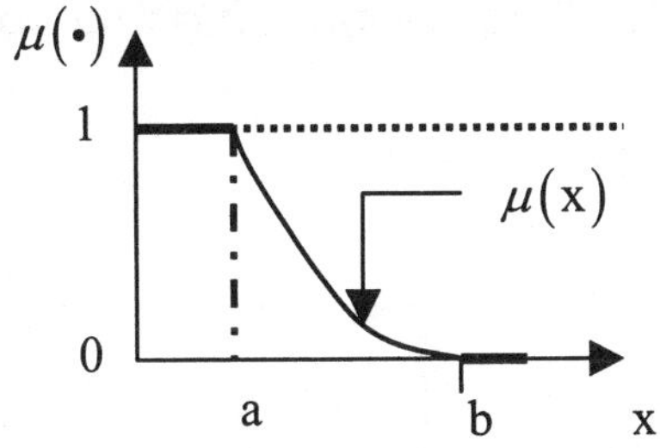

Figure 5.5.1.3.4 Geometry of a Z-Membership Function.

equation. to Figure 5.5.1.3.4

$$\mu(x)\begin{cases} = 1 \ , x \in [0, a) \\ = \frac{1}{2}\left\{1 - \sin\frac{\pi}{b\text{-}a}\left[x - \frac{1}{2}(a+b)\right]\right\} \ , x \in [a,b] \\ = 0 \ , \ x \in (b, \infty) \end{cases}$$

5.5.2 Membership Functions for "x is large"

Let us now turn our attention to fuzzy numbers that tend to express the numerical idea of largeness, such as "a is large". Just as the linguistic variable "x is small" we can express largeness in varying degrees such "medium large", "very large" and others depending on the problem at hand,

5.5.2.1 Step Membership function for "x is large"

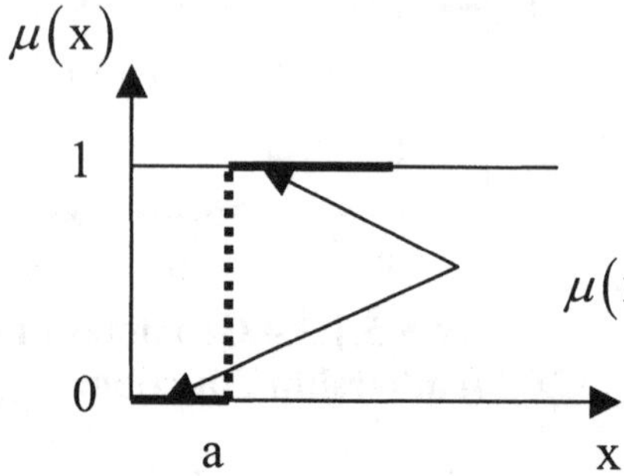

$$\mu(x)\begin{cases} =0 \ , \ x \in [0\,,a) \\ =1 \ , \ x \in [a\,,\infty) \end{cases}$$

Figure 5.5.2.1.1 Geometry of Step-Membership Function for "x is large"

5.5.2.2 Exponential Membership Functions for "x is large"

$$\mu(x)\begin{cases} 0\,,\ x \in [0,\,a] \\ 1\text{-}e^{-k(x\text{-}a)},x>a,\ k>0 \end{cases}$$

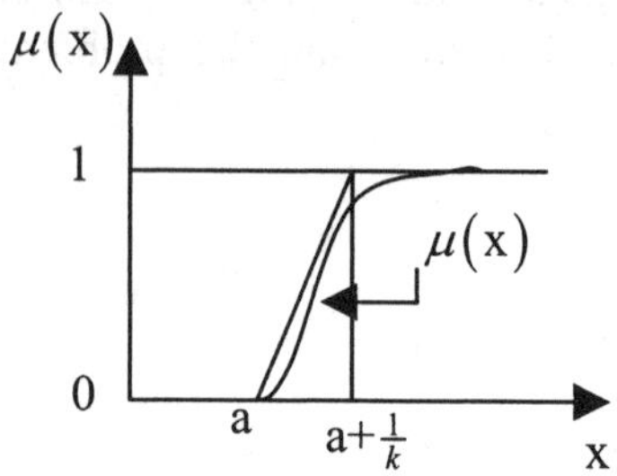

Figure 5.5.2.2.1 Geometry of an Exponential Membership Function

$$\mu(x)\begin{cases} 0,\ x \in [0,a] \\ 1\text{-}e^{-k(x\text{-}a)^2},\ x>a,\ k>0 \end{cases}$$

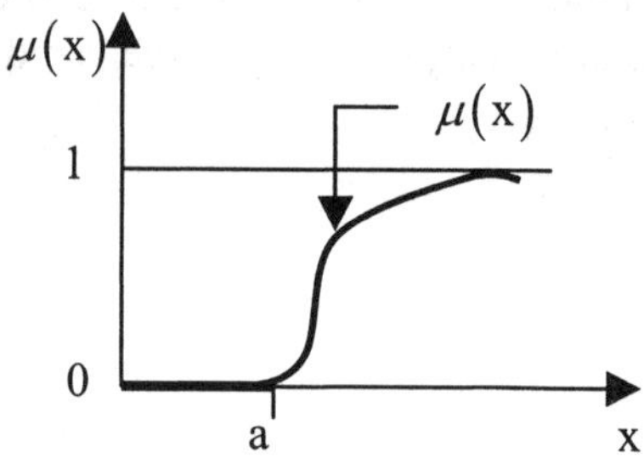

Figure 5.5.2.2.2 Geometry of an Exponential Membership Function

5.5.2.3 S- and Ramp Membership Function for "x is Large"

Let us now give examples of functional structure of membership functions of S-type and ramp type. Keep in mind that these membership functions can be approximated by either TFN (triangular) or TZFN (trapezoidal). The functional example, geometric example and corresponding equations are given in Figures 5.5.2.3.1-5.5.2.3.4.

$$\mu(x)\begin{cases} 0 \\ = a(x\text{-}a)^k, a < x < a + a^{-\frac{1}{k}} \\ = 1 \;,\; x \geq a + a^{-\left(\frac{1}{k}\right)} \end{cases}$$

$$\mu(x)\begin{cases} 0 \;,\; x \in [0, a_1] \\ \frac{x\text{-}a_1}{a_2\text{-}a_1} \;, x \in (a_1, a_2) \\ 1 \;, x \in [a_2, \infty) \end{cases}$$

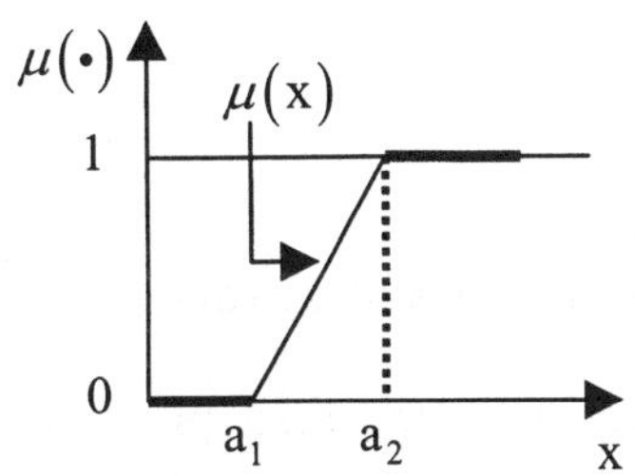

Figure 5.5.2.3.1 Geometry of an S-Membership Function

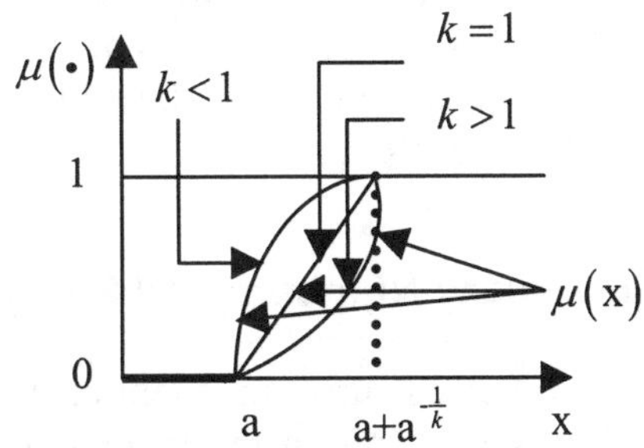

Figure 5.5.2.3.2 Geometry of a Ramp Membership Function

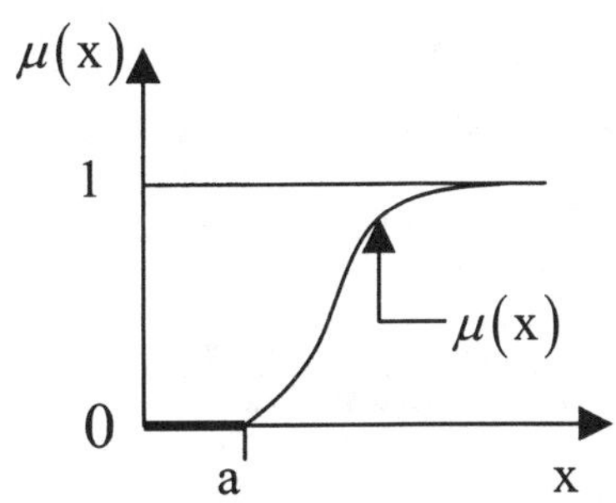

Figure 5.5.2.3.3 Geometry of an S-Membership Function

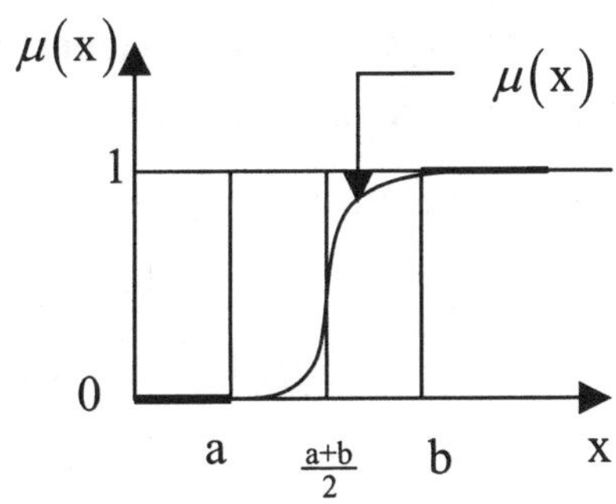

Figure 5.5.2.3.4 Geometry of an S-Membership Function

$$\mu(x)\begin{cases} = 0 \;,\; x \in [0, a] \\ = \frac{k(x-a)^2}{1+k(x-a)^2} \;,\; x \in (a, \infty] \end{cases}$$

Equation to Figure 5.5.2.3.3

$$\mu(x)\begin{cases} 0 \quad x \in [0, a] \\ \frac{1}{2}\left\{1 + \sin\frac{\pi}{b\text{-}a}\left[x - \frac{1}{2}(a+b)\right]\right\}, \\ \qquad\qquad x \in (a, b) \\ = 1 \;,\; x \in [a, \infty) \end{cases}$$

Equation to Figure 5.5.2.3.4

5.6 Basic Arithmetic of Fuzzy Numbers

In section 5.5 we introduced classes of membership functions that may be used to model fuzzy phenomena in many areas of scientific research. These functions tend to establish a range of fuzzy numbers. In this section we shall simply present the essential properties and the arithmetic of fuzzy numbers. Because of their properties for easy applications we shall use triangular and trapezoidal fuzzy numbers in the analysis that follows. What is being produced here are the basic elements of fuzzy arithmetic. For detailed treatment one is referred to [739] [748] [1150], [1152b].

5.6.1 Fuzzy Numbers and Fuzzy Confidence Intervals

Let us turn our attention to fuzzy numbers, their properties and their possible uses in the modeling of fuzzy phenomena. A question that arises is what is a fuzzy number?

Definition 5.6.1.1

An element, a, belonging to a set of real numbers, R, $(\text{tha is } a \in \mathsf{R})$ is said to be a fuzzy number if it is covered by a fuzzy set, $\mathbb{A}$, with a membership function $\mu_{\mathbb{A}}(x)$ such that

$$\mu_{\mathbb{A}}(x)\begin{cases} =1 \ , \ \text{if } x=a \\ \in(0,1) \ , \ \text{if } x \neq a \text{ and } x \in \mathbb{A} \\ =0 \ , \ \text{if } x \notin \mathbb{A} \end{cases}$$

$\mathbb{A}$ is said to be a fuzzy covering of "a".

The value "a" is said to be an ordinary real number if $\mu_{\mathbb{A}}(a)=1$. The membership value, $\mu_{\mathbb{A}}(x)$, may be viewed as *fuzzy confidence coefficient*. It measures the *degree of precision or exactness* to which "x" is "a". The value $\left[1-\mu_{\mathbb{A}}(x)\right]$ is the risk of *imprecision or inexactness* attached to the statement that "x" is close to "a". We shall refer to it as some measure of simple *fuzzy risk* and represented as, $\rho(x|a)$ where $x \in \mathbb{A}$ and $a \in \mathsf{R}$.

Note:

This definition of fuzzy number is special and will be modified to define triangular (TFN) and trapezoidal (TZFN) fuzzy numbers. The geometric structures and the topological properties of these numbers are given in figures (5.6.2.3.1.1 and 5.6.2.4.1.) and Sections (5.6.2.3)

Definition 5.6.1.2

A function, $\rho(x|a)$ is said to be a fuzzy risk function of $x \in \mathbb{A}$ around $a \in R$ with the property that

$$\rho(x|a) = \begin{cases} 1 \Rightarrow \text{ sure risk} \\ 1 - \mu_{\mathbb{A}}(x) \Rightarrow \text{ some risk} \\ 0 \Rightarrow \text{ no risk} \end{cases}$$

Statement 5.6.1.1

Every fuzzy covering, $\mathbb{A}$, of a fuzzy number has a lower limit of a real number, "a" and upper limit of a real number, "b" such that if $\mathbf{I} = [a,b]$ with a<b then

$$\mu_{\mathbb{A}}(x) \begin{cases} = 0 \text{ , if } x \leq a \\ \in (0,1) \text{ , if } a < x < b \\ = 1 \text{ , if } x > b \end{cases}$$

Statement 5.6.1.2

If "a" is a fuzzy number in the set of real numbers, R and $\mathbb{A}$ is its fuzzy covering then

a) $\mathbb{A}$ is a normal fuzzy set.

b) The α – level cuts, $\mathbb{A}_\alpha$, is a closed interval for every $\alpha \in (0,1]$.

c) The supports of $\mathbb{A}$, $\mathbb{A}^{0+}$ are bounded

d) Each fuzzy number may be represented by its α-level cuts

Definition 5.6.1.3

Given a fuzzy set, $\mathbb{A}$, with a membership function, $\mu_{\mathbb{A}}(x) \in [0,1]$ and $a_1 < a_2 \in R$ then $[a_1 , a_2]$ is a fuzzy confidence interval of R with $\mu_{\mathbb{A}}(x)$ as its confidence coefficient. The weighted index of total fuzzy confidence coefficient is a measure of fuzzy risk coefficient with a *fuzzy risk covering*.

Statement 5.6.1.2

1) Every fuzzy confidence interval $[a,b]$ is a fuzzy number and hence a fuzzy set.
2) Every fuzzy set may be uniquely represented by its α-level cuts where $\alpha \in [0,1]$
3) The α-level cuts of fuzzy sets and hence fuzzy confidence interval are closed interval of real numbers for all $\alpha \in (0,1]$

5.6.2 Basic Fuzzy Arithmetic on Fuzzy Numbers

There are two methods of performing fuzzy arithmetic on fuzzy numbers. They are methods of *interval arithmetic* and *extension principles.* The extension principle involves a process where arithmetic operations on the set of real numbers, R, are extended to fuzzy numbers through α-level cuts (α-cuts) where the fuzzy set coverings are normal and convex.

5.6.2.1 Basic definitions

We shall now offer some basic working definitions for the arithmetic operations on the fuzzy numbers.

Definition 5.6.2.1.1

If $\mathbb{A}=[a_1,a_3]$, $a_1, a_3 \in \mathbb{R}$ is a fuzzy number with $\mu_{\mathbb{A}}(\cdot) \in [0,1]$ as its membership function then $\mathbb{A}$ is normal if $\exists\, x \in \mathbb{R} \ni \bigvee_{x} \mu_{\mathbb{A}}(x)=1$ and $\mathbb{A}$ is convex if its $\alpha-$cuts, $\mathbb{A}_\alpha$ relative to the horizontal axis is nested in the sense that if $\alpha_1 > \alpha_2$ then for upper-level cut (lower-level cut)

$\mathbb{A}_{\alpha_1} \subset \mathbb{A}_{\alpha_2}$ $(\mathbb{A}_2 \subset \mathbb{A}_1)$. Generally if $\alpha_1 > \alpha_2 > \cdots > \alpha_n$ then $\mathbb{A}_{\alpha_i} \underset{i<j}{\subset} \mathbb{A}_{\alpha_j}$.

5.6.2.2 Fuzzy-Interval Arithmetic

Let us now consider real numbers a_1, a_2, b_1 and $b_3 \in R$. Define fuzzy intervals $\mathbb{A}=[a_1,a_2]$ and $\mathbb{B}=[b_1,b_2]$. Let (+) = addition, ($-$) = subtraction, ($\bullet$) = multiplication and (/) = division then:

i) Fuzzy Addition

$$\mathbb{A}+\mathbb{B} = [a_1,a_2]+[b_1,b_2]=[a_1+b_1 \ , a_3+b_3]$$

(5.6.2.2.1)

ii) Fuzzy Subtraction

$$\mathbb{A} - \mathbb{B} \ = \ [a_1,a_3]-[b_1,b_3]=[a_1 - b_3 \ , a_3 - b_1]$$

(5.6.2.2.2)

iii) Fuzzy Multiplication

$$\mathbb{A}\bullet\mathbb{B} = [a_1,a_3]\bullet[b_1,b_3]$$

$$=\left[(a_1\bullet b_1 \wedge a_1\bullet b_3 \wedge a_3\bullet b_1 \wedge a_3\bullet b_3) \ , \ (a_1\bullet b_1 \vee a_1\bullet b_3 \vee a_3\bullet b_1 \vee a_3\bullet b_3)\right]$$

(5.6.2.2.3)

iv) Fuzzy Inverse

$$\mathbb{A}^{-1} = [a_1,a_3]^{-1} = \left[\left(\frac{1}{a_1} \wedge \frac{1}{a_3}\right), \left(\frac{1}{a_1} \vee \frac{1}{a_3}\right)\right]$$

(5.6.2.2.4)

except when $a_1 \leq 0 \leq a_2$

v) Fuzzy Division

$$\left(\mathbb{A}/\mathbb{B}\right) = [a_1,a_3]/[b_1,b_3]$$

$$=\left[\left(\frac{a_1}{b_1} \wedge \frac{a_1}{b_3} \wedge \frac{a_3}{b_1} \wedge \frac{a_3}{b_3}\right), \left(\frac{a_1}{b_1} \vee \frac{a_1}{b_3} \vee \frac{a_3}{b_1} \vee \frac{a_3}{b_3}\right)\right]$$

(5.6.2.2.5)

vi) Fuzzy Minimum ($\wedge$)

$$\mathbb{A} \wedge \mathbb{B} = [a_1, a_3] \wedge [b_1, b_3] = [a_1 \wedge b_1 , a_3 \wedge b_3]$$

(5.6.2.2.6)

vii) Fuzzy Maximum

$$\mathbb{A} \vee \mathbb{B} = [a_1 , a_3] \vee [b_1 , b_3] = [(a_1 \vee b_1) , (a_3 \vee b_3)]$$

(5.6.2.2.7)

5.6.2.3 Illustrative Example

Let the fuzzy interval $\mathbb{A}$ and $\mathbb{B}$ be defined as $\mathbb{A} = [3,5]$ and $\mathbb{B} = [-2, 8]$ then by applying the respective fuzzy operators we can write

i) Addition

$$\mathbb{A} + \mathbb{B} = [3,5] + [-2,8] = [3 + (-2) , 5 + 8] = [1,13]$$

ii) Subtraction

$$\mathbb{A} - \mathbb{B} = [3,5] - [-2,8] = [3 - 8 , 5 - (-2)] = [-5,7]$$

iii) Multiplication

$$\mathbb{A} \bullet \mathbb{B} = [3,5][-2,8]$$

$$= [\{(3\bullet(-2)) \wedge (3\bullet 8) \wedge (5\bullet(-2)) \wedge (5\bullet 8)\},$$

$$\{(3(-2)) \vee (3\bullet 8) \vee (5\bullet(-2)) \vee (5\bullet 8)\}]$$

$$= [(-6) \wedge 24 \wedge (-10) \wedge 40, (-6) \vee 24 \vee (-10) \vee 40]$$

$$= [-10 , 40]$$

iv) Inverse

$$\mathbb{A}^{-1} = [3 , 5]^{-1} = \left[\left(\frac{1}{3} \wedge \frac{1}{5}\right) , \left(\frac{1}{3} \vee \frac{1}{5}\right)\right] = \left[\frac{1}{5} , \frac{1}{3}\right]$$

v) Division

$$\left(\mathbb{A}/\mathbb{B}\right) = \left\{[3,5]/[-2,8]\right\} = \left[\left(\frac{3}{-2} \wedge \frac{3}{8} \wedge \frac{5}{-2} \wedge \frac{5}{8}\right), \left(\frac{3}{-2} \vee \frac{3}{8} \vee \frac{5}{-2} \vee \frac{5}{8}\right)\right]$$

$$\left[\frac{5}{-2} , \frac{5}{8}\right] = [-2.5 , 0.625]$$

vi) Minimum

$$\mathbb{A} \wedge \mathbb{B} = [3 , 5] \wedge [-2 , 8] = [\{3 \wedge (-2)\} , \{5 \wedge 8\}] = [-2 , 5]$$

vii) Maximum

$$\mathbb{A} \vee \mathbb{B} = [3\ ,\ 5] \vee [-2\ ,\ 8] = [\{3 \vee (-2)\}\ ,\ \{5 \vee 8\}] = [3,8]$$

viii) Symmetric Image

$$-\mathbb{A} = -[3\ ,\ 5] = [-5\ ,\ -3]$$

5.6.2.4 $\alpha -$ Cut Algebra and Arithmetic (Extension Principle)

In section 5.3 we discussed $\alpha -$ level sets and their relationship to the decomposition theorem. In this section we shall discuss $\alpha -$ cut in terms of algebraic and arithmetic operations. First recall from section 5.3, Definition 5.3.2 that an $\alpha -$ cuts are $\alpha -$ level sets that are ordinary sets, $\mathbb{A}_\alpha$ whose members for upper-level cut are of the form:

$$\mathbb{A}_\alpha = \{(x, \mu_{\mathbb{A}}(x)) \mid \mu_{\mathbb{A}}(x) \geq \alpha, \alpha \in [0,1] \text{ and } x \in \mathsf{R}\}$$

such that if $\alpha_1 \leq \alpha_2$ then $\mathbb{A}_{\alpha_2} \leq \mathbb{A}_{\alpha_1}$, $\alpha_1, \alpha_2 \in [0,1]$ and R is the set of real numbers. The $\alpha -$ cuts of fuzzy confidence interval where the α's are fuzzy confidence coefficients may be written as $\mathbb{A}_\alpha = [a_1, a_2]_\alpha$, $\mathbb{B}_\alpha = [b_1, b_2]_\alpha$, $\alpha \in [0,1]$ and $a_i, b_i \in \mathsf{R}$. To illustrate the operational mechanism of the $\alpha -$ cut arithmetic we shall use two important class of fuzzy numbers of triangular fuzzy number **(TFN)** and trapezoidal fuzzy number **(TZFN).**

5.6.2.3.1 ***Triangular Fuzzy Number (TFN)***

In this section we redefine a triangular fuzzy number as has been stated in eqn. (5.5.3.4) and definition (5.5.1.1)

Definition 5.6.2.3.1.1

Let $a_1, a_2, a_3 \in \mathsf{R}$ be ordered real numbers such that $a_1 < a_2 < a_3$ then a_2 is a triangular fuzzy number if its $\varepsilon -$ neighborhood , $N_\varepsilon(a_2)$ contains the fuzzy intervals $\mathbb{A}_1 = [a_1\ ,\ a_2]$ and $\mathbb{A}_2 = [a_2\ ,\ a_3]$ with a membership function, $\mu_{\mathbb{A}}(x)$ defined as

$$\mu_{\mathbb{A}}(x)\begin{cases} = 0\ , \text{if } x \le a_1 \text{ and } x \ge a_3 \\ = \dfrac{x - a_1}{a_2 - a_1}\ , \text{if } a_1 < x \le a_2 \\ = \dfrac{a_3 - x}{a_3 - a_2}\ , \text{if } a_2 \le x < a_3 \end{cases}$$

(5.6 2.3.1.1.1)

Equation (5.6.2.3.1.1.1) is equivalent to eqn. (5.5.3.4) by simply defining $a_3 = a_2 + a_1$ and $a = a_2$ and with slight restructuring. Note that a_1 is the lower limit and a_3 is the upper limit of the set $N_\varepsilon(a_2)$. The triangular fuzzy number is geometrically shown in Figure 5.6.2.3.1.1. The number a_2 is the *core* of the TFN with $\mu_{\mathbb{A}}(x) = 1$

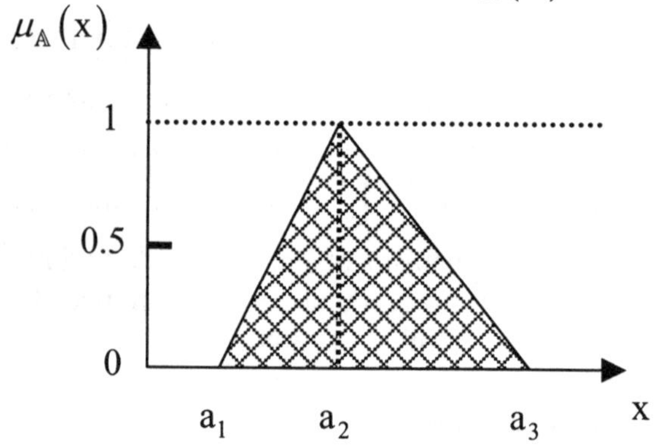

Figure 5.6.2.3.1.1TriangularFuzzyNumber

$\mathbb{A} = [a_1, a_2, a_3] = N_\varepsilon(a_2)$

Where a_2 is the Core

Definition 5.6.2.3.1.2

If $a_1, a_2 \in \mathsf{R}$ such that $a_1 < a_3$ then $\forall \alpha \in [0,1]$ the fuzzy interval $\mathbb{A} = [a_1, a_3]$ is a triangular fuzzy number if there $\exists\ a_2 \in \mathsf{R}$ such that the $\varepsilon-$neighborhood of a_2 represented as $N_\varepsilon(a_2)$ may be specified by an $\alpha-$cut of $\mathbb{A} = [a_1\ , a_3]$ in the form

$$\mathbb{A}_\alpha = [a_1, a_3]_\alpha = [a+(a_2 - a_1)\alpha \ , a_3-(a_3 - a_2)\alpha]$$

Example 5.6.2.3.1.2

Let $\mathbb{A} = [a_1 \ , a_2 \ , a_3] = [-3 \ , \ 2 \ , \ 4]$ be a TFN as shown in the fuzzy number geometry below.

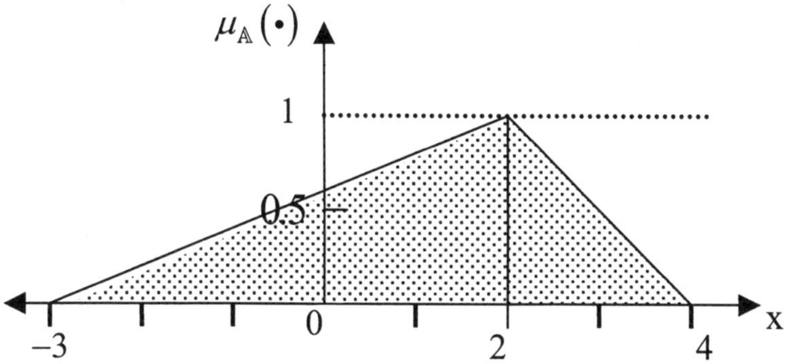

Figure 5.6.2.3.1.2 Triangular Fuzzy Number $\mathbb{A} = [-3, \ 2, \ 4]$

By Definition 5.6.2.3.1.2 we can write

$$\mu_{\mathbb{A}}(x) \begin{cases} = 0 \ , \ \text{if } x \le -3 \text{ and } x > 4 \\ = \dfrac{x+3}{5} \ , \ \text{if } -3 < x \le 2 \\ \dfrac{4-x}{2} \ , \ \text{if } 2 \le x < 4 \end{cases}$$

where (-3) is the lower limit and (4) is the upper limit of the fuzzy confidence interval and $[-3,4] \subseteq N_\varepsilon(2)$. Alternatively, by Definition 5.6.2.3.1.2 we can specify the triangular fuzzy number $\mathbb{A} = [-3 \ , \ 2 \ , \ 4]$ using interval of confidence or neighborhood intervals defined by $\alpha - \text{cut}, \ \alpha \in [0 \ , \ 1]$ as $\mathbb{A}_\alpha = [-3 \ , \ 4]_\alpha = [-3 + 5\alpha \ , 4 - 2\alpha]$ and if $\alpha = 0.75 \ , 0.50 \ , 0.25$ then

$\mathbb{A}_{\alpha=0.75} = [-0.75\ ,\ 1.25], \mathbb{A}_{\alpha=0.50} = [-0.5,\ 3]$ and $\mathbb{A}_{\alpha=1} = [2\ ,\ 2]$

5.6.2.3.2.2 ***An α – cut Arithmetic on TFN***

Let us consider two triangular fuzzy numbers a_2 and b_2 with fuzzy interval coverings $\mathbb{A} = [a_1\ ,\ a_3]$ and $\mathbb{B} = [b_1\ ,\ b_3]$ then from the Definition 5.6.2.3.1.2 we can write their α – cuts for $a_1, a_2, a_3, b_1, b_2, b_3 \in \mathsf{R}$ as

$$\mathbb{A}_{\alpha} = [a + (a_2 - a_1)\alpha\ ,\ a_3 + (a_2 - a_3)\alpha]$$

(5.6.2.3.2.2.1)

$$\mathbb{B} = [b_1 + (b_2 - b_1)\alpha\ ,\ b_3 + (b_2 - b_3)\alpha]$$

(5.6.2.3.2.2.2)

As a numerical example let $a_1 = -3,\ a_2 = 2,\ a_3 = 4$, $b_1 = -2$, $b_2 = 1$, $b_3 = 3$ where $\mathbb{A} = [-3\ ,\ 4]$ and $\mathbb{B} = [-2\ ,\ 3]$ and the α – cut representation may be written as:

$$\mathbb{A}_{\alpha} = [-3 + 5\alpha\ ,\ 4 - 2\alpha]\ ,\quad \mathbb{B} = [-2 + 3\alpha\ ,\ 3 - 2\alpha]$$

i) Addition

$$\mathbb{C}_{\alpha} = \mathbb{A}_{\alpha} + \mathbb{B}_{\alpha} = [-3 + 5\alpha\ ,\ 4 - 2\alpha] + [-2 + 3\alpha\ ,\ 3 - 2\alpha]$$
$$= [(-3 + 5\alpha - 2 + 3\alpha)\ ,\ (4 - 2\alpha + 3 - 2\alpha)]$$
$$= [(-5 + 8\alpha)\ ,\ (7 - 4\alpha)]$$

The geometry of addition of two fuzzy numbers is shown in Figure 5.6.2.4.2.2.1

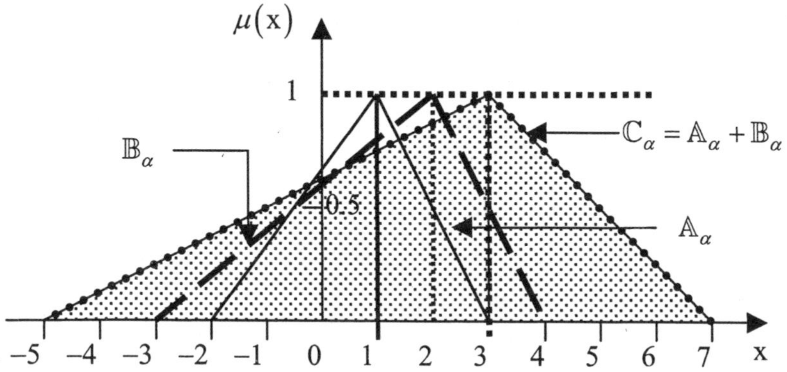

Figure 5.6.2.3.2.2.1 A Geometry of Addition of Two TFN $(\mathbb{A} + \mathbb{B})$

Note:

The lower and upper limits are obtained by setting $\alpha = 0$; the core or the middle number is obtained by setting $\alpha = 1$.

ii) Subtraction

$$\mathbb{C}_\alpha = \mathbb{A}_\alpha - \mathbb{B}_\alpha = [-3+5\alpha\ ,\ 4-2\alpha] - [-2+3\alpha\ ,\ 3-2\alpha]$$
$$= [(-3+5\alpha)-(-2+3\alpha)\ ,\ (4-2\alpha)-(3-2\alpha)]$$
$$= [-1+2\alpha\ ,\ 1]$$

The geometry of subtraction is shown in Figure 5.6.2.3.2.2.2

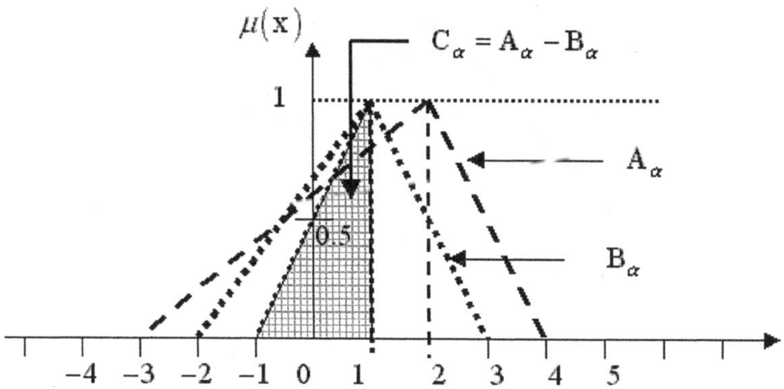

Figure 5.6.2.3.2.2.2 The Geometry of Subtraction of two TFN $(\mathbb{A} - \mathbb{B})$

Let us turn our attention to multiplication, division and inverse. Some difficulties arise with multiplication, division and inverse operations with the interval triplets. The $\alpha-\text{cuts}$ may be used for these operations. However care must be exercised with the min-max process when the triplets have a mixture of negative and positive numbers. In other words the fuzzy computing become more complex with these operations when $\mathbb{A}$ and $\mathbb{B} \subset \mathrm{R}$ appear as triangular fuzzy numbers. For simple computations as illustrative examples let $\mathbb{A}$ and $\mathbb{B} \subset \mathrm{R}^+$ where $\mathbb{A} = [2\ ,\ 4\ ,\ 6]$ and $\mathbb{B} = [1\ ,\ 5\ ,\ 8]$. The $\alpha-\text{cut}$ intervals of fuzzy confidence may be specified as

$$\mathbb{A} = [2+2\alpha\ ,\ 6-2\alpha]\ ,\quad \mathbb{B} = [1+4\alpha\ ,\ 8-3\alpha]$$

iii) Multiplication (using eqn. 5.6.2.2.3)

$$\begin{aligned}\mathbb{C}_\alpha = \mathbb{A}_\alpha \bullet \mathbb{B}_\alpha &= [2+2\alpha\ ,\ 6-2\alpha] \bullet [1+4\alpha\ ,\ 8-3\alpha] \\ &= [(2+2\alpha)(1+4\alpha)\ ,\ (6-2\alpha)(8-3\alpha)] \\ &= \left[\left(8\alpha^2+10\alpha+2\right),\ \left(6\alpha^2-34\alpha+48\right)\right]\end{aligned}$$

Note: $\mathbb{C}_{\alpha=0} = [2\ ,\ 48]$ and $\mathbb{C}_{\alpha=1} = [20,20] = 20$

The geometry of the multiplication $\mathbb{C}_\alpha = \mathbb{A}_\alpha \bullet \mathbb{B}_\alpha$ is represented in Figure 5.6.2.3.2.2.3.

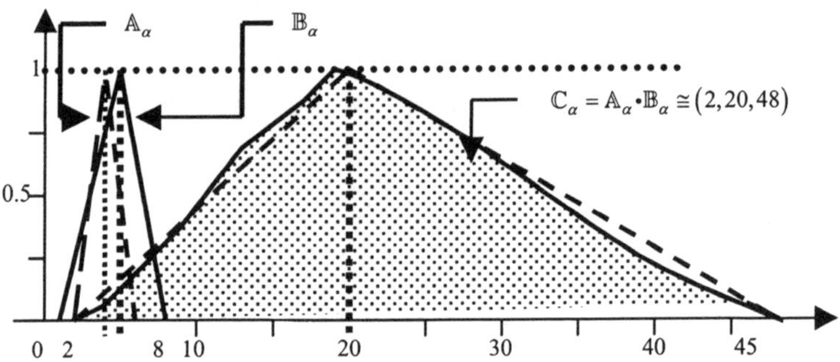

Figure 5.6.2.3.2.2.3 Multiplication of Two TFN, $\left(\mathbb{C}_\alpha = \mathbb{A}_\alpha \bullet \mathbb{B}_\alpha\right)$

The rest of the operations on $\mathbb{A}_\alpha$ and $\mathbb{B}_\alpha$ are quite involved. They, however, can be computed with $\alpha-\text{cut}$ and using the interval arithmetic formulae of (eqns. 5.6.2.2.1-5.6.2.2.6). It may also be noted that the straight lines joining the triplet of the TFN, the straight lines joining the triplet of the TFN, $\mathbb{C}_\alpha = \mathbb{A}_\alpha \bullet \mathbb{B}_\alpha = (2,\ 20,\ 48)$ are approximations. For discussions of the use of triangular fuzzy numbers (TFNs) as approximations to other complex fuzzy numbers, see [739] [1526] and for detailed treatment of arithmetic and basic algebra of fuzzy numbers see [1150].

It may be noted that if $\mathbb{A}$ and $\mathbb{B}$ are TFNs then $\mathbb{A} + \mathbb{B}$ and $\mathbb{A} - \mathbb{B}$ are also TFN's. Multiplication, inverse, division, maximum and minimum operations on two fuzzy numbers $\mathbb{A}$ and $\mathbb{B}$ do not necessarily yield fuzzy numbers.

5.6.2.4 Trapezoidal Fuzzy Numbers (TZFN)

Let us turn our attention to trapezoidal fuzzy numbers and corresponding arithmetic. Let us first define the trapezoidal fuzzy number (TZFN)

Definition 5.6.2.4.1

Let $a_1, a_2, a_3, a_4 \in \mathsf{R}$ be ordered numbers such that $a_1 < a_2 < a_3 < a_4$ then the interval $[a_2, a_3]$ is said to be trapezoidal fuzzy number if its neighborhood $\mathsf{N}_\varepsilon[a_2, a_3]$ contains fuzzy intervals $\mathbb{A}_1 = [a_1, a_2]$ and $\mathbb{A}_2 = [a_3, a_4]$ with a membership function, $\mu_{\mathbb{A}}(x)$ that is defined as

$$\mu_{\mathbb{A}}(x) \begin{cases} = 0 & \text{if } x \in (-\infty\ ,\ a_1] \text{ or } x \in [a_4\ ,\ \infty) \\ = \dfrac{x - a_1}{a_2 - a_1}, & \text{if } x \in (a_1\ ,\ a_2) \\ = 1\ , & \text{if } x \in [a_2\ ,\ a_3] \\ = \dfrac{a_4 - x}{a_4 - a_3}, & \text{if } x \in (a_3\ ,\ a_4) \end{cases}$$

(5.6.2.4.1)

The set $\mathbb{K} = [a_2\ ,\ a_3] \neq \varnothing$ is called the *core* of TZFN. If the cardinality of $\mathbb{K}$ is one then the TZFN becomes TFN (that is $\#\mathbb{K} = 1 \Rightarrow$ TZFN = TFN). The implication is that TFN is a special case

ply to the TZFN. An example of the geometry of a TZFN is given in Figure 5.6.2.4.1

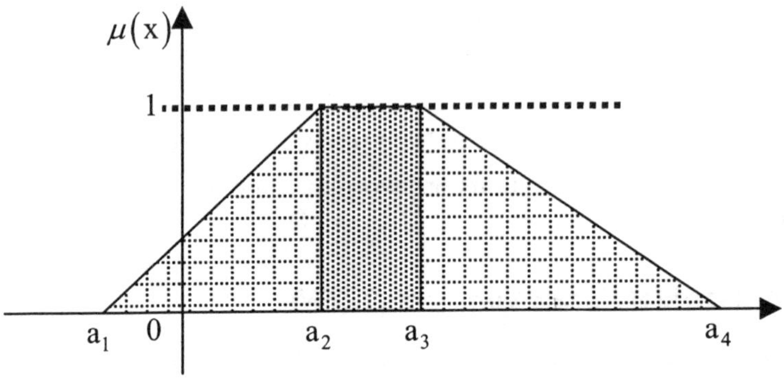

Figure 5.6.2.4.1 A Geometry of TZFN where $\mu_{\mathbb{A}}(x)\in[0,1]$ **and** $\mathbb{A}=[a_1, a_2, a_3, a_4]$

Before doing some arithmetic operations on TZFN let us give a general statement about characterization of fuzzy numbers.

Statement 5.6.2.4.1

Let $\mathbb{A}$ be a member of a family of all closed intervals of real numbers whose membership function, $\mu_{\mathbb{R}}(x)\in[0\ ,\ 1]$ then $\mathbb{A}$ is a fuzzy number if there is a non-empty closed interval, $[a_1\ ,\ a_2]\neq\varnothing$ such that

$$\mu_{\mathbb{A}}(x)\begin{cases}=\rho(x) & ,\ \text{if } x\in(-\infty\ ,\ a_1)\\ =1 & ,\ \text{if } x\in[a_1\ ,\ a_2]\\ =\varphi(x) & ,\ \text{if } x\in(a_2\ ,\ \infty)\end{cases}$$

(5.6.2.4.2)

where $\rho(x)\in(0,1)$ and $\varphi(x)\in[0,1)$, $\rho'>0$ and $\varphi'<0$ with $\rho'=\frac{d\rho}{dx}$ and $\varphi'=\frac{d\varphi}{dx}$.

Note:

The TFN and TZFN may easily be shown to follow equation (5.6.2.4.2) For example, let us visit equation (5.6.2.4.1) and notice that the system may be decomposed to represent a TFN as

$$\rho(x) = \frac{x - a_1}{a_2 - a_1} \;,\; \text{if } x \in (-\infty \;,\; a_2)$$

$$\varphi(x) = \frac{a_3 - x}{a_4 - a_3} \;,\; \text{if } x \in (a_3 \;,\; \infty)$$

and $[a_1 \, , a_2] = a_2$ with $\mu_{\mathbb{A}}(a_2) = 1$. Equation (5.6.2.4.2) may graphically be represented as in Figure 5.6.2.4.2.

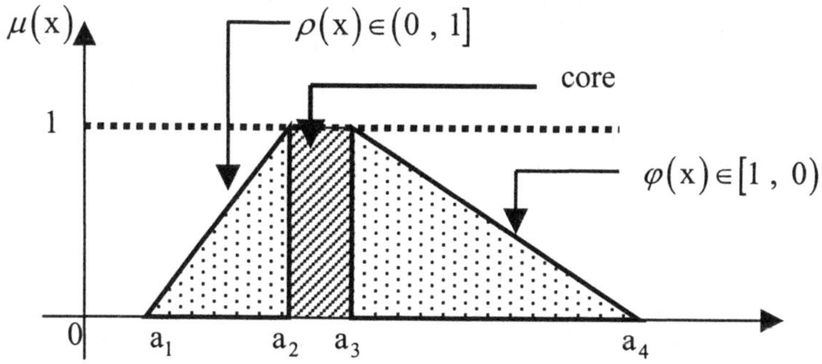

Figure 5.6.2.4.2 General Geometry of TZFN

5.6.2.4.1 ***Arithmetic Operations on Trapezoidal Fuzzy Numbers (TZFN)***

Consider two TZFNs $\mathbb{A} = (a_1, a_2, a_3, a_4)$ and $\mathbb{B} = (b_1, b_2, b_3, b_4)$ with membership functions $\mu_{\mathbb{A}}(x),\; x \in [a_1, a_4]$ and $\mu_{\mathbb{B}}(x),\; x \in [b_1, b_4]$ where the value $\mu_{\mathbb{A}}(x) = 1$ if

$x \in [a_2 \;,\; a_3]$ and $\mu_{\mathbb{B}} = 1$ if $x \in [b_2 \;,\; b_3]$. The structure of any TZFN is such that it can be decomposed into *left right-angled* TFN (LTFN) and *central rectangular* fuzzy number (RFN) and *right right-angled* TFN (RTFN). In the case of $\mathbb{A}$ and $\mathbb{B}$, the corresponding intervals of decomposition may be written as LTFN $= [a_1, a_2]_T$

$RFN=[a_2, a_3]_R$, $LTFN=[b_1, b_2]_T$, $RFN=[b_2, b_3]_R$ and $RTFN=[b_3, b_4]_T$ where T = triangular and R = rectangular.

Definition 5.6.2.4.2

Let $\mathbb{A} \subset \mathbb{R}$ then $\mathbb{A}$ is said to be a rectangular fuzzy number (RFN) if there exist a closed interval $[a_2, a_3]$ such that $\mu_{\mathbb{A}}(x)=1$ if $x \in [a_2, a_3]$ and $\mu_{\mathbb{A}}(x) \in (0, 1]$ whenever $x \notin [a_2, a_3]$.

Note:

It is useful to note that the complementary statements of TFN, TZFN and RFN lead to inverted fuzzy numbers such that $\mu_{\mathbb{A}}(x)=0$ if $x \in [a_2, a_3]$. Examples of inverted fuzzy numbers involve statements such as x is either negative or positive small (|x| is small); similarly x is either negative or positive big (that is, |x| is big). Graphical examples are shown in Figures 5.6.2.4.3 (a, b, c, d, e, f) [see also section 5.5.1]

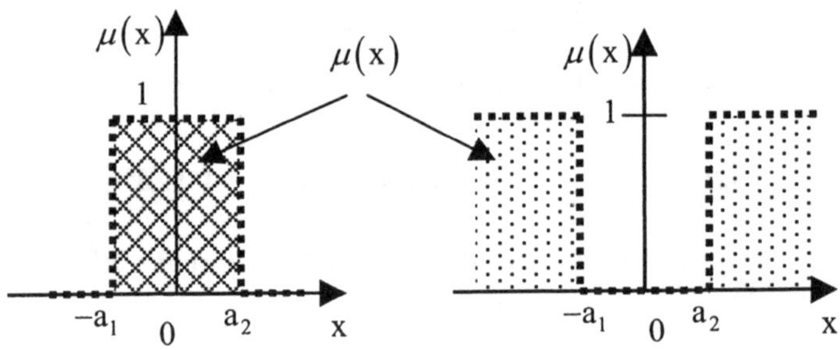

a) RFN for |x| is small **b) Inverted RFN for |x| is large**

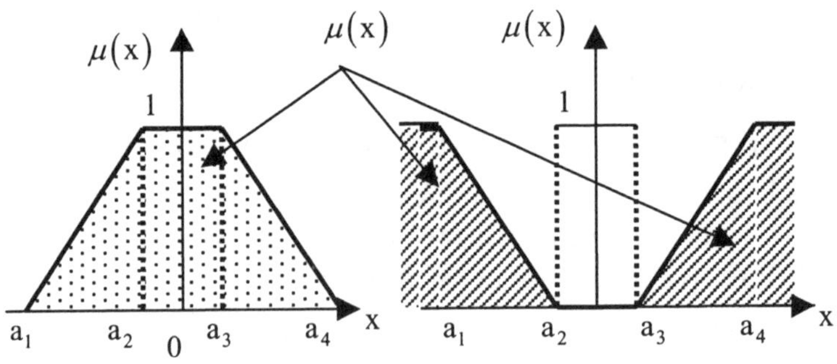

c) A TZFN for |x| is Small **d) An Inverted TZFN for |x| is Large**

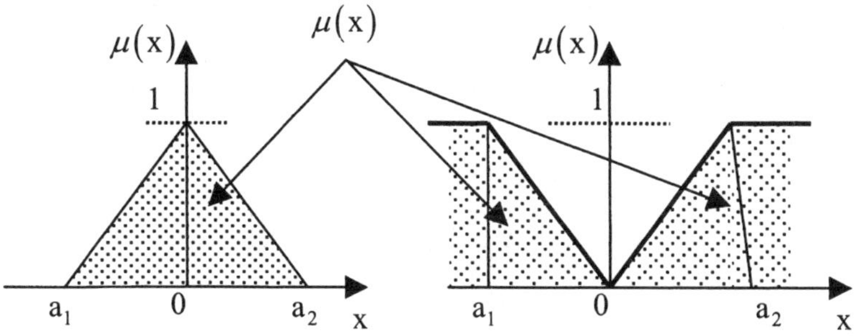

e) A TFN for |x| Small **f) An Inverted TFN for |x| is large**

Figure 5.6.2.4.3 Geometry of TFNs and TZFNs

The fuzzy statements negative or positive small or large may be replaced by statements "x is close to a" or "x is far away from a".

Let us return to the fuzzy arithmetic operations on TZFN's $\mathbb{A}=(a_1, a_2, a_3, a_4)$ and $\mathbb{B}=(b_1, b_2, b_3\ b_4)$. Corresponding to these sets let numerical examples be represented as $[a_1=2, a_2=4, a_3=7, a_4=9]$ $[b_1=1, b_2=3]$, $[b_3=5, b_4=8]$ and hence

$$\left.\begin{array}{l}\mathbb{A}=(a_1,a_2,a_3,a_4)\\ \mathbb{A}=(2,\ 4,\ 7,\ 9)\end{array}\right\}$$

(5.6.2.4.2)

$$\left.\begin{array}{l}\mathbb{B}=(b_1,\ b_2\ b_3\ b_4)\\ \mathbb{B}=(1\ ,3\ ,5\ ,8)\end{array}\right\}$$

(5.6.2.4.3)

The corresponding $\alpha-\text{level}$ cuts may be written as

$$\left.\begin{array}{l}\mathbb{A}_\alpha=\left[a_1+(a_2-a)\alpha\ ,\ a_4+(a_3-a_4)\alpha\right]\\ \mathbb{A}_\alpha=\left[2+2\alpha\ ,\ 9-2\alpha\right]\end{array}\right\}$$

(5.6.2.4.3)

$$\left.\begin{array}{l}\mathbb{B}_\alpha=\left[b_1+(b_2-b_1)\alpha\ ,b_4+(b_3-b_4)\right]\\ \mathbb{B}_\alpha=\left[1+2\alpha\ ,\ 8-3\alpha\right]\end{array}\right\}$$

(5.6.2.4.4)

We may now compute the following numbers by their indicated operations

i) Addition

$$\mathbb{C} = \mathbb{A} + \mathbb{B} = (a_1, a_2, a_3, a_4) + (b_1, b_2, b_3\ b_4)$$

(5.6.2.4.5)

$$= (2, 4, 7, 9) + (1, 3, 5, 8)$$
$$= (a_1+b_1\ , a_2+b_2\ , a_3+b_3\ , a_4+b_4)$$
$$= (3\ , 7\ , 12\ , 17)$$

ii) Subtraction

$$\left.\begin{aligned} \mathbb{C} = \mathbb{A} - \mathbb{B} &= (a_1, a_2, a_3, a) - (b_1,b_2,b_3, b_4) \\ &= (2\ , 4\ , 7\ , 9) - (1, 3, 5, 8) \\ &= (a_1 - b_1\ , a_2 - b_2\ , a_3 - b_3\ , a_4 - b_4) \\ &= (1\ , 1\ , 2\ , 1) \end{aligned}\right\}$$

(5.6.2.4.6)

iii) Symmetric Image

$$\left.\begin{aligned} -(\mathbb{A}) &= -(a_1, a_2, a_3, a_4) = (-a_1, -a_2, -a_3, -a_4) \\ -(\mathbb{B}) &= -(b_1, b_2, b_3, b_4) = (-b_1, -b_2, -b_3, -b_4) \end{aligned}\right\}$$

(5.6.2.4.7)

These arithmetic operations can also be performed by using the interval of confidence at $\alpha - \text{levels}$ as specified by eqs. (5.6.2.4.3) and (5.6.2.4.4). Let us work with the $\alpha - \text{level}$ algebra and arithmetic.

Addition:

$$\mathbb{C}_\alpha = \mathbb{A}_\alpha + \mathbb{B}_\alpha = [2 + 2\alpha\ , 9 - 2\alpha] + [1 + 2\alpha\ , 8\text{-}3\alpha]$$
$$= [3 + 4\alpha\ , 17 - 5\alpha]$$

where $\alpha = 0 \Rightarrow \mathbb{C}_\alpha = [3\ , 17]$ and $\alpha = 1 \Rightarrow \mathbb{C}_\alpha = [7\ , 12]$. Thus

$$\mathbb{C}_\alpha = \mathbb{A}_\alpha + \mathbb{B}_\alpha = (3, 7, 12, 17)$$

as is obtained at the beginning example. The geometry of the operation is shown in Figure 5.6.2.4.4.

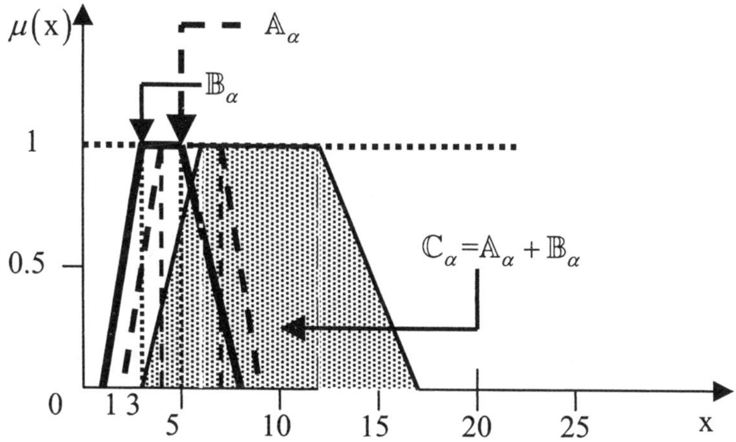

Figure 5.6.2.4.4 Geometry of the Sum of Two TZFN's

$$\mathbb{C}_\alpha = \mathbb{A}_\alpha + \mathbb{B}_\alpha$$

Multiplication:

$$\mathbb{C}_\alpha = \mathbb{A}_\alpha \bullet \mathbb{B}_\alpha = \left[(2+2\alpha)(1+2\alpha)\ ,\ (9-2\alpha)(8-3\alpha)\right]$$

$$= \left[\left(2+6\alpha+4\alpha^2\right)\ ,\ \left(72-43\alpha+6\alpha^2\right)\right]$$

where $\alpha = 0 \Rightarrow \mathbb{C}_\alpha = [2\ ,\ 72]$ and $\alpha=1 \Rightarrow \mathbb{C}_\alpha = [12\ ,\ 35]$ and hence

$$\mathbb{C}_\alpha = \mathbb{A}_\alpha \bullet \mathbb{B}_\alpha \cong [2\ ,\ 12\ ,\ 35\ ,\ 72]$$

The geometry of the product of two TZFN's is shown in Figure 5.6.2.4.5.

Note:

It may be noted that it is computationally easy to use the TFN's in applied problems. As such it is helpful and useful to use TFN's to approximate other fuzzy numbers. This approximation simplifies substantially the needed algebra and computation. For example, the exponential fuzzy (EFN) may be approximated by TFN or sometimes by TZFN. For extensive discussion on TFN approximations to other fuzzy numbers see [739], [1150] and [1152b].

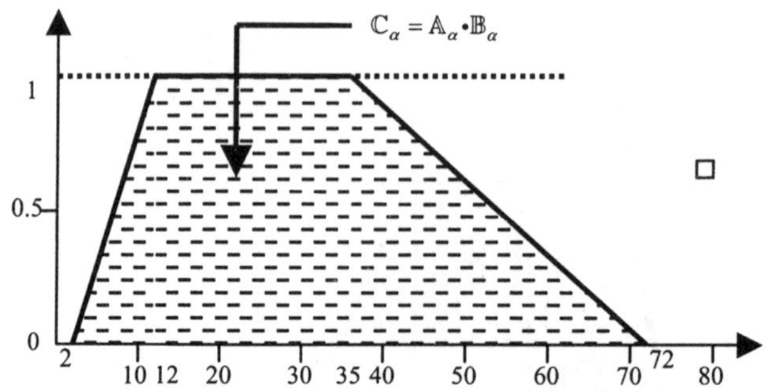

Figure 5.6.2.4.5 Geometry of the Product of two TZFN's

$$\mathbb{C}_\alpha = \mathbb{A}_\alpha \bullet \mathbb{B}_\alpha$$

6 The Basic Theory of Fuzzy Decisions

In Chapter 5 we presented some basic and essential characteristics of fuzzy mathematics that has been used in the development of cost-benefit analysis. In the process the theory of fuzzy decision was woven into the cost benefit-analysis to create a theory of soft computing of collective decision. The cost-benefit analysis is seen as an approach to collective decision. The objective is to combine cost-benefit rationality and fuzzy rationality to deal with the essential problem of democratic decision making. The fuzzy rationality requires the development of the theory of fuzzy decisions.

6.1 The Fuzzy Decision Problem

In this chapter we shall be concerned only with the analytical representations of the essential elements of fuzzy decisions and their properties. This representation and the corresponding solution should lead to morphological structure of optimal fuzzy decision rules that have been used to supplement and complement cost-benefit analysis. Central to the theory of fuzzy decisions are the notions of vagueness, imprecision, ill-definedness, subjective approximations, cognitive constraints on information processing etc. The building blocks of the theory are the fuzzy constraints and fuzzy goals where such fuzzy goals and constraints are analytically represented by notions of fuzzy sets and subsets as have been discussed in Chapter 5. The objective here is to discuss some important ideas of effective decision modeling in fuzzy environment. This will allow the reader to be familiar with the basic tools of fuzzy modeling and computing applicable to cost-benefit analysis of any decision. The idea is to assemble for quick reference the tools and basic logic of the theory of fuzzy decisions and the applicable mathematics that we have utilized in the framework of cost-benefit analysis and the applications to which it may be put. For full discussions on the theory of fuzzy decisions see [661] [682] [694] [709] [732][735][739][806][809].To this end, a number of definitions are required.

It must be pointed out that the decision problem and the optimal solution function represent a family of families of decision problems. The specific family of the decision problems would depend on the type of decision environment assumed. Such a decision environment would be described by the structural parameters on the basis of which the information support of the decision is obtained. The specific algorithm designed to realize an optimal decision rule would then depend on the family of the decision. Given the type of assumed environment, each decision problem may be classified according to conditions of consistency and similarity problems [694] [756] [760][780b]. The nature of the decision environment may be illustrated in a Venn-like diagram. Let $\mathbb{I}$ be the general environment of decision-making that is partitioned into certainty space, $\mathbb{C}$, and uncertainty space, $\mathbb{C}'$. We may now impose on the partition the fuzzy space, $\mathbb{F}$, and non-fuzzy space, $\mathbb{F}'$. The partitions lead to collectively exhaustive and mutually exclusive sub-spaces that are presented in Figure 6.1.1

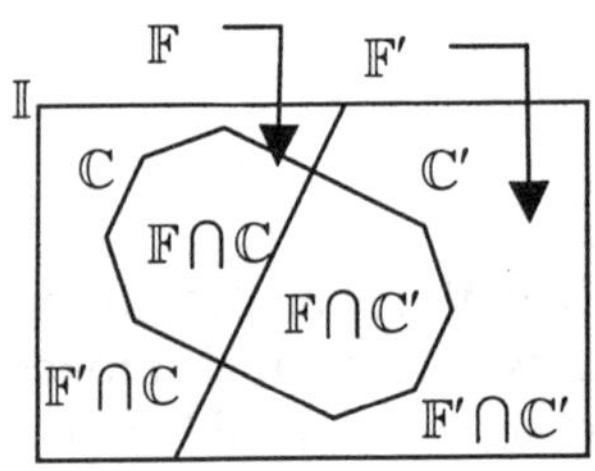

SPACES	$\mathbb{C}$	$\mathbb{C}'$
$\mathbb{F}'$	NON-FUZZY AND CERTAIN $\mathbb{F}'\cap\mathbb{C}$	NON-FUZZY AND UNCERTAIN $\mathbb{F}'\cap\mathbb{C}'$
$\mathbb{F}$	FUZZY AND CERTAIN $\mathbb{F}\cap\mathbb{C}$	FUZZY AND UNCERTAIN $\mathbb{F}\cap\mathbb{C}'$

Figure 6.1.1 Certain, Uncertain, Fuzzy and Un-fuzzy Information and Decision Spaces

Note:

$$\mathbb{I}=\mathbb{F}\cup\mathbb{F}'=\mathbb{C}\cup\mathbb{C}',\ \mathbb{F}=(\mathbb{F}\cap\mathbb{C})\cup(\mathbb{F}\cap\mathbb{C}'),\ \mathbb{F}'=(\mathbb{F}'\cap\mathbb{C})\cup(\mathbb{F}'\cap\mathbb{C})$$

$$\mathbb{I}=(\mathbb{F}\cap\mathbb{C})\cup(\mathbb{F}'\cap\mathbb{C})\cup(\mathbb{F}'\cap\mathbb{C}')\cup(\mathbb{F}\cap\mathbb{C}')$$

$$\mathbb{C}=(\mathbb{F}\cap\mathbb{C})\cup(\mathbb{F}'\cap\mathbb{C}),\ \mathbb{C}'=(\mathbb{F}\cap\mathbb{C}')\cup(\mathbb{F}'\cap\mathbb{C}')$$

Definition 6.1.1

An environment for a decision is called *fuzzy* if the relevant information characteristics about the structural properties of the decision are either known and vague or unknown and vague.

Definition 6.1.2

A fuzzy environment for a decision is said to be *fuzzy-certainty* if it combines vagueness and full information in arriving at an optimal fuzzy decision rule. It is said to be *fuzzy uncertainty* if it combines less than full information and vagueness in the arrival of the optimal decision rule. It is said to be *non-fuzzy-certainty* if it combines full and exact information.

Observation 6.1.1

A decision problem in a fuzzy environment is generally called fuzzy decision problem. It is qualified by either certainty or uncertainty depending on whether all information regarding the decision is known or not, prior to the decision. A number of fuzzy decision problems can be identified in economics. Such examples may include the problems of social choice or decision [926][938], the consumer decision or choice[905a][954], problems of welfare (normative) economics[906], risk analysis [903][959]. The problems of fuzzy certainty and fuzzy uncertainty have been referred to as β-class and η-class respectively by Dompere [694]. If the environment of decision is non-fuzzy then it might be either certain or non-certain generating classes of decision problems that we may refer to as α-class (non-fuzzy and certain) and σ-class (non-fuzzy and uncertain) respectively. Thus the partition of the information space induces a partition on the general decision space creating four classes of decision problems as shown in Figure 6.1.2.

	$\mathbb{C}$	$\mathbb{C}'$
$\mathbb{F}'$	α-CLASS $\mathbb{F}\cap\mathbb{C}$	σ-CLASS $\mathbb{F}'\cap\mathbb{C}'$
$\mathbb{F}$	β CLASS $\mathbb{F}\cap\mathbb{C}$	η-CLASS $\mathbb{F}\cap\mathbb{C}'$

Figure 6.1.2 Families of Theoretical Decision Problems

Let Ω be a general set of goals and Γ a general set of constraints. In this chapter we shall treat only the β-class of decision problems (fuzzy and certain). The β-class of decision problems may be distinguished from other classes by the following characteristics: 1) All the relevant information regarding the decision is given to the decision agent; 2) The information on decision is, however imprecise, the concepts and notions defined over the decision space may be vague or ill-defined while the measurements of relevant variables are either inexact or approximation; 3) The information is summarized by fuzzy structural parameters. Such a class of decision problems may be referred to as *no-data-fuzzy-decision* problem. It is no-data-problem because all the relevant information regarding the decision is known to the decision agent. It is fuzzy because the concepts, notions and the measurements are either vague, inexact, approximation, linguistic or ill-defined.

The formal representation of this class of decision problems is similar to that of the α -class except that there is an added problem. The added problem involves designing methods and techniques that will allow proper inclusions of fuzzy ends and means in the decision-making process under conditions of certainty [710] [715b] [768]. Since the basic concept, measurements and related mathematical structure have been presented in Chapter 5 we shall deal directly with the basic structure of fuzzy decisions.

Definition 6.1.3

The set of fuzzy goals, $\mathbb{A}$, is a subset of the space of goals, Ω , and is equipped with a membership function $\mu_{\mathbb{A}}(x)$. It is written as

$$\mathbb{A} = \{(x, \mu_{\mathbb{A}}(x)) \mid x \in \Omega \text{ and} \mu_{\mathbb{A}}(x) \in \mathbb{M}\}$$

(6.1.1)

where $\mathbb{M} = \{\mu_{\mathbb{A}}(x) \mid x \in \mathbb{A} \text{ and } 0 \leq \mu_{\mathbb{A}}(x) \leq \mu_{\Omega}(x) \leq 1\}$ is a set of membership functions where $\mathbb{A}$ is a fuzzy subset of Ω.

Example 6.1

Consider a socioeconomic system and suppose that its objective is to undertake all yearly social projects whose average social benefit over output is substantially greater than a specified value, say "a". This is a fuzzy goal and the set of this fuzzy goal may be written as

$$\mathbb{A} = \left\{(x, \mu_{\mathbb{A}}(x)) \mid x \in \Omega,\ \mu_{\mathbb{A}}(x) = \left[k|1-x|e^{-(1-x)^2} + 1\right]^{-1},\ k > 0\right\}$$

(6.1.2)

Graphically, eqn. (6.1.2) may be represented as in Figure 6.1.1. This is also an example of E-membership function of eqn. 5.4.2.3 for "a" is large.(E= exponential)

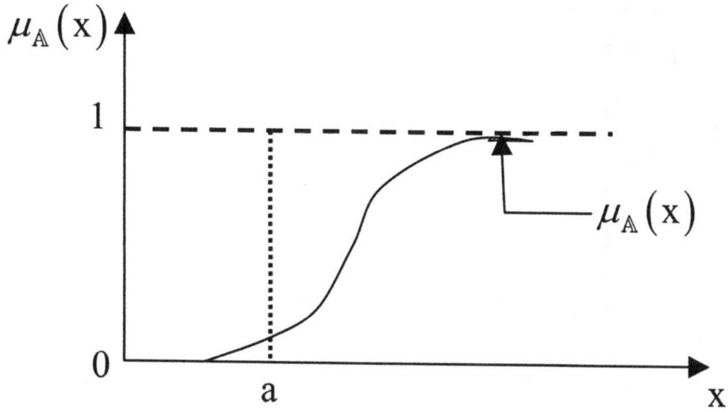

Figure 6.1.1 A Fuzzy Social Goal

Definition 6.1.4

A fuzzy constraint $\mathbb{B}$ is a subset of the general constraint set, Γ, that is equipped with a membership function $\mu_{\mathbb{B}}(x)$. It may be written as

$$\mathbb{B} = \left\{ \left(x, \mu_{\mathbb{B}}(x) \right) \mid x \in \Gamma \text{ and } \mu_{\mathbb{B}}(x) \in \mathbb{M} \right\}$$

(6.1.3)

where $\mathbb{M}$ is a set of membership functions whose members are constrained between zero and one.

Example 6.1.2

Continue with example 6.1.1. The fuzzy constraint on the social decision may be specified as; the average social cost over output should not be far greater than a specified value say b. An example of this constraint fuzzy set may be written as

$$\mathbb{B} = \left\{ \left(x, \mu_{\mathbb{B}}(x) \right) \mid x \in \Gamma,\ \mu_{\mathbb{B}}(x) = e^{-h\|x-b\|^2} \right\},\ h = \begin{cases} 0, \text{ if } x \le b \\ \frac{b}{x}, \text{ if } x > b \end{cases}$$

(6.1.4)

where $\|\bullet\|$ denotes the usual Euclidean norm. Equation (6.1.4) is graphically represented in Figure. 6.1.2. It is another example of E-membership function.

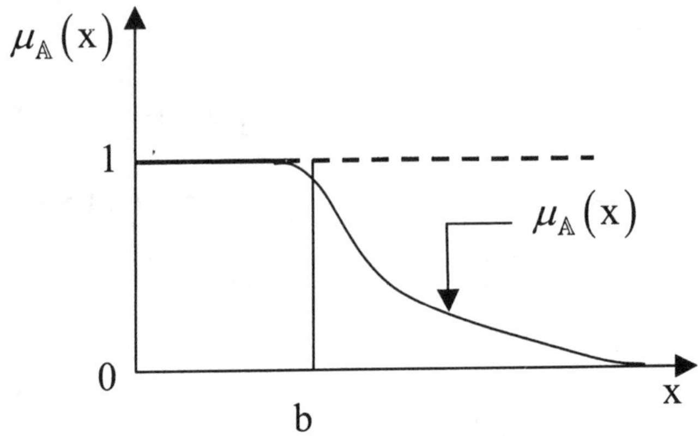

Figure. 6.1.2 A Fuzzy Social Constraint

Observation 6.1.2

The class of membership functions that one may select from is large. The selected functional form will depend on the nature of the problem, the linguistic representation of the goals and constraints. The decision rule that would emerge will be influenced by the structure and form of the membership functions of the fuzzy goal and the constraint which will also affect the quality of such a decision. Most of the time it is useful to use either the triangular or the trapezoidal membership as approximation [1153a] [1153b] and also see Chapter 5, a Volume on fuzzy value theory [931b] of a companion book for some explicit applications.

Definition 6.1.4

Let $\mathbb{A}$ be a fuzzy set of goals and $\mathbb{B}$ a fuzzy set of constraints, each of which is equipped with their respective membership function of $\mu_{\mathbb{A}}(x)$ and $\mu_{\mathbb{B}}(x)$ where $\mathbb{A} \subset \Omega$ and $\mathbb{B} \subset \Gamma$. The intersection of these fuzzy subsets defines a new fuzzy set which is the critical region of fuzzy decision, Δ, where

$$\Delta = \{\mathbb{A} \cap \mathbb{B} \mid \mathbb{A} \subset \Omega \text{ and } \mathbb{B} \subset \Gamma\} \tag{6.1.5}$$

that may further be specified with a membership function as

$$\Delta = \left\{ \left(x, \mu_{\Delta}(x) \right) \mid \mu_{\Delta}(x) = \mu_{\mathbb{A}}(x) \wedge \mu_{\mathbb{B}}(x),\ x \in \mathbb{A} \cap \mathbb{B} \right\}$$

(6.1.6)

Given the critical region of the decision space, a question arises as to what would be the decision rule that would lead to the selection of the most preferred goal. In other words, what is the optimal fuzzy decision principle or what is the rule for implementing the social decision? To answer this question we need to construct defuzzification decision engine through an appropriate set of algorithms. Hence the following definitions are offered.

Definition 6.1.5

If Δ is a fuzzy decision set with a membership function $\mu_{\Delta}(x)$ then a fuzzy maximal decision set is a fuzzy set $\mathbb{M}$ defined as

$$\mathbb{M} = \left\{ \left(x^*, \mu_{\Delta}(x^*) \right) \mid \mu_{\Delta}(x^*) = \sup_{x \in \Omega} \left[\mu_{\mathbb{A}}(x) \wedge \mu_{\mathbb{B}}(x) \right] \right\}$$

(6.1.7)

where $\mu_{\mathbb{A}}(x)$ and $\mu_{\mathbb{B}}(x)$ are as specified in equations (6.1.2) and (6.1.3).

Definition 6.1.6

An element $\hat{x} \in \Omega$ is said to be an optimal decision element if.

$$\hat{x} \in \mathbb{M} \Leftrightarrow \mu_{\Delta}(\hat{x}) = \sup_{x \in \Omega} \mu_{\Delta}(x)$$

(6.1.8)

Observation 6.1.3

Definitions (6.1.5) and (6.1.6) provide us with an optimal fuzzy decision rule. In its bare essentials, the decision rule requires one to choose an element in the goal-constraint confluence that has the largest membership grade in the critical region. The concept of critical region of decision is crucial to all classes of decision problems whether fuzzy or non-fuzzy. Furthermore, the fuzzy optimal decision problem may be stated as a mathematical programming problem that may be written as

$$\left.\begin{array}{l}\sup\limits_{x\in\Omega} \mu_{\Delta}(x) \\ \text{s.t.} \quad x \in \mathbb{A} \cap \mathbb{B}\end{array}\right\}$$

(6.1.9)

We may represent equations (6.1.6), (6.1.7) and (6.1.8) as in Figure 6.1.3. It may be kept in mind that the intersection point of the membership functions of the fuzzy goals and constraints needs not be the optimal decision element as in Figure. 6.1.3. The point of the optimal decision will depend on the curvatures of the membership functions used and the problem under analysis for choice.

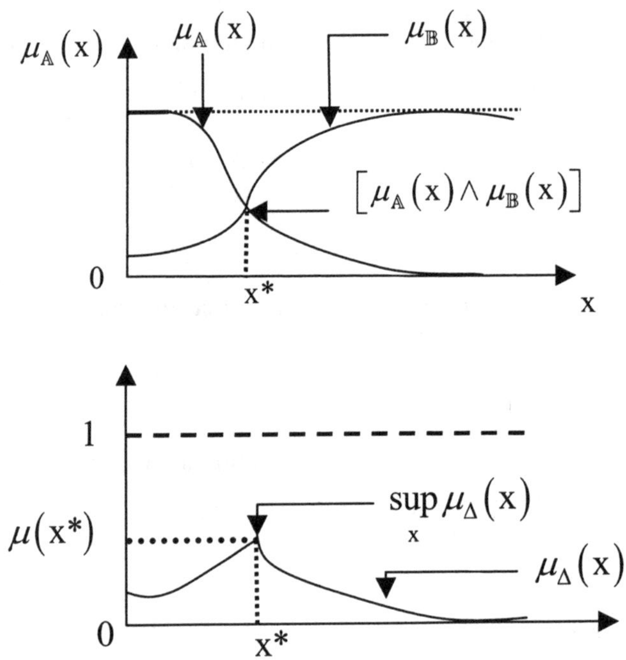

Figure. 6.1.3. A Fuzzy Optimal Decision

Example 6.1.3

Consider examples (6.1.1) and (6.1.2) where the basic decision problem is to determine those projects that must be undertaken. The qualified projects must satisfy the goals and the constraints. The fuzzy decision problem is obtained by combining equations (6.1.2) and (6.1.4) where the fuzzy decision set may be written as

$$\Delta = \left\{ \left(x, \mu_\Delta(x) \right) \mid \mu_\Delta(x) = \left[(k|1-x|e^{-(1-x)^2} + 1) - e^{-h\|x-b\|} \right] \right\}$$

$$\text{where } k > 0, \ h \begin{Bmatrix} 0, \ x < b \\ \frac{x}{b}, \ x > b \end{Bmatrix}$$

(6.1.10)

6.2 The Fuzzy Decision Problem Extended

So far we have only considered a decision problem with only one fuzzy goal and one fuzzy constraint. In many decision problems we are usually confronted with sets of fuzzy goals and constraints. Such decision problems include collective, multicriteria and multiattribute and multiobjective decision problems. The framework that has been established may be extended to account for the presence of many goals and many constraints. Let $\mathbb{I}$ be an index set for fuzzy goals and $\mathbb{J}$ a corresponding index set for fuzzy constraints. Furthermore, let $\mathbb{A}_i$, $i \in \mathbb{I}$, be the family of fuzzy goals and $\mathbb{B}_j$, $j \in \mathbb{J}$, that of fuzzy constraints. The fuzzy decision space Δ with multiple goals and constraints is a set

$$\Delta = \left(\bigcap_{i \in \mathbb{I}} \mathbb{A}_i \right) \cap \left(\bigcap_{j \in \mathbb{J}} \mathbb{B}_j \right)$$

(6.2.1)

with a membership function that may be specified as

$$\mu_\Delta(x) = \left(\bigwedge_{i \in \mathbb{I}} \mu_{\mathbb{A}_i}(x) \right) \bigwedge \left(\bigwedge_{j \in \mathbb{J}} \mu_{\mathbb{B}_j}(x) \right)$$

(6.2.2)

It has been said that the fuzzy optimal decision is the element which has the largest membership value, in the critical region of decision. If $x^* \in \mathbb{M}$ is the element, then x^* is the element that solves the fuzzy optimization problem

$$\mu_{\Delta}(x^*) = \sup_{x\in\Omega}\left[\left(\bigwedge_{i\in\mathbb{I}} \mu_{\mathbb{A}_i}(x)\right)\wedge\left(\bigwedge_{j\in\mathbb{J}} \mu_{\mathbb{B}_j}(x)\right)\right]$$

(6.2.3)

Generally, the element x^* may not be unique and hence a definition of optimal fuzzy decision set will be analytically useful. Let $\breve{\mathbb{M}}$ be such a set then we may write

$$\breve{\mathbb{M}} = \left\{\left(x^*, \mu_{\Delta}(x)\right) \mid \mu_{\Delta}(x^*) = \sup_{x\in\Delta\subset\Omega} \mu_{\Delta}(x)\right\}$$

(6.2.4)

If $\breve{\mathbb{M}}$ happens to contain no element or one element then no problem arises. On the other hand, if $\breve{\mathbb{M}}$ contains more than one optimal fuzzy element that solves the fuzzy decision problem then a question arises as to what choice must be made among the optimal elements in the set. This question may be dealt with in many ways. One of such ways is the *center of gravity method* [1128, 1153a, 1153b, 1165] to find x^{**} by computing

$$x^{**} = \frac{\sum_{x^*\in\breve{\mathbb{M}}} x^*\left[\bigwedge_{i\in\mathbb{I}} \mu_{\mathbb{A}_i}(x^*)\right]}{\sum_{x^*\in\breve{\mathbb{M}}}\left[\bigwedge_{i\in\mathbb{I}} \mu_{\mathbb{A}_i}(x^*)\right]}$$

(6.2.5)

for a discrete case. If the membership functions are continuous then we may write

$$x^{**} = \frac{\int_{x^*\in\breve{\mathbb{M}}} x^*\left[\bigwedge_{i\in\mathbb{I}} \mu_{\mathbb{A}_i}(x^*)\right] dx^*}{\int_{x^*\in\breve{\mathbb{M}}}\left[\bigwedge_{i\in\mathbb{I}} \mu_{\mathbb{A}_i}(x^*)\right] dx^*}$$

(6.2.6)

An alternative approach is to establish a new fuzzy set $\widehat{\mathbb{M}}$ with a membership function $\mu_{\widehat{\mathbb{M}}}(x^*)$ established over the optimal fuzzy decision set, $\breve{\mathbb{M}}$, and then define a fuzzy decision problem whose solution may be specified as

$$x^{**} = \begin{cases} \sup\limits_{x^* \in \mathbb{M}} \mu_{\widehat{\mathbb{M}}}(x^*) \\ \text{s.t.} \begin{cases} \breve{\mathbb{M}} = \left\{ (x, \mu_\Delta(x^*)) \mid \mu_\Delta(x^*) = \sup\limits_{x \in \Omega} \mu_\Delta(x) \right\} \\ \Delta = \left(\bigwedge\limits_{i \in \mathbb{I}} \mu_{\mathbb{A}_i}(x) \right) \bigwedge \left(\bigwedge\limits_{j \in \mathbb{J}} \mu_{\mathbb{B}_j}(x) \right) \end{cases} \end{cases}$$

So far we have presented the elements of the theory of fuzzy decision where fuzzy goals and constraints are assumed to have equal importance. Similarly, it assumes equal importance for differential goals and constraints in the case of multiple goals and constraints. In the practice of decision, however, the relative importance of different goals, of different constraints, and between constraints and goals may vary depending on the nature of the decision problem. The problem of relative importance may be accounted for by using convex-fuzzy decision structure.

Definition 6.2.1

Let $\mathbb{E} = [0,1]$ be a proportionality set. A convex-fuzzy decision of families of goals $\mathbb{A}_i' \mathrm{s}$ and constraints $\mathbb{B}_i' \mathrm{s}$ is a new fuzzy set, Δ, formed whose membership function is specified as

$$\mu_\Delta(x) = \sum_{i \in \mathbb{I}} \alpha_i \mu_{\mathbb{A}_i}(x) + \sum_{j \in \mathbb{J}} \beta_j \mu_{\mathbb{B}_j}(x)$$

(6.2.8)

where

$$\sum_{i \in \mathbb{I}} \alpha_i + \sum_{j \in \mathbb{J}} \beta_j = 1$$

(6.2.9)

Observation 6.2.1

It may be shown that if $\mu_{\Delta}(x)$ is a convex-fuzzy decision as specified in equation (6.2.8) then

$$\left(\underset{i\in\mathbb{I}}{\wedge}\mu_{\mathbb{A}_i}(x)\right)\wedge\left(\underset{j\in\mathbb{J}}{\wedge}\mu_{\mathbb{B}_j}(x)\right)\leq\mu_{\Delta}(x)\leq\left(\underset{i\in\mathbb{I}}{\vee}\mu_{\mathbb{A}_i}(x)\vee\left(\underset{j\in\mathbb{J}}{\vee}\mu_{\mathbb{B}_j}(x)\right)\right)$$

This inequality will become useful when combinations of expert information or judgment or social consensus is needed (see for example Chapters 3 and 4 of this book).

Observation 6.2.2

An important problem concerning the definition of fuzzy decision as intersections of fuzzy sets of goals and constraints must be noted. It is possible that a problem of absolute independence between fuzzy goals and fuzzy constraints may arise in the construction of the fuzzy decision. The decision problem may be characterized by goals-constraint interdependence.

1. A fuzzy decision, Δ, with a membership function, $\mu_{\mathbb{A}}(x)$ is said to be fuzzy independent of *constraint set* if $\mu_{\mathbb{A}}(x)\leq\mu_{\mathbb{B}}(x),\ \forall x\in\Omega$
2. A fuzzy decision, Δ, with a membership function, $\mu_{\Delta}(x)$, is said to be fuzzy independent of *goals* if $\mu_{\mathbb{B}}(x)\leq\mu_{\mathbb{A}}(x),\ \forall x\in\Omega$
3. A fuzzy decision, Δ, with a membership function , $\mu_{\Delta}(x)$, is said to be completely fuzzy interdependent if $\mu_{\mathbb{A}}(x)\not\leq\mu_{\mathbb{B}}(x)$ and $\mu_{\mathbb{B}}(x)\not\leq\mu_{\mathbb{A}}(x)$, in other words $\Delta\neq\varnothing$.

The problem of fuzzy independence of goals and constraints when it occurs may be dealt with by utilizing a fuzzy product decision.

Definition 6.2.2

A product-fuzzy decision, Δ ,of a family of fuzzy constraints and goals of $\mathbb{B}_j'$s and $\mathbb{A}_i'$s is a new fuzzy set $\Delta=\mathbb{A}\bullet\mathbb{B}$ formed with a membership function $\mu_{\Delta}(x)$, that may be specified as

$$\mu_{\Delta}(x)=\left[\prod_{i\in\mathbb{I}}\mu_{\mathbb{A}_i}(x)\right]\bullet\left[\prod_{j\in\mathbb{J}}\mu_{\mathbb{B}_j}(x)\right]$$

(6.2.11)

Observation 6.2.2

The product-fuzzy decision, Δ, may be modified to account for weights that may be technical or social or both such that they would allow differential importance to be attached to the goals and constraints as seen fit by the decision makers. Equation (6.2.11) may then be modified as

$$\mu_{\Delta}(x)=\left[\prod_{i\in\mathbb{I}}\alpha_i\mu_{\mathbb{A}_i}(x)\right]\bullet\left[\prod_{j\in\mathbb{J}}\beta_j\mu_{\mathbb{B}_j}(x)\right],\ \alpha,\beta\in[0,1]$$

(6.2.12)

6.3 The Fuzzy Mathematical Programming (FMP)

From the fuzzy decision problems and the optimal decision rule an important question arises as to what are the algorithms that would allow the computations of the optimal values. The algorithms that would lead to the realization of the optimal decision rule fall under what has come to be known as fuzzy mathematical programming. Two approaches have been suggested for the development of the fuzzy mathematical programming algorithms. One of the approaches is by the way of fixed-level set and the other through the fuzzy utility function and fuzzy sets of non-dominating alternatives. Because of its immediate applications to the cost-benefit analysis we shall consider the FMP.

6.3.1 The Formulation of the Fuzzy Mathematical Programming Problem

First it is observed that the optimization problem of the membership function of the decision space may be transformed through the use of appropriate set partitioning. Consider a set Ω on which an objective function $\phi:\Omega\rightarrow\mathsf{R}$ is defined. One may take $\Omega=\mathsf{R}^{n}$. Furthermore, let $\mathbb{B}$ be a fuzzy constraint in Ω and characterized by a membership function of the form $\mu_{\mathbb{B}}:\Omega\to[0,1]$. The fuzzy mathematical programming problem is to find $\sup_{x\in\Omega}\phi(x)$ on $\mathbb{B}$.

Suppose $\phi(\bullet)$ is bounded from above in the form $\phi(x)\leq \mathrm{u}$. Define the fuzzy goal as a fuzzy set in the form

$$\mathbb{A}=\{(x,\mu_{\mathbb{A}}(x))\,|\,x\in\Omega \text{ and } \mu_{\mathbb{A}}(x)=\phi(x)/\mathrm{u}\}$$

(6.3.1.1)

The critical region of the decision may then be specified as

$$\Delta=\{(x,\mu_{\Delta}(x))\,|\,x\in\Omega,\ \mu_{\Delta}(x)=\mu_{\mathbb{A}}(x)\wedge\mu_{\mathbb{B}}(x)\}$$

(6.3.1.2)

and the optimization problem is

$$\sup_{x\in\Omega}\mu_{\Delta}(x)=\sup_{x\in\Omega}\left[\mu_{\mathbb{A}}(x)\wedge\mu_{\mathbb{B}}(x)\right]$$

(6.3.1.2)

6.3.1.1 A Note on Decomposition of the Fuzzy Decision

To find the optimal solution to the fuzzy decision problem we decompose the fuzzy decision space into fixed-level sets of optimization that should exhaust the decision space.

Lemma 6.3.1.1

$$\sup_{x\in\Omega}\mu_{\Delta}(x)=\sup_{\alpha\in[0,1]}\left[\alpha\wedge\sup_{x\in\mathbb{B}_{\alpha}}\mu_{\mathbb{A}}(x)\right]$$

where $\mathbb{B}_{\alpha}=\{(x,\mu_{\mathbb{B}}(x))\,|\,x\in\Omega \text{ and } \mu_{\mathbb{B}}(x)\geq\alpha\}$ *the* α *-level set of* $\mathbb{B}$ *that induces a crisp set,* $\mathbb{B}_{\alpha}$ *in* $\mathbb{B}$.

Proof

Consider the fuzzy constraint set , $\mathbb{B}$. By the theorem of decomposition of fuzzy set (See chapter 5 of this volume) we may write

$$\left.\begin{aligned}\mu_{\mathbb{B}}(x) &= \bigvee_{\alpha\in[0,1]}\left(\alpha\bullet\mu_{\mathbb{B}_\alpha}(x)\right)\\ &= \bigvee_{\alpha\in[0,1]}\left(\alpha\wedge\mu_{\mathbb{B}_\alpha}(x)\right)\end{aligned}\right\}$$

Theo.5.1 and Cor. 5.1

Define

$$\mu_{\mathbb{B}}(x)=\begin{cases}1, \text{ if } x\in\mathbb{B}_\alpha\\ 0, \text{ if } x\notin\mathbb{B}_\alpha\end{cases}$$

We may write

$$\begin{aligned}\mu_\Delta(x) &= \mu_{\mathbb{A}}(x)\wedge\mu_{\mathbb{B}}(x)\\ &= \left[\bigvee_\alpha\left(\alpha\wedge\mu_{\mathbb{B}_\alpha}(x)\right)\right]\wedge\mu_{\mathbb{A}}(x)\\ &= \bigvee_\alpha\left[\alpha\wedge\mu_{\mathbb{B}_\alpha}(x)\wedge\mu_{\mathbb{A}}(x)\right]\end{aligned}$$

and hence

$$\begin{aligned}\bigvee_{x\in\Omega}\mu_\Delta(x) &= \bigvee_{x\in\Omega}\bigvee_\alpha\left[\alpha\wedge\mu_{\mathbb{B}_\alpha}(x)\wedge\mu_{\mathbb{A}}(x)\right]\\ &= \bigvee_\alpha\bigvee_{x\in\Omega}\left[\alpha\wedge\mu_{\mathbb{B}_\alpha}(x)\wedge\mu_{\mathbb{A}}(x)\right]\\ &= \bigvee_{\alpha\in[0,1]}\left\{\alpha\wedge\left[\bigvee_{x\in\Omega}\left(\mu_{\mathbb{B}_\alpha}(x)\wedge\mu_{\mathbb{A}}(x)\right)\right]\right\}\end{aligned}$$

However,

$$\begin{aligned}\bigvee_{x\in\Omega}\left[\mu_{\mathbb{B}_\alpha}(x)\wedge\mu_{\mathbb{A}}(x)\right] &= \bigvee_{x\in\mathbb{B}_\alpha}\left[\mu_{\mathbb{B}_\alpha}(x)\wedge\mu_{\mathbb{A}}(x)\right]\bigvee\left\{\bigvee_{x\in\mathbb{B}_\alpha}\left[\mu_{\mathbb{B}_\alpha}(x)\wedge\mu_{\mathbb{A}}(x)\right]\right\}\\ &= \bigvee_{x\in\mathbb{B}}\mu_\Delta(x)=\sup_{x\in\mathbb{B}_\alpha}\mu_\Delta(x)\end{aligned}$$

hence

$$\sup_{x\in\Omega}\mu_\Delta(x)=\sup_{\alpha[0,1]}\left\{\alpha\wedge\sup_{x\in\mathbb{B}_\alpha}\mu_{\mathbb{A}}(x)\right\}\qquad\square$$

Lemma 6.3.1.2

Let $\mathbb{B}_j$, $j \in \mathbb{J}$, *be a family of fuzzy constraint sets where* $\mathbb{J}$ *is an index set in* Ω *and* $\mathbb{A}$ *a fuzzy goal as specified in equation (6.3.1) then*

$$\sup_{x\in\Omega} \mu_{\Delta}(x) = \sup_{\alpha_j\in[0,1]} \left[\left(\bigwedge_{j\in\mathbb{J}} \alpha_j \right) \bigwedge \left(\sup_{x\in\bigcap_{j\in\mathbb{J}} \mathbb{B}_{\alpha_j}} \mu_{\mathbb{A}}(x) \right) \right]$$

where

$$\Delta = \begin{cases} (x, \mu_{\Delta}(x)) \mid x \in \Omega \, , \mu_{\Delta}(x) = \left[\bigwedge_{j\in\mathbb{J}} \mu_{\mathbb{B}_j}(x) \right] \bigwedge \mu_{\mathbb{A}}(x) \text{ and } \mathbb{B}_{\alpha_j} \\ \text{are } \alpha\text{-level sets} \end{cases}$$

Proof

The proof of the above lemma begins by examining the α -cut decomposition of the family of the constraint sets and optimizing them over the α -cuts. From the α -cuts we can write both the j-constraint set and the fuzzy decision as

$$\mu_{\mathbb{B}_j}(x) = \bigvee_{\alpha\in[0,1]} \left[\alpha \wedge \mu_{\mathbb{B}_{\alpha_j}}(x) \right] , j \in \mathbb{J}$$

It thus follows that

$$\mu_{\Delta}(x) = \bigvee_{\alpha} \left[\alpha \wedge \mu_{\mathbb{B}_{1\alpha}}(x) \right] \wedge \cdots \wedge \bigvee_{\alpha} \left[\alpha \wedge \mu_{\mathbb{B}_{j\alpha}}(x) \right] \wedge \mu_{\mathbb{A}}(x)$$

Now let $\alpha_j \in [0,1]$ and $j \in \mathbb{J}$ be the $\alpha -$cuts of the individual fuzzy constraint sets, $\mathbb{B}_j{}'$s then we may write the fuzzy decision membership function as

$$\mu_{\Delta}(x) = \bigvee_{[\alpha_j | j\in\mathbb{J}]} \left[\left(\bigwedge_{j\in\mathbb{J}} \alpha_j \right) \left(\bigwedge_{j\in\mathbb{J}} \mu_{\mathbb{B}_{j\alpha_j}}(x) \right) \left(\mu_{\mathbb{A}}(x) \right) \right]$$

But

$$\mu_{\Delta}(x)=\left(\mu_{\bigcap_{j\in\mathbb{J}}\mathbb{B}_j}(x)\right)\wedge\mu_{\mathbb{A}}(x)=\bigvee_{\alpha}\left[\left\{\alpha\wedge\mu_{\bigcap_{j\in\mathbb{J}}\mathbb{B}_{j\alpha}}(x)\right\}\wedge\mu_{\mathbb{A}}(x)\right]$$

By taking the suprenum and using the definition of α −level set we obtain

$$\sup_{x\in\Omega}\mu_{\Omega}(x)=\sup_{x\in\Omega}\left\{\bigvee_{\alpha\in[0,1]}\left[\left(\alpha\wedge\bigcap_{j\in\mathbb{J}}\mu_{\mathbb{B}_{j\alpha}}(x)\right)\wedge\mu_{\mathbb{A}}(x)\right]\right\}$$

$$=\sup_{\alpha\in[0,1]}\left[\alpha\wedge\left(\sup_{x\in\bigcap_{j\in\mathbb{J}}\mathbb{B}_{j\alpha}}\mu_{\mathbb{A}}(x)\right)\right]$$

$$=\sup_{[\alpha_j|j\in\mathbb{J}]}\left[\left(\bigwedge_{j\in\mathbb{J}}\alpha_j\right)\wedge\left(\sup_{\bigcap_{j\in\mathbb{J}}\mathbb{B}_{j\alpha}}\mu_{\mathbb{A}}(x)\right)\right] \quad \square$$

6.3.2 The α -level Function of FMP

One may view the problem of finding the maximal decision element through the α -level cut in the critical region of fuzzy decision space as simply a problem of finding an optimal element to a scalar function induced by α -rization process. Let $\xi:[0,1]\rightarrow[0,1]$ such that

$$\xi(\alpha)=\sup_{x\in\mathbb{B}_{\alpha}}\mu_{\mathbb{A}}(x)$$

(6.3.2.1)

Furthermore, define a function $\psi:[0,1]\rightarrow[0,1]$ such that

$$\psi[\alpha]=\alpha\wedge\xi(\alpha)$$

(6.3.2.2)

By lemma 6.3.1.1 we can write suprenum of $\mu_{\Delta}(x)$ as

$$\sup_{x\in\Omega}\mu_{\Delta}(x)=\sup_{\alpha\in[0,1]}\psi(\alpha)$$

(6.3.2.3)

Let the support(supp.) of a fuzzy measure on Ω be denoted by

$$\operatorname{supp}\mu\ (x)=\overline{\{(x,\mu_{\mathbb{B}}(x))\mid\mu_{\mathbb{B}}(x)\neq 0\}}$$

where the closure in $\Omega=\mathbb{R}^{n}$ is with ordinary topology. The properties of the scalar function $\xi(\alpha)$ may be stated as a lemma.

Lemma 6.3.2.1

The scalar function $\xi(\alpha)$ *has the properties*

a) $\xi(0)=\sup_{x\in\Omega}\mu_{\mathbb{A}}(x)$

b) $\alpha_1\leq\alpha_2\in[0,1]\Rightarrow\xi(\alpha_1)\geq\xi(\alpha_2)$

Proof

a) If $\alpha=0$ then $\mathbb{B}_0=\Omega$ and hence by equation (5.2.1)

$$\xi(0)=\sup_{x\in\mathbb{B}_0}\mu_{\mathbb{A}}(x)=\sup_{x\in\Omega}\mu_{\mathbb{A}}(x)$$

b) By the construct of $\alpha-$level sets if

$\alpha_1\leq\alpha_2$ then $\mathbb{B}_{\alpha_2}\subseteq\mathbb{B}_{\alpha_1}$ (See Lemma 5.1, Chapter 5)

$$\Rightarrow\xi(\alpha_2)=\sup_{x\in\mathbb{B}_{\alpha_2}}\mu_{\mathbb{A}}(x)\leq\sup_{x\in\mathbb{B}_{\alpha_1}}\mu_{\mathbb{A}}(x)=\xi(\alpha_1)\ \square$$

Observation 6.3.2.1

If one defines $\mathfrak{B}([0,1])$ as the space of all continuous real-valued functions on set [0,1] then it may be observed that $\mathfrak{B}([0,1])$ is a Banach space with a norm value, λ

$$\lambda\in\mathfrak{B}([0,1])\Rightarrow\|\lambda\|=\sup_{\alpha\in[0,1]}|\lambda(\alpha)|$$

(6.3.2.4)

The norm measure, λ, generates the uniform convergence in the space $\mathfrak{B}([0,1])$. If we assume that the membership functions, $\mu_{\mathbb{A}}(x)$ and $\mu_{\mathbb{B}}(x)$ defined on the set of goals and constraints are continuous then it may immediately be observed that

$$\xi(\alpha) = \left\| \mu_{\mathbb{B}_\alpha}(x) \bullet \mu_{\mathbb{A}}(x) \right\|$$

(6.3.2.5)

where

$$\mu_{\mathbb{B}_\alpha}(x) \begin{cases} = 1 & \text{if } \mu_{\mathbb{B}}(x) \geq \alpha \\ = 0 & \text{if } \mu_{\mathbb{B}}(x) < \alpha \end{cases}$$

(6.3.2.6)

Lemma 6.3.2.2.

If ξ is a continuous mapping on [0,1] then $\exists\ \alpha_0 \in [0,1]$ such that $\xi(\alpha) = \alpha_0$..

Proof

The lemma is easily proven by utilizing one of the fixed point theorems on unit simplex [1035] [1059] [1106] [1112].

Note. From the lemma 6.3.2.2 $\exists\ \alpha_0 \in [0,1]$ such that $\sup_{x \in \mathbb{B}_{\alpha 0}} \mu_{\mathbb{A}}(x) = \alpha$

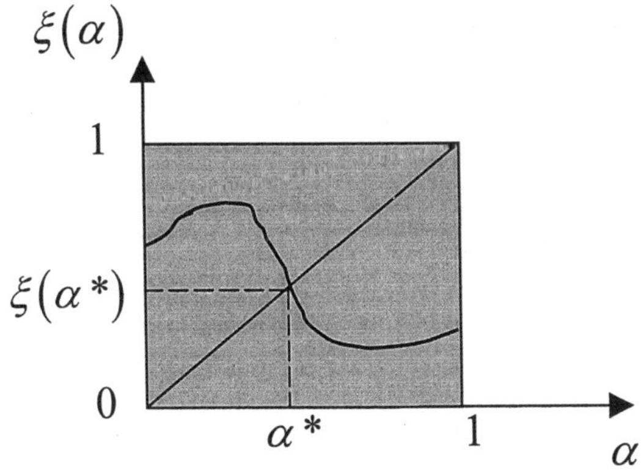

Figure 6.3.2.1 An Illustration of the Fixpoint Mapping

Lemma 6.3.2.3

If $(\mathbb{A}, \mu_{\mathbb{A}}(x))$ and $(\mathbb{B}, \mu_{\mathbb{B}}(x))$ *defined over decision space* Ω *are fuzzy goal and constraint respectively with* $\alpha \in [0,1]$ *as fixed level crisp set generator then*

$$\sup_{x\in\Omega}\left[\mu_{\mathbb{A}}(x) \wedge \mu_{\mathbb{B}}(x)\right] = \sup_{x\in\Omega} \mu_{\Delta}(x) = \alpha_0 \in [0,1]$$

Proof

From equation (6.3.2.3)

$$\sup_{x\in\Omega} \mu_{\Delta}(x) = \sup_{\alpha\in[0,1]} \psi(\alpha)$$

By equation (6.3.2.2)

$$\psi(\alpha) = \alpha \wedge \xi(\alpha)$$

It must be shown that $\sup_{\alpha\in[0,1]} \psi(\alpha) = \alpha_0$.

Take $\alpha_0 \in [0,1]$ and hence by substitution

$$\psi(\alpha_0) = \alpha_0 \wedge \xi(\alpha_0) = \alpha_0$$

since $\xi(\alpha_0) = \alpha$ by lemma (6.2.2).We have to show that for all $\alpha \in [0,1]$, $\psi(\alpha) \le \psi(\alpha_0)$ if α_0 is a maximum.

Suppose $\alpha < \alpha_0$ then $\xi(\alpha) \ge \xi(\alpha_0) = \alpha_0$.

But $\alpha_0 > \alpha \Rightarrow \psi(\alpha) = \alpha < \alpha_0 = \psi(\alpha_0)$. On the other hand if $\alpha > \alpha_0$,then $\xi(\alpha) \le \xi(\alpha_0) = \alpha_0 < \alpha$. This implies that $\psi(\alpha) = \xi(\alpha) \le \alpha_0 = \psi(\alpha_0)$. Thus

$$\sup_{x\in\Omega} \mu_{\Delta}(x) = \sup_{x\in\mathbb{B}_{\alpha_0}} \mu_{\mathbb{A}}(x) = \alpha$$

Theorem 6.3.2.1

If $(\mathbb{A}, \mu_{\mathbb{A}}(x))$ and $(\mathbb{B}, \mu_{\mathbb{B}}(x))$ *are fuzzy goal and fuzzy constraint respectively with a fuzzy decision defined as* $(\Delta = \mathbb{A} \cap \mathbb{B},\ \mu_{\Delta}(x))$ *over a decision space* Ω *then*

$$\sup_{x\in\Omega}\mu_\Delta(x)=\begin{cases}\sup\limits_{x\in\mathbb{R}}\mu_{\mathbb{A}}(x)\\ \text{s.t}\quad \mathbb{R}=\{x\in\Omega\mid\mu_{\mathbb{B}}(x)-\mu_{\mathbb{A}}(x)\geq 0\}\end{cases}$$

Proof

To prove this theorem we need to show that the α -cut of the constraint $\mathbb{B}_\alpha\subset\mathbb{R}$. Consider $\mathbb{B}_\alpha$ and suppose $x_0\in\mathbb{B}_\alpha$ then $\mu_{\mathbb{B}}(x_0)\geq\alpha_0$. We must show that $x_0\in\mathbb{A}$. But

$$\alpha_0=\sup_{x\in\mathbb{B}_{\alpha_0}}\mu\ (x)\qquad\text{(Lemma 6.2.2)}$$

$$\Rightarrow\mu_{\mathbb{B}}(x_0)\geq\mu_{\mathbb{A}}(x_0)\Rightarrow\mu_{\mathbb{B}}(x_0)-\mu_{\mathbb{A}}(x_0)\geq 0$$

Hence

$x_0\in\mathbb{R}.$

But $\mathbb{R}=\left(\mathbb{R}\cap\ \text{supp}\mu_{\mathbb{B}}(x)\right)\cup\left(\mathbb{R}\cap\ \text{supp}\mu_{\mathbb{B}}(x)\right)$

Furthermore,

$$\sup_{x\in\Omega}\mu_\Delta(x)=\sup_{x\in\text{supp}\mu_{\mathbb{B}}(x)}\left[\mu_{\mathbb{A}}(x)\wedge\mu_{\mathbb{B}}(x)\right]$$
$$\geq\sup_{x\in\mathbb{R}\text{supp}\mu_{\mathbb{B}}(x)}\left[\mu_{\mathbb{A}}(x)\wedge\mu_{\mathbb{B}}(x)\right]$$

But

$$\sup_{x\in\mathbb{R}}\left[\mu_{\mathbb{A}}(x)\wedge\mu_{\mathbb{B}}(x)\right]=\max\begin{bmatrix}\sup\limits_{x\in\mathbb{R}\cap\text{supp}\mu_{\mathbb{B}}(x)}\{\mu_{\mathbb{A}}(x)\wedge\mu_{\mathbb{B}}(x)\},\\ \sup\limits_{x\in\mathbb{A}\cap[\text{supp}\mu_{\mathbb{B}}(x)]'}\{\mu_{\mathbb{A}}(x)\wedge\mu_{\mathbb{B}}(x)\}\end{bmatrix}$$
$$=\sup_{x\in\mathbb{R}\cap\text{supp}\mu_{\mathbb{B}}(x)}\left[\mu_{\mathbb{A}}(x)\wedge\mu_{\mathbb{B}}(x)\right]$$

It immediately follows that

$$\sup_{x\in\Omega} \mu_{\Delta}(x) \geq \sup_{x\in\mathbb{R}} \{\mu_{\mathbb{A}}(x) \wedge \mu_{\mathbb{B}}(x)\}$$

$$= \sup_{X\in\mathbb{R}} \mu_{\mathbb{A}}(x) \geq \sup_{x\in\mathbb{B}_\alpha} \mu_{\mathbb{A}}(x)$$

Hence

$$\sup_{x\in\Omega} \mu_{\Delta}(x) = \sup_{x\in\mathbb{R}} \mu_{\mathbb{A}}(x)$$

It may be also useful to see [694] [1165] [1259] for more treaties on FMP.

Observation 6.3.2.1

The basic requirement for $\xi(\alpha) = \sup_{x\in\mathbb{B}} \mu_{\mathbb{A}}(x)$ to be reduced to solving the problem $\sup_{x\in\mathbb{R}} \mu_{\mathbb{A}}(x)$ with properly defined set, $\mathbb{R}$ is that $\xi(\alpha) = \sup_{x\in\mathbb{B}_\alpha} \mu_{\mathbb{A}}(x)$ is continuous in the relevant domain. To check the continuity either in practice or in theory requires us to know the sufficient conditions for the existence of such a continuous mapping.

Definition 6.3.2.1

Let Ω be a fuzzy set with a membership function $\mu_\Omega(x) \in [0,1]$ that is defined in its entire domain. The set Ω is said to be a fuzzy convex if for any $\lambda \in [0,1]$ and $x, y \in \Omega$

$$\mu_\Omega(\lambda x + (1-\lambda) y) \geq \mu_\Omega(x) \wedge \mu_\Omega(y)$$

for x $\neq$ y [one may set $\Omega = \mathbb{R}^n$.

Lemma 6.3.2.4

A fuzzy set Ω with a membership function $\mu_\Omega(x) \in [0,1]$ is convex if and only if all its α -level sets $\{\Omega_\alpha, \alpha \in [0,1]\}$ are convex.

Proof

Suppose Ω is convex, implying that, for $\lambda \in [0,1]$, $z = \lambda x + (1-\lambda) y \in \Omega$, x, y $\in \Omega$. Define α -level set

$\Omega_\alpha = \left\{x \in \Omega \mid \mu_{\Omega_\alpha}(x) \geq \alpha\right\}$ and suppose $x, y \in \Omega_\alpha$. We must demonstrate that

$$z = \left[\lambda x + (1-\lambda) y\right] \in \Omega_\alpha .$$

But

$$\mu_\Omega(z) \geq \mu_{\Omega_\alpha}(x) \wedge \mu_{\Omega_\alpha}(y) \geq \alpha \Rightarrow z \in \Omega_\alpha .$$

Now suppose that $\Omega_\alpha{}'s$ are convex sets and $x, y \in \Omega$.. Let $\mu_\Omega(x) \leq \mu_\Omega(y)$ then by Lemma 5.1, Chapter 5 we have
$\Omega_{\mu_\Omega}(y) \subseteq \Omega_{\mu_\Omega}(x)$

This implies $x, y \in \Omega_{\mu_\Omega}(x)$. But $\Omega_{\mu_\Omega}(x)$ is convex by assumption; thus $z \in \Omega_{\mu_\Omega}(x)$
and $\mu_\Omega(z) \geq \mu_\Omega(x) \wedge \mu_\Omega(y)$ and $z \in \Omega$ by theorem of decomposition. The argument is similar if $\mu_\Omega(x) > \mu_\Omega(y)$.

Theorem 6.3.2.2

If the fuzzy constraint $\mathbb{B}$ with membership function $\mu_\mathbb{B}(x)$ is strictly fuzzy convex then the functional equality $\xi(\alpha) = \sup\limits_{x \in \mathbb{B}_\alpha} \mu_\mathbb{A}(x)$ is continuous where $\mathbb{A}$ a fuzzy set of goals is and $\mu_\mathbb{A}(x)$ is its corresponding membership function.

6.4 Fuzzy Multicriteria Decision Making

The theory of decision making in fuzzy decision space that we have presented so far is directly related to both individual and collective (group) decision making. In this framework the preference ordering is induced by a single criterion. In the case of collective decision, the preference ordering by a single criterion may be constrained by a fundamentally democratic principle that the preferences of all individuals in the group must count in the ranking process. There are some decisions, however, where the individual or the collective (group) is to make decisions about alternatives on the basis of more than one criterion, in other words in a multicriteria environment.

The analytical problem is to relate the criterion space of ranking measures so as to induce an order on the alternatives for choice. In such a multicriteria decision situation, each criterion may be different from others. The problem, then, is simply to create a logical process and mathematically acceptable method that may be used to aggregate the rankings of the different criteria. This type of decision problem is called simply as multicriteria decision problem. The complexity and the difficulty of the analysis of multicriteria decision problem increase in intensity and scope when we are dealing additionally with collective (multi-person) decision which is further complicated when we are in multi-attribute decision space.

6.4.1 Problem Definition

Let us recall from the previous chapters that we may simply define the decision space in many ways depending on the choice problem at hand. The specific definitions may be pulled together to construct a general definition.

Definition 6.4.1

A general decision space, Θ, is made up of sets of project alternatives, Π, objectives, Ω, attributes, $\mathbb{X}$, and criteria, $\mathbb{K}$, that may be specified as

$$\Theta = \Pi \otimes \Omega \otimes \mathbb{X} \otimes \mathbb{K}$$

$$= \left\{ \theta \mid \theta = (\omega, \pi, x, k), \text{ where } \pi \in \Pi, \omega \in \Omega, x \in \mathbb{X} \text{ and } k \in \mathbb{K} \right\}$$

The $\theta \in \Theta$ is called a decision element under analysis and choice. It is a vector of four related sub-elements that define the explicit nature of the decision problem. For a single attribute, x, with multiple criteria, the decision space reduces to

$$\Theta_x = \Pi \otimes \Omega \otimes \mathbb{K}$$

$$= \left\{ \theta \mid \theta = (\omega, \pi, k) \text{ where } \omega \in \Omega, \pi \in \Pi, \, k \in \mathbb{K} \text{ and } x \text{ is fixed in} \mathbb{X} \right\}$$

(6.4.1)

The element x is the common defining attribute of the elements in the set of alternatives. Similarly with a single criterion, k, and with multiple attributes the general decision space becomes

$$\Theta_k = \Pi \otimes \Omega \otimes \mathbb{X}$$

$$= \left\{ \theta \middle| \theta (\omega, \pi, x) \text{ where } \omega \in \Omega, \pi \in \Pi, \, x \in \mathbb{X} \text{ and } k \text{ is fixed in } \mathbb{K} \right\}$$

It may be noted that k is given and known criterion fixed in $\mathbb{K}$ for ranking all the alternatives. An example is measurable cost-benefit ratio in cost-benefit analysis. The general decision space, Θ, collapses to sets of project alternatives, Π, with a defining set of attributes, $\mathbb{X}$, and a set of objectives, Ω, where there is only one criterion that induces a preference ranking. It is useful to point out that in defining the decision space the personality characteristics and the information processing capability of the decision maker are given. If the attribute is one (that is, $\#\mathbb{X}=1$ and $\#\mathbb{K}\geq 1$) then the decision is embedded in the criterion field; while if the criterion is one (that is, $\#\mathbb{K}=1$ and $\#\mathbb{X}\geq 1$) then the decision is attribute embedded.

Let us consider a multicriteria decision problem in a decision space, Θ_x. Any given criterion, $k\in\mathbb{K}$, given the behavioral consistency of the decision agent induces a rank-order relation in either ascending or descending order of preferences on all $\pi\in\Pi$ that satisfies the comparability principles of choice. The decision structure in a multicriteria decision space reveals a family of sets of ordered elements that constitute a *criterion field* in permutation.

Definition 6.4.2

A criterion field $\mathfrak{f}_k(\Pi)$ is a family of sets of ordered elements of set, Π, with an index set, $\mathbb{I}$, and induced by the criterion space, $\mathbb{K}$, in the form

$$\mathfrak{f}_k(\Pi)=\left\{\Pi_{k_i}\mid \pi\in\Pi, k\in\mathbb{K}, i\in\mathbb{I}\right\}$$

such that

1. $\Pi\in\mathfrak{f}_k(\Pi)$
2. $\#\left(\Pi_{k_i}\right)=\#\left(\Pi_{k_{i+1}}\right)=\#\Pi=\#\mathbb{I},\ i\in\mathbb{I}$;

For all, $\Pi_{k_i}\in\mathfrak{f}_k(\Pi)$ if π_l,π_q and $\pi_j\in\Pi_{k_i}$ then either

3. $\pi_l\succ\pi_j$ or $\pi_l\sim\pi_j$ or $\pi_l\prec\pi_j$

   and that $\pi_j\succcurlyeq\pi_l$ and $\pi_l\succcurlyeq\pi_q\Rightarrow\pi_j\succcurlyeq\pi_q$, $i,l,q\in\mathbb{I}$
4. $\#\left(\mathfrak{f}_k(\Pi)\right)=(\#\Pi)!$

One way of viewing the multicriteria decision problem is to find or construct a single criterion, k^*, that is a functional dependent on all the relevant elements, $k \in \mathbb{K}$ such that the elements in Π can be well ordered through the decision process for choice. The criterion, k^*, is called the *global criterion* that is to be used to induce ordered ranking on the alternative elements in Π.

As a logical construct, the global criterion, k^*, may or may not be in the initial criterion set, $\mathbb{K}$. The first step in finding k^* is either to locate within the set, $\mathbb{K}$, a dominating criterion $k_i = k^*$ that meets the preference ordering of the decision agents as well as the essential elements of the criterion $k \in \mathbb{K}$. Alternatively, we may seek an aggregate k^* on the basis of some preferential weights that allow either aggregation or summability if a dominating criterion can not be found. Let the criterion space, $\mathbb{K}$, be finite with a corresponding index set $\mathbb{I}$ where $\#\mathbb{K} = \#\mathbb{I} = \text{m}$ and $\#\mathbb{J} = \#\Pi = \text{n}$. Let us suppose that we have been able to solicit from the decision agents or developed from the solicited information criterion weights, $w_{ij}, i \in \mathbb{I}, j \in \mathbb{J}$. The weights express the distribution of degree of importance that are attached to each criterion elements $k_i, i \in \mathbb{I}$ and for any fixed $j \in \mathbb{J}$ such that $k_i \in \mathbb{K}$ and $\pi_j \in \Pi$. Accordingly the set of distributional weights, $\mathbb{W}$, given a relation, $\mathcal{R}$, may be specified as

$$\mathbb{W} = \left\{ w_{ij} \mid w_{ij} = \mathcal{R}\left(k_i, \pi_j\right), \ \forall i \in \mathbb{I}, \text{ and } \forall j \in \mathbb{J} \right\}$$

(6.4.3)

It may be noted that for any fixed $i \in \mathbb{I}$ and $k_i \in \mathbb{K}$ a row vector $\mathbb{V}_{i\bullet}$ is generated as we traverse over the j's projects. Similarly, for any fixed $j \in \mathbb{J}$ and $\pi_j \in \Pi$ a column vector, $\mathbb{V}_{\bullet j}$ of weights are generated as we move over ith criteria. Thus we obtain a matrix of distributional weights of the form shown in Table 6.4.1.

Table 6.4.1 Distributional Weights of Importance between Criteria and Project Alternatives

$\mathcal{R}(k_i, \pi_j) = w_{ij}$

$k \backslash \pi$	π_1	π_2	$\cdots$	π_j	$\cdots$	π_n	
k_1	w_{11}	w_{12}	$\vdots$	w_{1j}	$\cdots$	w_{1n}	$\mathbb{V}_{1\bullet}$
k_2	w_{21}						
$\vdots$	$\vdots$	$\ddots$				$\vdots$	$\vdots$
k_i	w_{i1}			w_{ij}			$\mathbb{V}_{i\bullet}$
$\vdots$	$\vdots$				$\ddots$	$\vdots$	$\vdots$
k_m	w_{m1}		$\cdots$	w_{mj}	$\cdots$	w_{mn}	$\mathbb{V}_{m\bullet}$
	$\mathbb{V}_{\bullet 1}$		$\cdots$	$\mathbb{V}_{\bullet j}$	$\cdots$	$\mathbb{V}_{\bullet n}$	

The values of the table may be written in a matrix form as

$$\Pi_{k_i} = \begin{pmatrix} w_{11} & \cdots & w_{1n} \\ \vdots & w_{ij} & \vdots \\ w_{m1} & \cdots & w_{mn} \end{pmatrix}$$

(6.4.4)

These distributional weights are such that

1. $w_{ij} \in [0,1]$ and $\sum_{i \in \mathbb{I}} w_{ij} = 1 \; \forall j \in \mathbb{J}$
2. $\mathbb{V}_{\bullet j} \geq \mathbb{V}_{\bullet l} \Rightarrow \pi_j \succcurlyeq \pi_l,\; j,l \in \mathbb{J}$
3. $\mathbb{V}_{h\bullet} \geq \mathbb{V}_{i\bullet} \; \forall i \in \mathbb{I}$ and a fixed $h \in \mathbb{I} \Rightarrow k_h = k^*$ is the

dominating criterion that may be selected as the global criterion

We shall explore the conceptual meanings and decision usefulness of both the vertical and horizontal sums of the criterion-alternative distributional weights in later discussions. For any fixed $\pi_j \in \Pi$ in eqn. (6.4.3) the needed criterion, k_j* for each column vector, given the distributional weights may be expressed as

$$k_j* = \varphi\left(w_{ij}, k_i \mid i \in \mathbb{I}, j \in \mathbb{J}\right)$$

(6.4.5)

where $\varphi(\bullet)$ may be seen in terms of aggregation function. Such an aggregation function must meet certain conditions of aggregation which may not be the same as conditions of summability. The first simple approach is to adopt a weighted average model (WAM) where we compute a weighted average as:

$$k_{\bullet j}* = \varphi\left(w_{ij}, k_i\right) = \sum_{i \in \mathbb{I}} w_{ij} k_i$$

(6.4.6)

If w_{ij} 's are ordered in an ascending or descending order of magnitude then the obtained k_j* would qualify as *ordered weighted average* (OWA) [2] [76] [77][78][79].

There are some problems with the aggregation through the WAM model. The aggregation in terms of weighted summability can not be undertaken until we know the units in which the k_i's $\in \mathbb{K}$ are defined and measured. If the k_i's are in different units of measurement then k_j* would appear as a composite sum and can not be computed. This is another problem of aggregation of heterogeneous entities. The problem occurs in all multi-dimensional decision-making environments such as multi-attribute, multiobjective and multi-person (social or collective or group) decision environment. In all these cases, the analytical problem confronted by the practitioner is to design a set of algorithms that would allow a combination of n-dimensional information into a grand element whose values are comparable in a manner the will induce hierarchies of preference for selection where the first element has the maximum preference.

The values of the column vectors of Table 6.4.1 may have a number of interpretations as well as being obtained by different methods. One interpretation is to view w_{ij} as the level of decision importance that is attached to criterion k_i for any fixed project (decision) alternative, π_j. The decision importance may be social, collective or individual. Alternatively, one may interpret the values, w_{ij}, as the degree to which the fixed alternative, π_j, conforms to the decision *characteristics* (see Chapter 3 of this volume) of the criteria, k_i. For any fixed $\pi_j \in \Pi$, the vector $\left[w_{ij}\right]$ will generate a lattice for the criteria in the form $\left[\left(k_{1j}, w_{1j}\right), \left(k_{2j}, w_{2j}\right), \cdots, \left(k_{mj}, w_{mj}\right)\right]$. The same analytical situation holds logically when the concept of criterion is replaced by the concept of either attribute or differential groups. The degrees of decision importance or conformity may be subjective scores such that w_{ij}'s reflect subjective evaluations that must be obtained by a process of normalization.

The lattice of criteria (or for that matter the elements in question) may take the forms of Hasse diagrams for dimensional vector lattice of either A or B. For examples and possible hierarchical decision representation, see [1153a, 1154c]. Now let priority scores of two individuals in terms of preferences for nine criteria $\left[k_1, \cdots, k_i \cdots k_9\right]$ be given for a fixed alternative, π_j, as

$$k_1(a_1, b_1) = k_1(3,2) \quad , \quad k_2(a_2, b_2) = k_2(4,4)$$
$$k_3(a_3, b_3) = k_3(3,3) \quad , \quad k_4(a_4, b_4) = k_4(3,4)$$
$$k_5(a_5, b_5) = k_5(2,4) \quad , \quad k_6(a_6, b_6) = k_6(2,3)$$
$$k_7(a_7, b_7) = k_7(2,2) \quad , \quad k_8(a_8, b_8) = k_8(4,3)$$
$$k_9(a_9, b_9) = k_9(4,2)$$

The Hasse diagram of this lattice of criteria may be represented as in Figure 6.4.1.

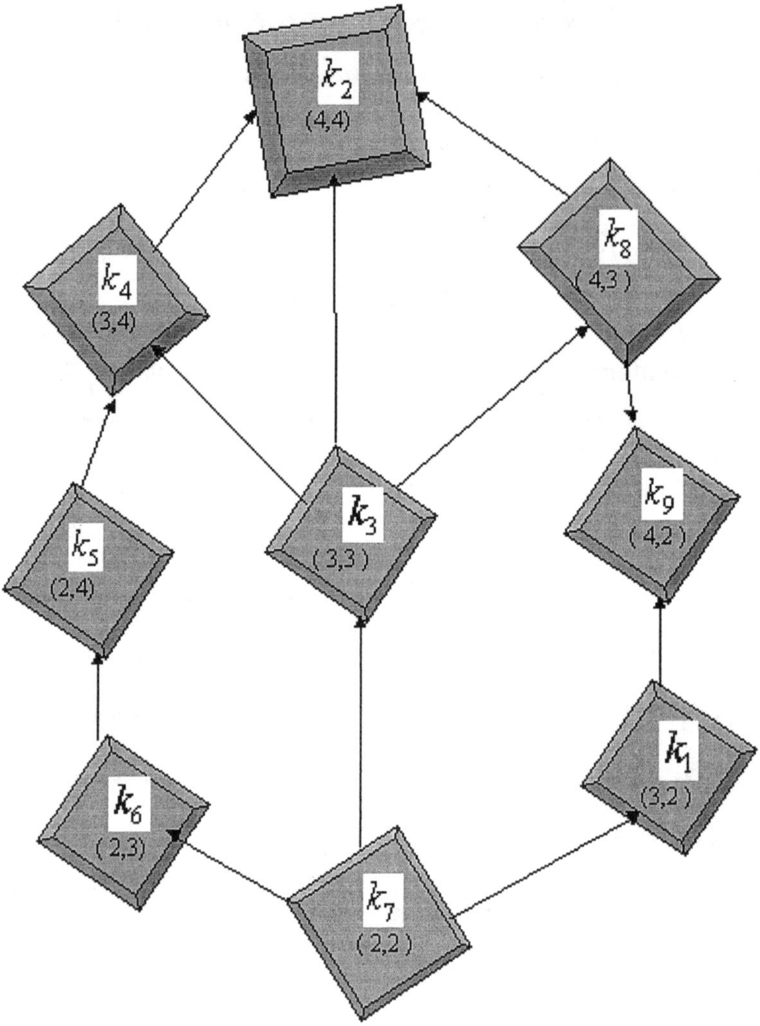

Figure 6.4.1 Hasse Diagram for Criterion Preferences

Alternatively, without the criterion (9) we can represent the Hasse diagram as in Figure 6.4.2

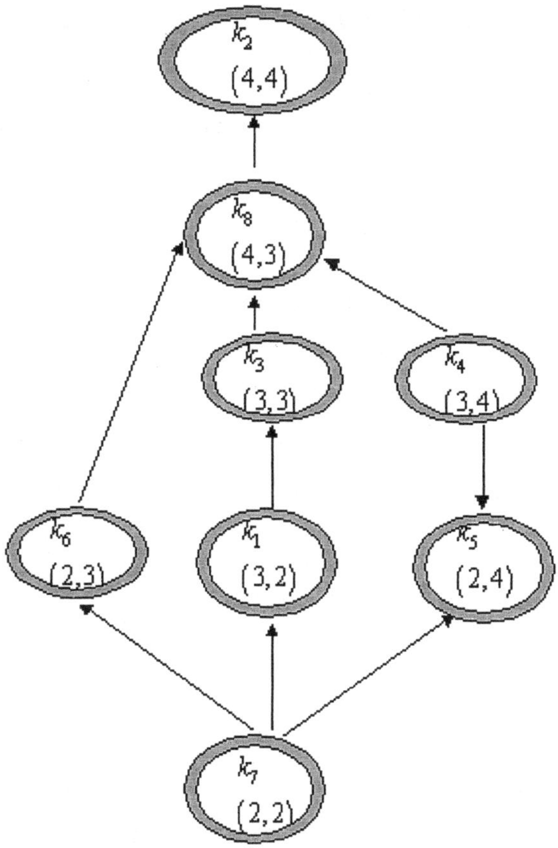

Figure 6.4.2 Hasse Diagram for Vector Lattice of Criterion Preferences

Note that while k_6 and k_1 are preferred to k_7; k_6 and k_1 are non-comparable. Both Figures 6.4.1 and 6.4.2 constitute vector lattices. The problem of multicriteria decision may be illustrated with few examples of 6.4.1 and 6.4.2 and 6.4.3 (for extensive discussions on the uses of Hasse diagrams in decision analysis see [463])

Example 6.4.1 : A candidate Selection Problem:

Consider an academic department in a university where a faculty position is to be filled. Suppose the short list consists of five candidates $\{\pi_1, \pi_2, \pi_3, \pi_4, \pi_5\}$. Let the selection criteria in accordance with the general requirements for competitive academic programs be made up of characteristics of candidates' profile $\mathbb{K} = \{k_1, k_2, k_3, k_4\}$ where k_1 = degree level, k_2 = publication record, k_3 = teaching experience, k_4 = number of courses that the candidate can teach (that is, flexibility of areas of teaching).

Example 6.4.2: Project selection problem

Consider a firm or a community confronted with the selection of a project from a set of five projects. Let us suppose that there is a set of five projects, $\{\pi_1, \pi_2, \pi_3, \pi_4, \pi_5\}$. Suppose further that the selection is to be made in accordance with a set of aggregate attributes that constitute the criteria, $\mathbb{K} = \{k_1, k_2, k_3, k_4\}$ for choice where k_1 = a cost-benefit measure, k_2 = payback period, k_3 = return on book value, k_4 = worth maximization rate.

Example 6.4.3: A Mixed Cost-Benefit Selection Problem

Consider a social decision on a set of five projects, $\Pi = \{\pi_1, \pi_2, \pi_3, \pi_4, \pi_5\}$, one of which must be selected on the basis of a set of three criteria, $\mathbb{K} = \{k_1, k_2, k_3\}$ that are to be used for evaluation , ranking and selection. The criterion set is such that, k_1 = a cost-benefit measure, k_2 = small value of measured physical damage (number of people killed) and k_3 = low degree of measured pain and suffering. All these examples may quickly be related to vector lattice of objectives where the structure of the lattice may be represented by a Hasse diagram.

The decision-choice process in all these cases depends on multiple criteria. If we introduce the concept and measure of *priority scores* such as W_i for the k_i where the W_i may have any of the two interpretations offered previously. The values W_i's may then be normalized to $w_i = \left(\mathrm{W}_i \big/ \sum_{i \in \mathbb{I}} \mathrm{W}_i\right)$ where $w_i \in [0,1]$ and $\sum_{i \in \mathbb{I}} w_i = 1$. Whatever the interpretation that one gives to the values and representation of, w_i, it carries with it imprecision and subjectivity. The priority scores and the ranking of the alternatives must be abstracted from information that is imprecise in one way or the other under the general term of fuzziness. Thus the priority scores may be viewed as fuzzy sets that will induce fuzzy equilibrium weights.

Let the priority scores corresponding to criteria, k_i and alternative, π_j be W_{ij}, with defined membership functions that is given by two classes of decision agents made up of policy makers and experts. These two groups are assumed for simplicity. The number of the groups may be increased depending on the selection problem in the sense that the priority assessments that induce group ranking of the set of criteria is a collective decision problem. For theoretical development at issue we shall consider a criterion set of three, $\mathbb{K} = \{k_1, k_2, k_3\}$ and project alternative set of four, $\Pi = \{\pi_1, \pi_2, \pi_3, \pi_4\}$ and two evaluators (1 and 2). The three vectors together generate a matrix, $\mathbb{W}$, of the form

$$\mathbb{W} = \mathbb{K} \otimes \Pi = \left\{ \mathrm{W}_{ij} = \mathcal{R}\left(k_i, \pi_j\right) \mid k \in \mathbb{K} \text{ and } \pi \in \Pi \right\}$$

(6.4.7)

The evaluators are to assign priority values to each criterion, The criterion values must reflect the decision importance attached to k_i for any fixed project, π_j. The decision importance must capture the degree to which the fixed alternatives $\pi_j \in \Pi$ conforms to the decision characteristics of criterion, $k_i \in \mathbb{K}$. In an explicit form the matrix, $\mathbb{W}$, may be written as

$$\mathrm{W}_{ij} = \mathcal{R}\left(k_i, \pi_j\right)$$

$$\mathbb{W} = \begin{array}{c|cccc} & \pi_1 & \pi_2 & \pi_3 & \pi_4 \\ \hline k_1 & \mathrm{W}_{11} & \mathrm{W}_{12} & \mathrm{W}_{13} & \mathrm{W}_{14} \\ k_2 & \mathrm{W}_{21} & \mathrm{W}_{22} & \mathrm{W}_{23} & \mathrm{W}_{24} \\ k_3 & \mathrm{W}_{31} & \mathrm{W}_{32} & \mathrm{W}_{33} & \mathrm{W}_{34} \end{array}$$

(6.4.9)

where the priority score satisfy $\mathrm{W}_{ij} \in [0,10]$ in that a score of 0 is the least importance and a score of 10 is the most important that must be assigned by the two decision makers.
Define the priority scores of an individual, l as a column vector for any given alternative where $[\cdots]^{\mathrm{T}}$ is a transpose.

$$\left.\begin{aligned} \mathbf{W}_l^1 &= [\mathrm{W}_{11l}, \mathrm{W}_{21l}, \mathrm{W}_{31l}]^{\mathrm{T}} \\ \mathbf{W}_l^2 &= [\mathrm{W}_{12l}, \mathrm{W}_{22l}, \mathrm{W}_{32l}]^{\mathrm{T}} \\ \mathbf{W}_l^3 &= [\mathrm{W}_{13l}, \mathrm{W}_{23l}, \mathrm{W}_{33l}]^{\mathrm{T}} \\ \mathbf{W}_l^4 &= [\mathrm{W}_{14l}, \mathrm{W}_{24l}, \mathrm{W}_{34l}]^{\mathrm{T}} \end{aligned}\right\}$$

(6.4.10a)
Generally we may write

$$\mathbf{W}_l^i = \left[\mathrm{W}_{ijl}\right]^{\mathrm{T}}, i \in \mathbb{I}, j \in \mathbb{J} \text{ and } l \in \mathbb{L}$$

(6.4.10b)

where $\mathbb{L}$ is an index set of evaluators or members of the decision collective.

We consider the priority scores as fuzzy sets, $\mathbb{W}_{ijl}$ where $\mu_{\mathbb{W}_{ijl}} : \mathbb{W}_{ijl} \to [0,1]$ and $\mu_{\mathbb{W}_{ijl}}(\bullet)$'s the corresponding membership functions. There are two sets of fuzzy sets that are definable over each element of the matrix, $\mathbb{W}$, of eqn. 6.4.9 for the specified example. The sets may be written as

$$\mathbb{W}_{ijl}^i = \left\{\left(\mathrm{W}_{ijl}, \mu_{\mathbb{W}_{ijl}^i}(\bullet)\right) \mid \mu_{\mathbb{W}_{ijl}^i}\left(\mathrm{W}_{ijl}\right) \in [0,1] \text{ and } l \in \mathbb{L}\right\}$$

(6.4.11)

To induce ranking over the set of alternatives, Π, using the multiple criteria $k_i \in \mathbb{K}$, $i \in \mathbb{I}$, we compute fuzzy optimal values for each cohort through a fuzzy decision process. For the two evaluator case, one of the evaluator's fuzzy set is taken as a fuzzy goal while the other's fuzzy set is considered as a fuzzy constraint. This goal-constraint approach allows checks and balances of overvaluation and undervaluation through the logic of fuzzy decision construct. The objective of the fuzzy decision process is to provide a computational system with logical consistency that allows checks and balances in terms of overvaluation and undervaluation that may characterize information elicitation under democratic decision making.

The fuzzy decision process of checks and balances to obtain fuzzy equilibrium weights associated with each criterion, k_i and a fixed alternative, $\pi_j, i \in \mathbb{I}$ and $j \in \mathbb{J}$ requires us to define a fuzzy decision, Δ, for the two individual decision makers in the form

$$\Delta_{ij} = \mathbb{W}^i_{ij1} \cap \mathbb{W}^i_{ij2}$$

(6.4.12a)

$$\mu_{\Delta_{ij}}\left(\mathrm{W}_{ij}\right) = \mu_{\mathbb{W}^i_{ij1}}\left(\mathrm{W}_{ij}\right) \wedge \mu_{\mathbb{W}^i_{ij2}}\left(\mathrm{W}_{ij}\right)$$

(6.4.12b)

To solve the problem defined by eqn.(6.4.12b) we simply select the membership function $l = 1$ as the fuzzy goal that is constrained by the membership function l=2. On the basis of this we may write the fuzzy decision problem of $\mu_{\Delta_{ij}}(\bullet)$ as a fuzzy optimization problem in the form

$$\max_{\mathrm{W}_{ij}} \mu_{\Delta_{ij}}\left(\mathrm{W}_{ij}\right) = \begin{cases} \operatorname*{opt}_{\mathrm{W}_{ij}} \mu_{\mathbb{W}^i_{ij1}}\left(\mathrm{W}_{ij1}\right) \\ \quad \text{s.t. a)}\ \mu_{\mathbb{W}^i_{ij1}}(\mathrm{W}) - \mu_{\mathbb{W}^i_{ij2}}(\mathrm{W}) \geq 0 \\ \quad \text{or b)}\ \mu_{\mathbb{W}^i_{ij1}}(\mathrm{W}) - \mu_{\mathbb{W}^i_{ij2}}(\mathrm{W}) \leq 0 \end{cases}$$

(6.4.13)

where constraint (a) holds if the initial values are such that $\mu_{\mathbb{W}^i_{ij1}}(\bullet) \geq \mu_{\mathbb{W}^i_{ij2}}(\bullet)$ and constraint (b) holds if the initial values are such that $\mu_{\mathbb{W}^i_{ij1}}(\bullet) \leq \mu_{\mathbb{W}^i_{ij2}}(\bullet)$. The solution to equation (6.4.13) yields values of the form

$$w_{ij}^{*} = \sup_{W_{ij}} \mu_{\Delta_{ij}}\left(W_{ij}\right), \quad w_{ij}^{*} \in [0,1]$$

(6.4.14)

Equation (6.4.14) yields optimal decision matrix of size $(\#\mathbb{I})\mathrm{x}(\#\mathbb{J})$. To use the values to rank the alternatives with multiple criteria we compute the *total criterion weight* for each project alternative, $\pi_j \in \Pi$ as

$$w_{\bullet j} = \sum_{i \in \mathbb{I}} w_{ij} = \sum_{i \in \mathbb{I}} \sup_{W_{ij}} \mu_{\Delta_{ij}}\left(W_{ij}\right), \quad w_{ij} \in [0,1]$$

(6.4.15)

From eqn.(6.4.14) we may obtain the optimal inverse values as

$$W_{ij}^{*} = \mu_{\Delta_{ij}}^{-1}\left(w_{ij}^{*}\right)$$

(6.4.16)

and

$$W_{\bullet j}^{*} = \sum_{i \in \mathbb{I}} W_{ij}^{*} = \sum_{i \in \mathbb{I}} \mu_{\Delta_{ij}}^{-1}\left(w_{ij}^{*}\right)$$

(6.4.17)

We may also compute horizontal sums of the form

$$w_{i\bullet}^{*} = \sum_{j \in \mathbb{J}} w_{ij}^{*} = \sum_{j \in \mathbb{J}} \sup_{W_{ij}}\left(W_{ij}\right)$$

(6.4.18)

and

$$W_{i\bullet}^{*} = \sum_{j \in \mathbb{J}} W_{ij}^{*} = \sum_{j \in \mathbb{J}} \mu_{\Delta_{ij}}^{-1}\left(w_{ij}^{*}\right)$$

(6.4.19)

Equationns.(6.4.18) and (6.4.19) give us the overall *optimal coefficient of importance* of each of the criteria being satisfied by all the project alternatives π_j's . The value, $w_{i\bullet}^{*}$, is the optimal aggregate coefficient of importance attached to criterion $k_i \in \mathbb{K}$ and induced by the set of alternatives, $\pi_j \in \Pi$; and with corresponding optimal aggregate priority score, $W_{i\bullet}^{*}$ Thus the values of $w_{i\bullet}^{*}$ and $W_{i\bullet}^{*}$ may be used to rank the elements in the criterion set, $\mathbb{K}$. The most important criterion $\widehat{k}$ is one that has the highest optimal aggregate coefficient of importance. Thus

$$\widehat{k} = \begin{cases} \sup\limits_{i\in\mathbb{I}} \sum\limits_{j\in\mathbb{J}} \max\limits_{\mathrm{W}_{ij}} \mu_{\Delta_{ij}}\left(\mathrm{W}_{ij}\right) \\ \text{or} \\ \max\limits_{i\in\mathbb{I}} \sum\limits_{j\in\mathbb{J}} \mu_{\Delta_{ij}}^{-1}\left(w_{ij}^{*}\right) \end{cases}$$

(6.4.20)

Similarly $w_{\bullet j}^{*}$ is the optimal aggregate coefficient of importance induced by the criterion set, $\mathbb{K}$, on alternative, $\pi_j \in \Pi$ while the corresponding optimal aggregate priority score is $\mathrm{W}_{\bullet j}^{*}$, in accordance with the decision makers' preferences. Thus the most preferred project alternative, $\pi^{*} \in \Pi$ may be obtained from eqn.(6,4.15) or (6.4.17) as

$$\pi^{*} = \begin{cases} \sup\limits_{j\in\mathbb{J}} \sum\limits_{i\in\mathbb{I}} \max\limits_{\mathrm{W}_{ij}} \mu_{\Delta_{ij}}\left(\mathrm{W}_{ij}\right) \\ \text{or} \\ \sup\limits_{j\in\mathbb{J}} \sum\limits_{i\in\mathbb{I}} \mu_{\Delta_{ij}}^{-1}\left(w_{ij}^{*}\right) \end{cases}$$

(6.4.21)

If the criteria are in the same unit of measurement then we can compute the global criterion, k^{*}, from eqn.(6.4.6) as

$$k^{*} = \sum_{i\in\mathbb{I}} k_i \left[\frac{\sup\limits_{\mathrm{W}_{ij}} \mu_{\Delta_{ij}}\left(\mathrm{W}_{ij}\right)}{\sum\limits_{i\in\mathbb{I}} \sup\limits_{\mathrm{W}_{ij}} \mu_{\Delta_{ij}}\left(\mathrm{W}_{ij}\right)} \right] = \sum_{i\in\mathbb{I}} k_i \left(\frac{w_{ij}^{*}}{\sum\limits_{i\in\mathbb{I}} w_{ij}^{*}} \right) = \sum_{i\in\mathbb{I}} \left(\frac{w_{ij}^{*} k_i}{w_{\bullet j}^{*}} \right)$$

(6.4.22)

Another approach that may be used to induce a multicriteria ranking of project alternatives is through the method of *optimal membership averaging*(OMA) and *minimum squared averaging* (MSA) or *absolute value-difference averaging* (AVA) . The procedure is first to compute the fuzzy mean $\overline{w}$ of the optimal membership values over all the project alternatives, $\pi_j \in \Pi$. Thus

$$\overline{w} = \left[\frac{\sum_{i \in \mathbb{I}} \sum_{j \in \mathbb{J}} \max_{W_{ij}} \mu_{\Delta_{ij}} \left(W_{ij} \right)}{(\#\mathbb{I})(\#\mathbb{J})} \right]$$

(6.4.23)

Now we specify a function $\sigma\left(w_{ij}^*\right)$ of the form

$$\sigma_j\left(w_{ij}^*\right) = \sqrt{\frac{\sum_{i \in \mathbb{I}} \left(w_{ij}^* - \overline{w}\right)^2}{\#\mathbb{I}}}$$

(6.4.24)

The project ranking is then induced by the minimum fuzzy standard deviation by minimizing eqn.(6.4.24) over $\pi_j \in \Pi$. Thus the optimal project alternative, π^* is

$$\pi^* = \min_{j \in \mathbb{J}} \sigma_j\left(w_{ij}\right) = \min_{j \in \mathbb{J}} \sqrt{\frac{\sum_{i \in \mathbb{I}} \left(w_{ij}^* - \overline{w}\right)^2}{\#\mathbb{I}}}$$

(6.4.25)

This method of ranking is referred to here as *fuzzy ordered preference by minimum variance* (FOPMV).

6.5 Acomputational Example of Fuzzy Multicriteria Decision Problem

Consider a decision problem of two alternatives, π_1 and π_2 and with three criteria made up of k_1, k_2 and k_3 and two decision makers, 1 and 2 (they may be groups or individuals) that must rank the alternatives π_1 and π_2. An experiment is conducted on criteria k_i to assess the degree of decision importance that the individual decision makers attach to each criterion given alternatives π_1 and π_2 respectively. The individual one is asked to state in scores between 0 and 10 the degree of importance (I) unimportance (U) that must be attached to each criterion (k) given the project alternative $\pi_1(\pi_2)$. The information elicitation situation is reversed for decision agent 2 in that he or she is to state the degree of importance (unimportance) attached to each criterion k with respect to π_1 and π_2 respectively. In general, the values of priority scores will range between a_{min} and a_{max} where the, a's , are the number scores. In continuous form the structure leads to a fuzzy decision $\mu_I(x),\ (d\mu_I/dx)\geq 0$ as a fuzzy set of importance that is taken to be the fuzzy goal and $\mu_U(x),\ (d\mu_U/dx)\leq 0$ as the fuzzy set of unimportance that is taken as fuzzy constraint set. Together the fuzzy goal and constraint lead to a fuzzy decision of the form $\mu_\Delta(x)=\mu_I(\bullet)\wedge\mu_U(\bullet)$. The geometric structure of the decision is presented in Figure 6.5.1

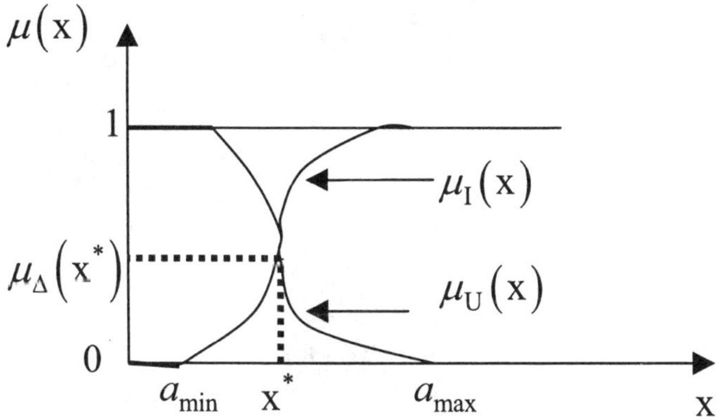

Figure 6.5.1 A Geometry of fuzzy Decision on Weight of Importance

Note that if $\mu_\Delta(x^*)=\alpha\in[0,1]$ then $x^*=\mu_\Delta^{-1}(\alpha)=\mu_\Delta^{-1}(\mu_\Delta(x^*))$

The task, after the information elicitation, is to select or construct appropriate membership functions to capture the concepts of importance and unimportance. Let W_{ij} be the values of the elicited priority scores as to the degree of importance and unimportance of criterion, k_i, for the alternatives, π_j. For each cohort we may specify two membership functions. One membership function is for the individual 1, and the other for the individual, 2. The structure of the decision process is such that there are six priority scores for each of the individual decision makers. The two elicited sets of priorities constitute two matrices of $\mathbb{W}^1$ and $\mathbb{W}^2$ for individual 1 and 2 respectively. They may be written as

$$\mathcal{R}^1(k_i,\pi_j)=\mathrm{W}_{ij}$$

$$\mathbb{W}^1 = \begin{array}{c|cc} & \pi_1 & \pi_2 \\ \hline k_1 & \mathrm{W}_{11}^{\mathrm{I}} & \mathrm{W}_{12}^{\mathrm{U}} \\ k_2 & \mathrm{W}_{21}^{\mathrm{I}} & \mathrm{W}_{22}^{\mathrm{U}} \\ k_3 & \mathrm{W}_{31}^{\mathrm{I}} & \mathrm{W}_{32}^{\mathrm{U}} \end{array}, \quad 0 \le \mathrm{W}_{ij} \le 10 \tag{6.5.1}$$

$$\mathcal{R}^2(k_i,\pi_j)=\mathrm{W}_{ij}$$

$$\mathbb{W}^2 = \begin{array}{c|cc} & \pi_1 & \pi_2 \\ \hline k_1 & \mathrm{W}_{11}^{\mathrm{U}} & \mathrm{W}_{12}^{\mathrm{I}} \\ k_2 & \mathrm{W}_{21}^{\mathrm{U}} & \mathrm{W}_{22}^{\mathrm{I}} \\ k_3 & \mathrm{W}_{31}^{\mathrm{U}} & \mathrm{W}_{32}^{\mathrm{U}} \end{array}, \quad 0 \le \mathrm{W}_{ij} \le 10 \tag{6.5.2}$$

The first column vectors of $\mathbb{W}^1$ and $\mathbb{W}^2$ represent the degree of importance and unimportance respectively. The second column vectors of $\mathbb{W}^1$ and $\mathbb{W}^2$ represent the degrees of unimportance and importance respectively. In both matrices $W_{ij} = \mathcal{R}\left(k_i, \pi_j\right)$ where $\mathcal{R}$ is a relation. Equations (6.5.1) and (6.5.2) may be combined and organized to define matrices of priority scores for degrees of importance (I) and unimportance (U) as $\mathbb{W}^{\mathrm{I}}$ and $\mathbb{W}^{\mathrm{U}}$.

$$\mathcal{R}^{\mathrm{I}}\left(k_i, \pi_j\right)$$

$$\mathbb{W}^{\mathrm{I}} = \begin{array}{c|c|c} & \pi_1 & \pi_2 \\ \hline k_1 & W_{11}^{\mathrm{I}} & W_{12}^{\mathrm{I}} \\ k_2 & W_{21}^{\mathrm{I}} & W_{22}^{\mathrm{I}} \\ k_3 & W_{31}^{\mathrm{I}} & W_{32}^{\mathrm{I}} \end{array}, \quad 0 \le W_{ij}^{\mathrm{I}} \le 10$$

(6.5.3)

$$\mathcal{R}^{\mathrm{U}}\left(k_i, \pi_j\right)$$

$$\mathbb{W}^{\mathrm{U}} = \begin{array}{c|cc} & \pi_1 & \pi_2 \\ \hline k_1 & W_{11}^{\mathrm{U}} & W_{12}^{\mathrm{U}} \\ k_2 & W_{21}^{\mathrm{U}} & W_{22}^{U} \\ k_3 & W_{31}^{\mathrm{U}} & W_{32}^{\mathrm{U}} \end{array}, \quad 0 \le W_{ij} \le 1$$

(6.5.4)

From equations (6.5.3) and (6.5.4) two matrices of fuzzy sets $\widehat{\mathbb{W}}^{\mathrm{I}}$ and $\widehat{\mathbb{W}}^{\mathrm{U}}$ may be organized as fuzzy information that must be processed for ranking of the alternatives. They are given in equations (6.5.5) and (6.5.6).

$$\mathcal{R}^{\mathrm{I}}\left(k_i, \pi_j\right)$$

$$\widehat{\mathbb{W}}^{\mathrm{I}} = \begin{array}{c|cc} & \pi_1 & \pi_2 \\ \hline k_1 & \mathbb{W}_{11}^{\mathrm{I}} & \mathbb{W}_{12}^{\mathrm{I}} \\ k_2 & \mathbb{W}_{21}^{\mathrm{I}} & \mathbb{W}_{22}^{\mathrm{I}} \\ k_3 & \mathbb{W}_{31}^{\mathrm{I}} & \mathbb{W}_{32}^{\mathrm{I}} \end{array}$$

(6.5.5)

$$\mathcal{R}^{\mathrm{U}}\left(k_i, \pi_j\right)$$

$$\mathbb{W}^{\mathrm{U}} = \begin{array}{c|c|c} & \pi_1 & \pi_2 \\ \hline k_1 & \mathbb{W}_{11}^{\mathrm{U}} & \mathbb{W}_{12}^{\mathrm{U}} \\ k_2 & \mathbb{W}_{21}^{\mathrm{U}} & \mathbb{W}_{22}^{\mathrm{U}} \\ k_3 & \mathbb{W}_{31}^{\mathrm{U}} & \mathbb{W}_{32}^{\mathrm{U}} \end{array}$$

(6.5.6)

where for $i \in \mathbb{I}$ and $j \in \mathbb{J}$ we have

$$\mathbb{W}_{ij}^{\mathrm{I}} = \left\{\left(\mathrm{W}_{ij}^{\mathrm{I}}, \mu_{\mathbb{W}_{ij}^{\mathrm{I}}}\left(\mathrm{W}_{ij}^{\mathrm{I}}\right)\right) \mid 0 \le \mathrm{W}_{ij}^{\mathrm{I}} \le 10,\ \mu_{\mathbb{W}_{ij}^{\mathrm{I}}}(\bullet) \in [0,1], \left(\mathrm{d}\mu^{\mathrm{I}} / \mathrm{dW}_{ij}^{\mathrm{I}}\right) \ge 0\right\}$$

(6.5.7)

$$\mathbb{W}_{ij}^{U} = \left\{ \left(W_{ij}^{U}, \mu_{\mathbb{W}_{ij}}^{U}\left(W_{ij} \right) \right) \mid 0 \le W_{ij} \le 10, \mu_{\mathbb{W}}^{U}(\bullet) \in [0,1], \left(d\mu^{U} / dW_{ij}^{U} \right) \le 0 \right\}$$

(6.5.8)

By combining eqns. (6.5.5) and (6.5.6) we obtain a computable matrix of optimal decision elements as

$$\mathbf{\Delta}_{ij} = \widehat{\mathbb{W}}^{I} \cap \widehat{\mathbb{W}}^{U} = \left\{ \left(\mathbb{W}_{ij}^{I} \cap \mathbb{W}_{ij}^{U} \right) \mid i \in \mathbb{I} \text{ and } j \in \mathbb{J} \right\}$$

(6.5.9)

where $\mathbb{W}_{ij}^{I}$ and $\mathbb{W}_{ij}^{U}$ are as defined in equations (6.5.7) and (6.5.8) respectively. The $\mathbf{\Delta}_{ij}$ defines a fuzzy decision set on the basis of which the optimal decision elements are to be found for each (*i-j*) cohort. The membership of the fuzzy decision function may then be written as

$$\mu_{\Delta_{ij}}\left(W_{ij} \right) = \left\{ \mu_{\mathbb{W}_{ji}^{U}}(\bullet) \wedge \mu_{\mathbb{W}_{ij}^{I}}(\bullet) \mid i \in \mathbb{I} \text{ and } j \in \mathbb{J} \right\}$$

(6.5.10)

The solution to eqn. (6.5.10) may be abstracted by maximization which may simply be written as

$$\underset{W_{ij}}{\text{opt}}\, \mu_{\Delta_{ij}}\left(W_{ij} \right) = \begin{cases} \underset{W_{ij} \in \mathbb{S}}{\text{opt}}\, \mu_{\mathbb{W}_{ij}^{U}}\left(W_{ij} \right) \\ \text{s.t. } \mathbb{S} = \left\{ W_{ij} \mid \left(\mu_{\mathbb{W}_{ij}^{I}}(\bullet) - \mu_{\mathbb{W}_{ij}^{U}}(\bullet) \right) \le 0 \right\} \end{cases}$$

(6.5.11)

The combined system of equations (6.5.9) – (6.5.11) yield optimal decision matrices of the form

$$\widehat{\mathbb{W}}_{\Delta_{ij}} = \begin{array}{c|c|c} & \pi_1 & \pi_2 \\ \hline k_1 & w_{11}^* & w_{12}^* \\ \hline k_2 & w_{21}^* & w_{22}^* \\ \hline k_3 & w_{31}^* & w_{32}^* \end{array}$$

(6.5.12a)

$$\widehat{\mathbb{W}} = \begin{array}{c|c|c} & \pi & \pi \\ \hline k_1 & \mathrm{W}_{11} & \mathrm{W}_{12} \\ k_2 & \mathrm{W}_{21} & \mathrm{W}_{22} \\ k_3 & \mathrm{W}_{31} & \mathrm{W}_{32} \end{array}$$

(6.5.12b)

where $w_{ij}^* = \sup \mu_{\Delta_{ij}}\left(\mathrm{W}_{ij}\right)$ and $\mathrm{W}_{ij}^* = \mu_{\Delta_{ij}}^{-1}\left(w_{ij}^*\right)$.

Let us now insert in explicit membership functions for fuzzy sets, $\mathbb{W}_{ij}$'s for degrees of unimportance and importance for decision makers (1) and (2).

$$\text{COHORT}_{11} \quad \mathbb{W}_{11} \quad \left\|\begin{array}{l} \text{Decision Agent 1 : } \textit{Degree of Unimportance} \\ \mu_{\mathbb{W}_{11}^{U}}(\mathrm{W}) = \begin{cases} e^{-k_1\left(\frac{\mathrm{W}}{10-\mathrm{W}}\right)} &, \quad k_1 > 0, \quad \mathrm{W} \in [0,10) \\ 0 &, \qquad\qquad \mathrm{W} \in [10,\infty) \end{cases} \\ \\ \text{Decision Agent 2: } \textit{Degree of Importance} \\ \mu_{\mathbb{W}_{11}^{I}}(\mathrm{W}) = \begin{cases} 0 &, \quad \mathrm{W} \in (0,3] \\ 1 - e^{-k_2(\mathrm{W}-3)} &, \quad k_2 > 0, \quad \mathrm{W} \in (3,10) \\ 1 &, \quad \mathrm{W} \in [0,\infty) \end{cases} \end{array}\right.$$

(6.5.13)

To solve the above system we optimize $\mu_{\Delta}(\bullet) = \mu_{\mathbb{W}^{U}}(\bullet) \wedge \mu_{\mathbb{W}^{I}}(\bullet)$ with respect to W by defining

$$\begin{aligned} &\max_{\mathrm{W}} \mu_{\mathbb{W}^{U}}(\mathrm{W}) \\ &\text{s.t.} \quad \left\{\mu_{\mathbb{W}^{U}}(\mathrm{W}) - \mu_{\mathbb{W}^{I}}(\mathrm{W})\right\} \le 0 \end{aligned}$$

(6.5.14)

The solution yields optimal, w_{11}^{*} and $\mathrm{W}_{11}^{*} = \mu_{\Delta}^{-1}\left(w_{11}^{*}\right)$

$$
\underset{\mathrm{W}_{21}}{\mathrm{COHORT}_{21}}
\left\|
\begin{array}{l}
\text{Decision Agent 1}: \textit{Degree of Unimportance} \\
\mu_{\mathrm{W}_{21}^{\mathrm{U}}}(\mathrm{W}) = \begin{cases} 1 & , \quad \mathrm{W} \in (0,2] \\ \dfrac{10-\mathrm{W}}{8} & , \quad \mathrm{W} \in (2,10) \\ 0 & , \quad \mathrm{W} \in [10,\infty) \end{cases} \\
\\
\text{Decision Agent 2}: \textit{Degree of Importance} \\
\mu_{\mathrm{W}_{21}^{\mathrm{U}}}(\mathrm{W}) = \begin{cases} 0 & , \quad \mathrm{W} \in (0,1] \\ \dfrac{\mathrm{W}-1}{9} & , \quad \mathrm{W} \in (1,10) \\ 1 & , \quad \mathrm{W} \in [0,\infty) \end{cases}
\end{array}
\right.
$$

(6.5.15)

The system of equation (6.5.15) may be solved by optimizing the fuzzy decision $\mu_{\Delta}(\bullet) = \mu_{\mathrm{W}^{\mathrm{U}}}(\bullet) \wedge \mu_{\mathrm{W}^{\mathrm{I}}}(\bullet)$ in the same manner as in equation (6.5.14). It may also be solved by either a method of elimination or substitution to obtain optimal, w_{21}^{*} as well as $\mathrm{W}_{21}^{*} = \mu_{\Delta}^{-1}\left(w_{21}^{*}\right)$. The next step is to specify the constrained-decision process of cohort (31).

$$
\text{COHORT}_{31}^{\mathbb{W}_{31}}
\left\|
\begin{array}{l}
\text{Decision Agent 1: } \textit{Degree of Unimpertance} \\
\mu_{\mathbb{W}^{\mathrm{U}}_{31}}(\mathrm{W}) = \begin{cases} 1 & , \ \mathrm{W} \in (-\infty, 0] \\ e^{-k_1\left(\frac{\mathrm{W}}{10-\mathrm{W}}\right)^2} & , \ \mathrm{W} \in (0, 10) \\ 0 & , \ \mathrm{W} \in [10, \infty) \end{cases} , \ k_1 > 0 \\
\\
\text{Decision Agent 2 : } \textit{Degree of importance} \\
\mu_{\mathbb{W}^{\mathrm{I}}_{31}}(\mathrm{W}) = \begin{cases} 0 & , \ \mathrm{W} \in [0, 3] \\ 1 - e^{-k_2\left(\frac{\mathrm{W}-3}{10-\mathrm{W}}\right)^2} & , \ \mathrm{W} \in (3, 10) \\ 1 & , \ \mathrm{W} \in [10, \infty) \end{cases}
\end{array}
\right.
$$

(6.5.16)

The set of the system of three equations above provides us with values of column vector one. We now turn our attention to column vector two and specify the system of equations that will yield the corresponding values.

$$
\text{CORHOT}_{12}^{\mathbb{W}_{12}}
\left\|
\begin{array}{l}
\text{Decision Agent 2 : } \textit{Degree of Unimportance} \\
\mu_{\mathbb{W}^{\mathrm{U}}_{12}}(\mathrm{W}) = \begin{cases} e^{-k_2\left(\frac{\mathrm{W}}{10-\mathrm{W}}\right)} & , \ \mathrm{W} \in [0, 10) \ , \ k_2 > 0 \\ 0 & , \ \mathrm{W} \in [10, \infty) \end{cases} \\
\\
\text{Decision Agent 1 : } \textit{Degree of Importance} \\
\mu_{\mathbb{W}^{\mathrm{I}}_{12}}(\mathrm{W}) = \begin{cases} 0 & , \ \mathrm{W} \in (0, 2] \\ 1 - e^{-k_1(\mathrm{W}-2)} & , \ \mathrm{W} \in (2, 10) \ , \ k_1 > 0 \\ 1 & , \ \mathrm{W} \in [10, \infty) \end{cases}
\end{array}
\right.
$$

(6.5.17)

$$\underset{\mathbb{W}_{22}}{\text{COHORT}_{22}}\left\|\begin{array}{l}\text{Decision Agent 2 : } \textit{Degree of Unimportance} \\ \mu_{\mathbb{W}_{22}^{U}}(\mathrm{W})=\begin{cases}1 & , \ \mathrm{W}\in(0,1] \\ \dfrac{10-\mathrm{W}}{9} & , \ \mathrm{W}\in(1,10) \\ 0 & , \ \mathrm{W}\in[10,\infty)\end{cases} \\ \\ \text{Decision Agent 1 : } \textit{Degree of Importance} \\ \mu_{\mathbb{W}_{22}^{I}}(\mathrm{W})=\begin{cases}0 & , \ \mathrm{W}\in(0,3] \\ \dfrac{\mathrm{W}-3}{7} & , \ \mathrm{W}\in(3,10) \\ 1 & , \ \mathrm{W}\in[10,\infty)\end{cases}\end{array}\right.$$

(6.5.18)

$$\underset{\mathbb{W}_{32}^{U}}{\text{COHORT}}\left\|\begin{array}{l}\text{Decision Agent 2: } \textit{Degree of Unimportance} \\ \mu_{\mathbb{W}_{32}^{U}}(\mathrm{W})=\begin{cases}\mathrm{e}^{-k_2\left(\frac{\mathrm{W}}{10-\mathrm{W}}\right)^2} & , \ \mathrm{W}\in[0,10) \ , \ k_2>0 \\ 0 & , \ \mathrm{W}\in[0,\infty)\end{cases} \\ \\ \text{Decision Agent 1: } \textit{Degree of Importance} \\ \mu_{\mathbb{W}_{32}^{I}}(\mathrm{W})=\begin{cases}0 & , \ \mathrm{W}\in(0,2] \\ 1-\mathrm{e}^{-k_1\left(\frac{\mathrm{W}-2}{10-\mathrm{W}}\right)} & , \ \mathrm{W}\in(2,10) \ , \ k_1>0 \\ 1 & , \ \mathrm{W}\in[10,\infty)\end{cases}\end{array}\right.$$

(6.5.19)

From the solutions to equations (6.5.13) - (6.5.19) the matrix of optimal priority scores may be obtained as

$$\hat{\hat{\mathbb{W}}}^* = \begin{bmatrix} 4.20 & 3.80 \\ 5.76 & 6.00 \\ 5.80 & 5.50 \end{bmatrix} = \begin{bmatrix} W_{11}^* & W_{12}^* \\ W_{21}^* & W_{22}^* \\ W_{31}^* & W_{32}^* \end{bmatrix}$$

(6.5.20)

The corresponding matrix of optimal membership values may be written as

$$\breve{\mathbb{W}}^* = \begin{bmatrix} 0.64 & 0.62 \\ 0.53 & 0.44 \\ 0.32 & 0.30 \end{bmatrix} = \begin{bmatrix} w_{11}^* & w_{12}^* \\ w_{21}^* & w_{22}^* \\ w_{31}^* & w_{32}^* \end{bmatrix}$$

(6.5.21)

It may be observed that the vertical sum of each vector is greater than one. The horizontal sum of each of the row vectors is either greater than or less than one. In general $w_{i\bullet} \lesseqgtr 1$ and $w_{\bullet j} \gtreqless 1$. In the above problem the overall optimal membership weights in an aggregative for π_1, $w_{\bullet 1} = 1.49$ and for π_2, $w_{\bullet 2} = 1.36$. Similarly we can compute the horizontal aggregation of the optimal membership values in a manner as in the vertical aggregation where for

k_1, $w_{1\bullet} = 1.26$, for k_2, $w_{2\bullet} = 0.97$ and for k_3, $w_{3\bullet} = 0.62$. The optimal priority aggregates are for π_1 we have $\mathrm{W}_{\bullet 1} = 15.76$ and for π_2 we have $\mathrm{W}_{\bullet 2} = 15.3$. In terms of criterion values the distribution is such that for k_1, k_2, and k_3 we have $\mathrm{W}_{1\bullet} = 8.0$, $\mathrm{W}_{2\bullet} = 11.76$ and $\mathrm{W}_{3\bullet} = 11.30$ respectively.

The optimal membership values may be used as distributional weights of importance as subjectively assessed and then combined with the optimal priority scores. To do this the column vector values must be normalized such that the sum is unity for each column. From eqn.(5.5.20) we can compute $\overset{\leftrightarrow}{w}_{ij}^{*} = \left(w_{ij}^{*} / w_{\bullet j}^{*} \right)$, $j = 1$ and 2. The computation yields a matrix of the form

$$\overset{\leftrightarrow}{\mathbb{W}} = \begin{bmatrix} \overset{\leftrightarrow}{w}_{11}^{*} & \overset{\leftrightarrow}{w}_{12}^{*} \\ \overset{\leftrightarrow}{w}_{21}^{*} & \overset{\leftrightarrow}{w}_{22}^{*} \\ \overset{\leftrightarrow}{w}_{31}^{*} & \overset{\leftrightarrow}{w}_{32}^{*} \end{bmatrix} = \begin{bmatrix} 0.43 & 0.46 \\ 0.36 & 0.32 \\ 0.21 & 0.22 \end{bmatrix}$$

(6.5.22)

We may now combine eqn.(6.5.21) with eqn.(6.5.19) and compute ordered weighted aggregate(OWA) to obtain

$$\overset{\leftrightarrow}{\overset{\leftrightarrow}{\mathbb{W}}} = \overset{\leftrightarrow}{\mathbb{W}} \widehat{\mathbb{W}}^{*T} = \begin{bmatrix} 0.43 & 0.46 \\ 0.36 & 0.32 \\ 0.21 & 0.22 \end{bmatrix} \begin{bmatrix} 4.20 & 4.76 & 5.80 \\ 3.80 & 6.00 & 5.50 \end{bmatrix}$$
$$= [4.74, 4.88]$$

(6.5.23)

In this case $\pi_2 \succcurlyeq \pi_1$. By the use of vertical weighted sum of the membership we have $w_{\bullet 1} > w_{\bullet 2} \Rightarrow \pi_1 \succcurlyeq \pi_2$ and by the use of aggregates of optimal priority sum $\mathrm{W}_{\bullet 1} > \mathrm{W}_{\bullet 2}$. $\Rightarrow \pi_1 \succcurlyeq \pi_2$. The weighted sum seems to override the two ranking approach. To use the method of fuzzy ordered preference by minimum variance approach as presented in eqns. (6.4.23) ,(6.4.24) and (6.4.25) we compute

$$\overline{w} = \frac{\sum_{i=1}^{3} \sum_{j=1}^{2} w_{ij}^{*}}{6} = \frac{2.85}{6} = 0.48$$

We now compute the FOPMV as $\sigma_1\left(w_{i1}^{*}\right) = 0.063$ and $\sigma_2\left(w_{i2}^{*}\right) = 0.052$ and since $\sigma_1(\bullet) > \sigma_2(\bullet)$ we conclude that $\pi_2 \succcurlyeq \pi_1$ by FOPMV method. The FOPMV gives the same ranking as the weighted sum of the priority scores.

6.6 Reflections on the General Fuzzy Decision Processes

The theory of fuzzy decision is a logical process which is made up of a number of sub-processes or modules with sub-algorithms that together constitute the organic algorithm. Given the general decision space, these modules form the chain of logical steps for fuzzy decision making. The logical steps are:

a) **Logical Step 1** : An analysis of conditions and presence of fuzziness;
b) **Logical Step 2** : Fuzzification process involving the decision environment, goals, Constraints and parameters;
c) **Logical Step 3:** Module of model representation of fuzzy equivalence composing of Selection of membership functions for the goals and constraints;
d) **Logical Step 4:** fuzzy decision module composed of an inference engine and supporting sub-inferences
e) **Logical Step 5**: Module of consistency test composed of examination, analysis and auditing of the decision module for any possible inconsistency;
f) **Logical Step 6**: Validation process consisting of checking for the validity and source of inconsistencies if any from the fuzzy decision model
g) **Logical Step 7**: Fuzzy optimization module that is made up of assembling together techniques and methods to abstract from the decision model optimal fuzzy values that may also define fuzzy equilibrium states;
h) **Logical Step 8**: Defuzzification process composed of approximate reasoning in interpreting the optimal fuzzy values;
i) **Logical Step 9**: Module of crisp value conversion composed of analytical structure of taking the defuzzified values and creating exact values and logic that become inputs into prescriptive rules of decision practice;
j) **Logical Step 10**: Module of prescriptive rules for decision action and implementation, composed of the optimal decision rules from the whole fuzzy decision process.

The general scheme of the fuzzy decision theory and the ten logical steps required for arriving at crisp decision rules for action and implementation is presented in Figure 6.6.1. Some analytical notes are

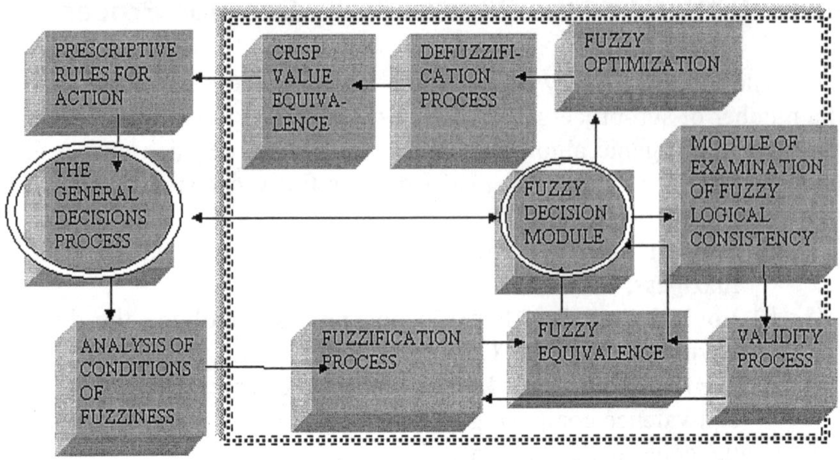

Figure 6.6.1 A General Logical Scheme of the Theory Fuzzy Decision

necessary and relevant to make clear the nature of the path of the theory of the fuzzy decision process. The analysis of the presence and characteristics of fuzziness in the logical step 1 allows one to determine whether a fuzzy decision model is needed and if so what kind of the model's structure would be appropriate for the problem at hand. Such a logical analysis provides us with conditions that become inputs into the fuzzification process of logical step (2).

The fuzzification module and the process that it entails allow one to determine what decision variables and parameters need to be fuzzified and in what manner should such fuzzification take place. The results of the fuzzification module provide explicit information as an input into module of fuzzy equivalence structure of logical step (3). The fuzzy equivalence module is a vehicle through which appropriate membership functions are either selected or constructed to represent the nature of the fuzzy decision variables. Here a determination is made as to whether the decision problem is collective, social, multiattribute, multicriteria or a combination of them. The results of logical step (3) become inputs into the fuzzy decision module of step (4).

The fuzzy decision module of logical step (4) pulls together all relevant elements of of the fuzzy decision problem into a mathematical or logical representation that must be analyzed for optimal solution. Before a solution is abstracted the fuzzy decision module of logical step (4) becomes an input into logical step (5) of critical examination module. In the logical step (5), a critical examination and auditing of the logical structure of the fuzzy decision model are conducted to derive conditions of consistency. If inconsistencies are discovered in the fuzzy decision model at the level of validity module, then the fuzzy decision module is sent to the fuzzification process for correction and logical refinement. On the other hand if the logical validity test is passed then the fuzzy decision model of logical step 4 is cleared of inconsistencies and becomes an input into the logical step 7 for fuzzy optimization.

The module of fuzzy optimization at the logical step 7 uses relevant mathematical algorithms to abstract fuzzy optimal membership values that satisfy both the membership functions of goals and constraints. The resulting optimal membership functional values then become inputs into the logical step of defuzzification process. The defuzzification process allows one to collect together all optimal fuzzy values that become translated into crisp values. The crisp values then become inputs into the logical step 9 of the module of crisp value process where crisp equivalencies are defined and checked to become inputs into the module of prescriptive rules for action and decision practice. For extensive and detailed usage of the logical steps of the theory of fuzzy decision see the companion volume on Fuzzy Value Theory [931b]. One may also consult [69], [694], [709] , [735], [739], [756], [809], [811],[1258].

References

On Aggregation, Distributional Weights, and Rankings

[1] Aczel, J. et al., "Procedures for Synthesizing Ratio Judgments," *Jour. Math. Psychology*, Vol. 27, pp. 93-102.

[2] Alsina, C. et al., "A Characterization of a Class of Aggregation Functions," *Fuzzy Sets and Systems*, Vol. 53 #1, 1993, pp. 33-38.

[3] Altrock, C.V. et al., "Multi-criteria Decision Making in German Automobile Industry Using Fuzzy Logic," *Fuzzy Sets and Systems*, Vol. 63, 1964, pp. 75-380.

[4] Baas, S.M., et al., "Rating and Ranking of Multiple Aspect Alternatives Using Fuzzy Sets," *Automatica,* Vol. 13, #1, 1977, pp. 47-58.

[5] Bardossy, A., et al., "Combination of Fuzzy Numbers Representing Expert Opinions," *Fuzzy Sets and Systems*, Vol. 57, 1993, pp. 173-181.

[6] Buckley, J.J., "The Multiple Judge, Multiple Criteria Ranking Problem Fuzzy Set Approach," *Fuzzy Sets and Systems*, Vol. 13, 1984, pp. 25-37.

[7] Buckley, J.J. et al., "A Fast Method of Ranking Alternatives using Fuzzy Numbers (Short Communications)," *Fuzzy Sets and Systems*, Vol. 30, 1989, pp. 337-339.

[8] Basile, L., et al., "Ordering for Classes of Aggregation Operators," *Inter. Jour. Uncertainty Fuzziness and Knowledge-Based Systems*, Vol. 4, #2, 1996 pp. 145-156.

[9] Bergin, J., "Common Knowledge with Monotone Statistics," *Econometrica*, Vol. 69, 2001, pp. 1315-1332.

[10] Bhat, K.V.S., "On the Notion of Fuzzy Consensus," *Fuzzy Sets and Systems*, Vol. 8, 1983, pp. 285-289.

[11] Billot, A., "Aggregation of Preferences: The Fuzzy Case," *Theory and Decision*, Vol. 30, #1, 1991, pp. 59-93.

[12a] Blackorby, C., et al., "Utility vs. Equity," *Jour. of Public Econ.,* Vol.7 1977, pp. 365-381.

[12b] Bodjanova, S., "Approximation of Fuzzy Concepts in Decision Making," *Fuzzy Sets and Systems*, Vol. 85, #1, 1997, pp. 23-30.

[13] Bye, B.V. et al., "A Fuzzy Algorithmic Approach to the Construction of Composite Indices: An Application to a Functional Limitation Index," *Intern. Jour. of General Systems*, Vol. 11, #2, 1985, pp. 163-172.

[14] Campos, L. et al, "Further Contributions to the Study of Average value for Ranking Fuzzy Numbers," *Intern. Jour. Approximate Reasoning*, Vol. 10, #2, 1994, pp. 135-153.

[15] Chen, C.–b., et al., " An Efficient Approach to Solving Fuzzy MADM Problem," Fuzzy Sets and Systems, Vol. 88, #1, 1997, pp. 51-68.

[16] Cheng, C. -H., "A New Approach For Ranking Fuzzy Numbers by Distance Method," *Fuzzy Sets and Systems*, Vol. 95, #3, 1998, pp. 307-318.

[17] Chichilinsky, G., "Social Aggregation and Continuity," *The Quart Jour. Econ.* Vol. 97, 1982, pp. 337-352.

[18] Cho, S. -B., "Fuzzy Aggregation of Modular Neural Networks with Ordered Weighted Averaging Operators," *Intern. Jour. Approximate Reasoning*, Vol. 13, #4, 1995, pp. 359-375.

[19] Cholewa, W., "Aggregation of Fuzzy Opinions-An Axiomatic Approach," *Fuzzy Sets and Systems*, Vol. 43, 1991, pp. 257-276.

[20] Choobineh, F., et al., "An Index for Ordering Fuzzy Numbers," *Fuzzy Sets and Systems*, Vol. 54, #3, 1993, pp. 287-294.

[21] Chu, A.T.W., et al., "A Comparison of two Methods for Determining the Weights of Belonging to Fuzzy Sets," *Jour. of Optimization Theory and Applications*, vol. 27, 1979, pp. 531-538.

[22] Cutello, V., et al., "Hierarchies of Intensity Preference Aggregations," *Intern Jour. Approximate Reasoning*, Vol. 10, #2, 1994, pp. 123-134.

[23] Delgado, M. et al., " On Aggregation Operations of Linguistic Labels," *International Jour. Intelligent Systems*, Vol. 8, 1993, pp. 351-370.

[24] Dockery, J.T., et al., "A Fuzzy Approach to Aggregating Military Assessments," *Intern. Jour. of Approximate Reasoning*, Vol. 1, #3, 1987, pp. 251-271.

[25] Dong, W.M. et al., "Fuzzy Weighted Averages and Implementation of the Extension Principle," *Fuzzy Sets and Systems*, Vol. 21, #2, 1987, pp.183-199.

[26] Dubois, D. et al., "Social Choice Axioms for Fuzzy Set Aggregation," *Fuzzy Sets and Systems*, Vol. 43, 1991, pp. 257-274.

[27] Dubois, D. et al., "A Review of Fuzzy Sets Aggregation Connectives "*Information Science*, Vol. 36, 1985, pp. 85-126.

[28] Farrell, M.J., "Some Aggregation Problems in Demand Analysis," *Review of Econ. Studies*, Vol. 21, 1953-54, pp. 193-203.

[29] Fie, J.C., "A Fundamental Theorem for the Aggregation Problem of Input-Output Analysis," *Econometrica*, Vol. 24, 1956, pp. 400-412.

[30] Frank, M. J., On the Simultaneous Associativity of $F(x,y)$ and $x + y - F(x,y)$," *Aequationes Math.,* Vol.19, 1979 pp137-152.

[31] Fisher, W.D., "Criteria for Aggregation in Input-output Analysis," *Rev. Economic Statistics*, Vol. 40, 1958, pp. 250-260.

[32] Fonck, P., et al., "An Application of Aggregation Procedures of Similarity Between Fuzzy Sets," *Fuzzy Sets and Systems*, Vol. 97, #1, 1998, pp. 67-74.

[33] Freeling, A.N.S. et al., "Combining Evidence in Evidentiary Value: Philosophical Judicial and Psychological Aspects of a Theory, Lund, 1983, pp. 58-74

[34] Gilardoni, G.L. et al., "On Reaching a Consensus Using DeGroot's Iterative Pooling," *Annals Statistics*, Vol. 21, 1993, pp. 391-401.

[35] Gorman, W.M., "Separable Utility and Aggregation," *Econometrica*, Vol.27, 1959, pp. 469-481.

[36] Gorman, W.M., "Additive Logarithmic Preference: A Further Note," *Rev. Econ. Studies*, Vol. 30, 1963, pp. 56-62.

[37] Green, J.H.A., *Aggregation in Economic Analysis: an Introductory Survey*, Princeton, N.J., Princeton Univ. Press, 1964.

[38] Green, E.J., "On the Difficulty of Eliciting Summary Information," *Jour. Economic Theory*, Vol. 32, 1984, pp. 228-245.

[39] Grunfield, Y. et al., "Is Aggregation Necessarily Bad," *Rev. Econ Statistics*, Vol. 42, 1960, pp. 1-13.

[40] Harberger, A.C., "On the Use of Distributional Weights in Social Cost-Benefit Analysis," *Jour. Political Economy*, Vol. 86, #2, 1978, pp. S87-S120.

[41a] Harberger, A.C., "Basic Needs Versus Distributional Weights in Social Cost-Benefit Analysis," *Economic Development and Cultural Change*, Vol.32, 1984, pp. 455-574.

[41b] Harberger, A. C. , *Taxation and Welfare*, Boston, Mass., Little, Brown and Co.,1974.

[42] Herrara, F., et al., "Direct Approach Processes in Group Decision Making Using Linguistic OWA Operators," *Fuzzy Sets and Systems*, Vol. 79, #2, 1996, pp. 175-190.

[43] Hong, D.H., et al., "On the Compositional Rule of Inference Under Triangular Norms," *Fuzzy Sets and Systems*, Vol. 66, 31, 1994, pp. 25-38.

[44] Houthaker, H.S., "Addictive Preference," *Econometrica*, Vol. 28, 1960, pp. 244-257.

[45] Houthaker, H. S., "Compensated Changes in Quantities and qualities Consumed," *Rev of Econ. Studies*, Vol.19, 1952, pp. 155-164.

[46] Hsu, H. -M. et al., "Aggregation of Fuzzy Opinions under Group Decision Making," *Fuzzy Sets and Systems*, Vol.79, #3, 1996, pp. 279-286.

[47] Hurwicz, L., "Aggregation in Macroeconomic Models," *Econometrica*, Vol. 20, 1952, pp. 489-490.

[48] Kaprzyk, J. et al., "'Soft' Measure of Consensus in the Setting of Partial Fuzzy Preferences," *European Jour. Operations Research*, Vol. 34, 1988, pp. 316-325.

[49] Keller, J.M., et al., "Evidence Aggregation Networks for Fuzzy Logic Inference," *IEEE Transactions on Neural Networks*, Vol. 3, #5, 1992, pp. 761-769.

[50] Koopmans, T.C., *Three Essays on the State of Economic Science*, New York, McGraw-Hill, 1957.

[51] Krishnapuram, R. et al., "Fuzzy Connective-based Hierarchical Aggregation Networks for Decision Making," *Fuzzy Sets and Systems*, Vol. 46, 1992, pp11-27.

[52] Kruse, R., "A Note On λ-Additive Fuzzy Measures," *Fuzzy Sets and Systems*, Vol. 8, #2, 1982, pp. 219-222.

[53] Kuncheva, L.I. et al., "Fuzzy Consensus Aggregation Operator," *Fuzzy Sets and Systems*, Vol. 79, #3, 1996, pp. 347-356.

[53a] Layard, R., "On the Uses of Distributional Weights in Social Cost-Benefit Analysis," *Jour. of Political Economy*, Vol. 88, #5, Oct. 1986.

[54] Lee, E.S. et al., "Comparison of Fuzzy Numbers based on the Probability Measures of Fuzzy Events," *Computer, Math, Applications*, Vol. 15, 1988, pp. 887-896.

[55] Leontief, W.W., "A Note on the Interrelation of Subsets of Independent Variables of a Continuous Function with Continuous First Derivatives," *Bulletin of the American Math Society*, Vol. 53, 1947, pp. 343-350.

[56a] Liou, T. et al., "Fuzzy Weighted Average: An Improved Algorithm, "*Fuzzy Sets and Systems*, Vol. 49, #3, 1992, pp. 307-315.

[56b] Lusk, E.J., "Priority Assignment: a Conditioned Sets Approach," *Fuzzy Sets and Systems*, Vol. 7, #1, 1982, pp. 43-56.

[57a] May, K. O., "Intransitivity, Utility and Aggregation in Preference Patterns," *Econometrica*, Vol. 22, #1, Jan. 1954, pp.1-13.

[57b] McKelvey, R.D. et al., "Common Knowledge Consensus and Aggregate Information," *Econometrica*, Vol. 54, 1986, pp. 109-127.

[58] Meier, K., "Methods for Decision Making with Cardinal Numbers and Additive Aggregation," *Fuzzy Sets and Systems*, Vol. 88, 1997, #2, pp. 135-159.

[59] Mizumoto, M., "Pictorial Representations of Fuzzy Connectives Part I: Cases of t-norms, t-co-norms, and Averaging Operators," *Fuzzy Sets and Systems*, Vol. 31, #2, 1989, pp. 217-242.

[60a] Montero, J., "Rational Aggregation Rules," *Fuzzy Sets and Systems,* Vol.63, #3, 1994, pp. 267-276.

[60b] Narasimham, R., "A Geometric Averaging Procedure for Constructing Supertransitive Approximation to Binary Comparison Matrices," *Fuzzy Sets and Systems*, Vol. 8, #1, 1982, pp. 53-62.

[61] Nielsen, L.T. et al., "Common Knowledge of Aggregation Expectations," *Econometrica*, Vol. 58, 1990, pp. 1235-1239.

[62] Park, K.S. et al., "A Note on the Fuzzy Weighted Additive Rule," *Fuzzy Sets and Systems*, Vol. 77, #3, 1996, pp. 315-320.

[63] Paun, G., "Impossibility Theorem for Indicators Aggregation," *Fuzzy Sets and Systems*, Vol.9, #2, 1983, pp. 205-210.

[64] Peston, M.H., "A View of the Aggregation Problem," *Review of Econ. Studies*, Vol. 27, 1959-1960, pp. 58-64.

[65] Ramakrishnan et al., "The Fuzzy Weighted Addictive Rule," *Fuzzy Sets and Systems*, Vol. 46, 1992, pp. 177-187.

[66] Saaty, T.L., "A Scaling Method For Priorities in Hierarchical Structures," *Jour. Math Psychology*, Vol. 15, #3, 19, pp. 234-281.

[67a] Schenkerman, S., "Use and Abuse of Weights in Multiple Objectives Decisions Support Models," *Decision Science*, Vol. 2, 1991, pp. 368-378.

[67b] Sen, A.K., "Interpersonal Aggregation and Partial Comparability," *Econometrica*, Vol. 38, 1970, pp. 393-409.

[68a] Sen, A.K., "On Weights and Measures: Information Constraints in Social Welfare Analysis," E*conometrica*, Vol. 45, 1977, pp. 1539-1572.

[68b] Shafer, G., "The Combination of Evidence," *International Jour. Of Intelligent Systems*, Vol. 1, 1986, pp. 155-179.

[68c] Simon, H.A. et al., "Aggregation of Variables in Dynamic Systems," *Econometrica*, Vol. 29, 1961, pp. 111-138.

[69] Skala, H.J., "Concerning Ordered Weighted Averaging Aggregation Operators," *Statistical Papers*, Vol. 32, 1991, pp. 35-44.

[70] Theil, H., *Linear Aggregation of Economic Relations*, Amsterdam, North-Holland, 1954.

[71] Theil, H., "The Aggregation Implications of Identifiable Structural Macro-relations," *Econometrica,* Vol. 27, 1959, pp. 14-29.

[72] Theil, H., *Statistical Decomposition Analysis*, Amsterdam, North-Holland, 1972.

[73] Triantaphyllou, E. et al., "Development and Evaluation of Five Fuzzy Multiattribute Decision-Making Methods," *Intern. Jour. Approximate Reasoning*, Vol. 14, #4, 1996, pp. 281-310.

[74] Yager, R.R., "MAM and MOM Operators for Aggregation," *Information* Science, Vol. 69, 1993, pp. 259-273.

[75] Yager, R.R., "Aggregation Operators and Fuzzy Systems Modeling," *Fuzzy Sets and Systems*, Vol. 69, 1994, pp. 125-145.

[76] Yager, R.R., "Constrained OWA Aggregation," *Fuzzy Sets and Systems*, Vol.81, 331, 1996, pp. 125-145.

[77] Yager, R.R., "On Ordered Weighted Averaging Aggregation Operators in Multicriteria Decision Making," *IEEE Transactions System, Man, Cybernetics*, Vol. 18, 1988, pp. 183-190.

[78] Yager, R.R., "Families of OWA Operators," *Fuzzy Sets and Systems*, Vol. 59, 1993, pp. 125-145.

[79] Yager, R.R., "Applications and Extension of OWA Aggregation," *International Jour. Man-Machine Studies,* Vol. 37, 1992, pp. 103-132.

[80] Yager, R.R., "On the RAGE Aggregation Method with Application to Neural Networks and Decision Making," *Intern. Jour. of Approximate Reasoning*, Vol. 11, #3, 1994, pp. 175-204.

[81] Yager, R.R., "Aggregating Fuzzy Sets Represented by Belief Structures," *Jour. of Intelligent and Fuzzy Systems*, Vol. 1, #3 pp. 215-224.

[82] Yager, R.R., "A General Approach to Rule Aggregation in Fuzzy Logic Control," *Applied Intelligence*, Vol. 2, #4, 1992, pp. 335-353.

On Economic Analysis, Benefits and Benefit Accounting

[83] Adams, R.M., S.A. Hamilton, and B.A. McCarl, "The Benefits of Pollution Control: The Case of Ozone and U.S. Agriculture," *American Jour. of Agricultural Economics*, Vol.68, no. 4, 1986, pp. 886-893.

[84] Adams, R.M., and B.A. McCarl, "Assessing the Benefits of Alternative Ozone Standards on Agriculture: The Role on Response Information," *Jour. of Environmental Economics and Management*, Vol.12, no. 3 1985, pp. 264-276.

[85] Anderson, R.C., and B. Ostro, "Benefits Analysis and Air Quality Standard," *Natural Resources Jour.*, Vol. 3, 1982, pp. 566-575.

[86] Bayless, M., "Measuring the Benefits of Air Quality Improvement: A Hedonic Salary Approach," *Jour. of Environmental Economics and Management*, Vol. 9, no. 1, 1982, pp. 81-99.

[87] Brady, G.L., B.T. Bower, and H.A. Lakhani, "Estimates of the National Benefits and Costs of Improving Ambient Air Quality," *Jour. of Environmental Management*, Vol.16, no. 3, 1983, pp. 191-210.

[88] Brookings Institution, *Can Benefits and Costs of Public Investments be Measured*, Brookings Research, Report #32, Washington, D.C., 1965.

[89] Bulter, J.R., and D.P. Doessel, "Measuring Benefits in Health: A Clarification," *Scottish Jour. of Political Economy*, Vol. 28, no. 2, 1981, pp. 196-205

[90] Crouch E. A. et al., *Risk/Analysis*, Cambridge, M.A., Ballinger, 1982.

[91] Cummings, R.G. et al., "General Methods for Benefit Assessment" in [548], 1986, pp. 161-191.

[92] Darling, A.H., "Measuring Benefits Generated by Urban Water Parks," *Land Economics*, Vol. 49, no. 1, 1973, pp. 22-34.

[93] DeSalvo, J.S., "Benefits and Costs of New York City's Middle Income Housing Programs," *Jour. of Political Economy*, Vol. 83, no. 4, 1975, 791-806

[94] Dorfman, R. (ed), *Measuring Benefits of Government Investments*, Washington, D.C., Brookings Institution, 1965.

[95] Dunn, R.M. Jr., "A Problem of Bias in Benefits of Government-Cost Analysis: Consumer Surplus Reconsidered," *Southern Economic Jour.*, Vol. 33, #35, 1967, pp. 337-342.

[96] Evans, R.G., and G.C. Robinson, "Surplus Day Care: Measurements of the Economic Payoff," *Canadian Medical Association Jour.*, Vol. 123, no. 9, 1980, pp. 873-880.

[97] Feldstein, M.S., "Net Social Benefit Calculation and the Public Investment Decision," *Oxford Economic Papers*, Vol. 16, #1, 1964, pp. 114-131.

[98] Feltham, G.A., "The Value of Information," *Accounting Rev.*, 1968, pp. 684-696.

[99] Fisher, G.W., "Willingness-to-Pay for Probabilistic Improvements in Functional Health Status: A Psychological Perspective," in S. Mushkin and D.W. Dunlop., (eds.).*Health: What Is It Worth? Measures of Health Benefits*. New York: Pergamon Press, 1979, pp. 167-200

[100] Foster, C.D. et al., "Estimating the Social Benefit of Constructing an Underground Railway in London," *Jour. of the Royal Stat. Society*, Vol. 126, part I, 1963, pp. 46-92.

[101] Freeman III, A.M., *The Benefit of Environmental Improvement*, Baltimore, MD, John Hopkins University press, 1979.

[102] Freenbury, D.F.,et al., *Measuring Benefits of Water Pollution Abatement*, New York, Academic Press, 1980.

[103] Greenly, D.A. et al., *Economic Benefits and Improved Water Quality*, Boulder, C.O., Westview Press1982.

[104] Grerking, S. et al., "What do We Know about Benefits of Reduced Mortality from Air Pollution Control," *Amer. Econ. Review*, Vol. 71, 1981, pp. 228-240.

[105] Hammond, C.M.H., *The Benefits of Subsidized Housing Programs: An Intertemporal Approach*. Cambridge: Cambridge University Press, 1987.

[106] Harrington, W., and P.R. Portney, "Valuing the Benefits of Health and Safety Regulation," *Jour. of Urban Economics*, Vol. 22, no. 1, 1987, pp. 101-112

[107] Joycc, T.J., M. Grossman, and F. Goldman, "An Assessment of the Benefits of Air Pollution Control: The Case of Infant Health," *Jour. of Urban Economics*, Vol. 25, no.1,1989, pp.32-51.

[108] Juster, F.T., *The Distribution of Economic Well-Being*, New York, National Bureau of Economic Research, 1977.

[109] Kraft, J., and E. O. Olsen, "The Distribution of Benefits from Public Housing," in F.T, Juster (ed.),.*The Distribution of Economic Well-Being,* New York: National Bureau of Economic Research, 1977. pp. 51-69.

[110] Krutilla, J.V., "Welfare Aspects of Benefit Cost Analysis," *Jour. of Political Econ.* Vol. 69, #3, 1961, pp. 226-235.

[111] Lave, L.B., *Quantitative Risk Assessment in Regulation*, Washington, D.C., Brookings Institution, 1982.

[112] Mack, R.P., et al., "Outdoor Recreation" in [94], 1965, pp71-100.

[113] Majid, J.J. et al., "Benefits Evaluation of Increments of Existing System of Public Facilities," *Land Economics*, Vol. 59, 1983, pp.377-392.

[114] Margolis, J., "Secondary Benefits, External Economies, and the Justification of Public Investment," *Review of Economics and Statistics*, Vol. 39, #3, 1957, pp. 284-291.

[115] Mushkin, S. et al. (eds.), *Health: What Is It Worth? Measures of Health Benefits*, New York Pergamon Press, 1979.

[116] Rhoads. S.E. (ed.), *Valuing Life: Public Policy Dilemmas*, Boulder Colorado, Westview Press, 1980.

[117] Rowe, R. et al., "An Experiment on the Economic Value of Visibility," *Jour. Of Environ. Economics and Management,* Vol. 7, 1980, pp. 1-19.

[118] Samuelson, P.A., "The Pure Theory of Public Expenditure," Rev of Econ Stat., Vol.36 Nov., 1954, pp. 357-389.

[119] Schulze, W.E. et al., "The Economic Benefits of Preserving Visibility in the National Parklands of Southwest," *Natural Resources Jour.* Vol 23, 1983, pp. 149-173.

[120] Starr, C., "Social Benefit versus Technological Risk," *Science,* Vol.165, 1969, pp. 1232-1238.

[121] Vaughn, W. et al., *The Benefits of Water Pollution Control: Fresh Water Recreational Fishing*, Baltimore, MD, Johns Hopkins Univ. Press, 1982.

[122] Weisbrod, B.A., "Benefit of Manpower Programs: Theoretical and Methodological Issues," in [647], 1969, pp. 3-15.

[123] Willig, R., "Consumer's Surplus without Apology," *Amer. Econ. Review,* Vol. 66, 1976, pp. 589-597.

On Economic Analysis, Costs and Costing

[124a] Akerlof, G. et al., "The Economic Consequences of Cognitive Dissonance," *Amer. Econ. Rev.*, Vol. 72, 1981, pp. 307-319.

[124b] Alchian, A., "Costs," *International Encyclopedia of Social Sciences*, Vol. 3, New York, Macmillan and C., 1968, pp. 404-415

[125] Andell, N., "Some Notes on Equating Private and Social Costs: Comment," *Southern Economic Jour.,* Vol. 33, #1, 1966, pp. 112.

[126] Berger, M.C., et al., "Valuing Changes in Health Risks: A Comparison of Alternative Measures," *Southern Economic Jour.*, Vol. 53, no. 4, 1987, pp. 967-984.

[127] Boyer, K., "The Costs of Price Regulation: Lessons from Railroad Deregulation," *Rand Jour. of Economics*, Vol.18, no. 3, 1987, pp.408-416.

[128] Breckner, N.V. et al., "Costing of Systems" in [138] 1967, pp. 42-59.

[129] Buchanan, J., *Cost and Choice: An Inquiry in Economic Theory*, Chicago, Markham, 1969.

[130] Bunker, J.P., B.A. Barnes and F. Mosteller, (eds). *Cost, Risks and Benefits of Surgery,* New York: Oxford University Press, 1977.

[131] Canning, John B. "Some Divergence of Accounting Theory form Economic Theory," *Accounting Review*, Vol. 4, Mar. 1929, pp. 1-8.

[132] Charles, S. "The Opportunity Cost of the Sale of Local Authority Rented Accommodation: A Comment." *Urban Studies*, Vol.19, 1982, p.8.

[132b] Cheung, S. N. S., *The Myth of Social Cost: A Critique of Welfare Economics and the Implications for Public Policy*, Lacing, Sussex, U.K. Goron Pro-Print Co. 1978.

[133] Cicchetti, C.J. et al., *The Cost of Congestion*, Cambridge, Mass, Ballinger, 1976.

[134] Coase, R.H., "The Problem of Social Cost," *Jour. of Law and Econ.* Vol. 3, 1960, pp. 1-44.

[135] Collier, V.C., and H. Rempel, "The Divergence of Private from Social Costs in Rural-Urban Migration: A Case Study of Nairobi, Kenya," *Jour. of Development Studies*, Vol.13, no. 3, 1977, pp.199-216

[136] Davis, O.A., et al., "Some Notes on Equating Private and Social Cost;" *South Econ. Jour.*, Vol. 32, #2, 1965, pp. 113-126.

[137a] Devine, E. J., "The treatment of Incommensurables in a Cost-Benefit Analysis," *Land Economics*, Vol. 42. #3, Aug., 1966, pp. 383-387.

[137b] Diamond, P.A., "The Opportunity Cost of Public Investment: Comment," *Quart. Jour. of Econ.*, Vol. 82, Nov. 1968, pp.682-688.

[138] Enke, S. (ed.), *Defense Management*, Englewood Cliffs, N.J., Prentice-Hall, 1967.

[139] Fabrycky, W.J. et al. *Life-Cycle Cost and Economic Analysis,* Englewood Cliffs, Prentice-Hall, 1991.

[140] Feldstein, M.S., "Opportunity Cost Calculations in Cost-Benefit Analysis," *Public Finance*, Vol. 19, #2, 1964, pp.117-139.

[141] Fisher, G.H., "Cost Functions and Budgets" in [158], 1970, pp. 231`-264.

[142] Fisher, G. "The Role of Cost-Utility Analysis" in D. Novick (ed.), *Program Budgeting*, New York, Rinehart and Winston, 1969 pp.61-79

[143] Hahn, R.W., and J.A. Hird, "The Costs and Benefits of Regulation: Review and Synthesis," *Yale Jour. of Regulation*, Vol. 8, no.1 (1991), pp. 233-278.

[144] Harberger, A., "On Measuring the Social Opportunity cost of Public Funds," in the *Discount Rate in Public Investment Evaluation*, Rep. #17, Conference Proc. From Committee on Economics of Water Resources Dev. of Western Agric. Econ. Reach. Coun. Denver, Dec.17-19, 1968

[145a] Harberger, A. C., "On Measuring the Social Opportunity Cost of Public Funds" in [145b], 1972, pp. 94-122.

[145b] Harberger, A. C., *Project Evaluation: Collected Papers*, Chicago, Chicago Press, 1972.

[146] Hartunian, N.S., C.N. Smart, and M.S. Thompson, "The Incidence and Economic Costs of Cancer, Motor Vehicle Injuries, Coronary Heart Disease and Stroke: A Comparative Analysis," *American Jour. of Public Health.* Vol.70, no.12, 1980, pp. 1249-1260.

[147] Hazilla, M., and R. Kop, "Social Cost of Environmental Quality Regulations: A General Equilibrium Analysis," *Jour. of Political Economy*, Vol.98, no. 4, 1990, pp. 853-873.

[148] Hufbauer, G.C., and K.A. Elliott, *Measuring the Cost of Protection in the United States*. Washington, DC: Institute for International Economics, 1994.

[149] Ippolito, R.A., "The Effect of Price Regulation in the Automobile Insurance Industry," *Jour. of Law and Economic*, Vol. 23, no. 1, 1979, pp. 55-90.

[150] Judy, R.W., "Costs: Theoretical and Methodological Issues" in [167],1969, pp. 16-29

[151] Jurgensen, H., "Private and Social Costs" *German Econ. Rev.,* Vol. 2, #4, 1964, pp. 273-288.

[152] Khatkhate, D.R., "Opportunity Costs and Its Application to Unemployment," *Social Research*, Vol. 28, #1, 1961, pp. 60-70.

[153] Knight, F.H., "some Fallacies in the Interpretation of Social Cost," in Stigler, et al, *Raedings in Price Theory*, Homewood Ill, Richard Irvin, 1952, pp.160-179.

[154] Krutilla, J. et al., "The Social Cost of Federal Financing" in *Multiple Purpose River Development*, John Hopkins Univ. Press, Baltimore, 1958, pp. 78-130.

[155] Ledebur, L.C., "The Problem of Social Cost," *American Jour .of Economics and Sociology*, Vol.26, #4, 1967, pp. 399-415

[156] Marglis, S.A., " The Opportunity Costs of Public Investment," *Quarterly Jour. of Econ.*, Vol. 77, 1963, pp. 274-249

[157] McCullough, J.D., *Cost Analysis for Planning Programming-Budgeting Cost-Benefit Studies*, Santa Monica, Rand Corp., 1966.

[158] Margolis, J. (ed.) *The Analysis of Public Output*, Washington, D.C National Bureau of Econ. Research, 1970.

[159] Mayeres, I., S. Ochelen, and S. Proost, "The Marginal External Costs of Urban Transport," *Transportation Reviews-D*, Vol.1, no. 2, 1996, pp.111-130.

[160] de Melo, J., and D. Tarr, "Welfare Costs of U.S. Quotas in Textiles, Steel and Autos," *The Review of Economics and Statistics*, Vol. 72, no. 3,1990, pp.489-197.

[161] Mishan, E.J., *The Cost of Economic Growth*, New York, Praeger, 1971.

[162] Nicholas, A., "The Opportunity Costs of Public Investments; Comment," *Quarterly Jour. of Econ.* May 1964, Vol. 78, pp. 499-505.

[163] Ritter, Jay. "The Costs of Going Public," *Jour. of Financial Economics* Vol.19, 1987, 269-281.

[164] Rosen, Harvey. "The Measurement of Excess Burden with Explicit Utility Functions," *Jour. of Political Economy* Vol. 86, April 1978 pp. s121-s135.

[165] Senauer, B., J.K. Kinsey, and T. Roe, "The Cost of Inaccurate Consumer Information: The Case of EPA Gas Mileage," *Jour. of Consumer Affairs*, Vol. 18, no. 2,1984, pp.193-212.

[166] Smith, A., and A.J. Venables, "Counting the Cost of voluntary Restraints in the European Car Market," in E. Helpman and A. Razin (eds.), *International Trade and Trade Policy*, Cambridge, MA: The MIT Press, 1991, pp.187-220 .

[167] Somers, G.G. et al., (eds.) *Cost-Benefit Analysis of Manpower Policies*, Proc. of North Amer. Conf., Industrial Relations Centre, Queens Univ., Kingston, Ontario, 1969

[168] Steiner, P.G., "The Role of Alternative Cost in Project Design and Selection," *Quarterly, Jour. of Econ.* Vol. 79, #3, 1965, pp. 417-430.

[169a] Steiner, P.G., *Measuring the Opportunity Costs of Government Investment*, Research Paper P-490, Institute of Defense Analysis, Washington, D.C., 1969.

[169b] Stiler, G. J., "The Cost of Subsistence," *Jour. of Farm Economics*, Vol. 27, 1945, pp. 303-314.

[170] Stober, W. et al., "Cost Bias in Benefit-Cost Analysis: Comment," *Southern Econ. Jour.*, Vol. 34, #4, 1968, pp. 563-568.

[171] Stuart, Charles. "Welfare Costs per Dollar of Additional Tax Revenue in the U.S.," *American Economic Review* , June 1984, pp. 352-362.

[172] Tullock, Gordon. "Rent-Seeking." In the new *Palgrave: A Dictionary of Economics*, London: Macmillan, 1987, p. 149.

[173] Tullock, Gordon. "The Welfare Costs of Tariffs, Monopolies and Theft." *Western Economic Jour.* Vol. 5 1968, pp. 224-232.

[174] Turvey, R., "On Divergences between Social Cost and Private Cost," *Economica*, Vol. 30, #119, 1963, pp. 309-313.

[175] Whitbread, M., "Measuring the Costs of Noise Nuisance From Aircraft," *Jour. of Transport Economics and Policy*, Vol.12, no. 2,1978, pp. 202-208.

[176] Worchester, Jr., D.A., "Pecuniary and Technological Externality Factor and Social Costs," *Amer. Econ. Rev.*, Vol. 59, 1969 pp. 873-885.

On Accounting Theory

[177] Belkaoui, A., *Accounting Theory*, New York, Harcourt Brace and Jovanich Pub. 1985.

[178] Churchill, N.C., et al., Accounting for Affirmative Action Programs: a Stochastic-flow Approach," *The Accounting Rev*, 1975, pp. 643-656.

[179] Dean, J., *Statistical Cost Estimation* Bloomington, IN, Indiana Univ. Press, 1976.

[180] Demski, J.S., et al., *Cost Determination: A Conceptual Approach*, Iowa, Iowa, State Univ. Press, IA, 1976

[181] Dittman, D.A., et al., "On the Existence of Unrecorded Human Assets: An Economic Perspective," *J. Accounting Rev.* (1976), pp. 498-464.

[182] Dompere, K.K., *On Accounting Information and the Theory of Economic Aggregates*, Washington, DC A Working Paper, Howard University, Department of economics, 1985.

[183] Dupuch N. et al., *Cost Accounting*, New York, Harcourt Brace Jovanovich Pub. 1982.

[184] M.J. Gordon, Postulates, principles and research in accounting, *The Accounting Rev.,* 1964, pp. 251-263.

[185] K.K. Humphreys (ed.), *Project and Cost Engineer's Handbook* ,Dekker Inc., New York, 2nd ed., 1984.

[186] Johnston J., *Statistical Cost Analysis*, New York, McGraw-Hill, 1972.

[187] Juster F. T., et al. (Eds.), *Social Accounting Systems,* New York, Academic Press, 1981.

[188] Lev, B. *Accounting and information theory*: studies in research #2, Sarasota, Fl., American accounting Association, Brownell Associates Inc., USA 1969

[189] Malstrom, E.M. (Ed.), *Manufacturing Cost Engineering Handbook,* New York, Dekker, 11984.

[190] Seidler ,L.J. et al., *Social Accounting: Theory, Issues and Cases, Los Angeles*, Marville Pub. Co., 1975.

[191] R.R. Sterling, "Decision-oriented financial accounting," *Accounting Business Res*. 1972, pp. 198-208.

[192] M.C. Wells, A revolution in accounting, *The Accounting Rev*. 1976, pp. 471-482.

On Economic Analysis, Cost-Benefit, and Investment Criteria I

[193] Arrow, K.J. and M. Kurz, *Public Investment, The Rate of Return, and Optimal Fiscal Policy*, Baltimore, John Hopkins Univ. Press, 1970.

[194] Arrow, K.J. and R.C. Lind, "Uncertainty and the Evaluation of Public Investment Decisions," *Amer. Econ. Rev. June* 1970, pp. 364-378.

[195] Arrow, K.J., "Uncertainty and the Evaluation of Public Investment Decisions: Reply," *Amer. Econ. Rev*. March, 1972, pp. 171-172.

[196] Bailey, M.J., "Formal Criteria for Investment Decision," *Jour. of Political Econ*. Vol. 67, #5, Oct. 1959, pp. 476-488.

[197] Chenery, H. B., The Interdependence of Investment Decisions," in M. Abramovitz et al. (eds.), *The allocation of Economic Resources*, Stanford, CA, Stanford University Press, 1969, pp. 82-120. Reprinted in[402], 1969, pp.336-371

[198] Clarkson, K.W., *Intangible Capital and Rates of Return*, A Publication of Washington, D.C., American Enterprise Institute, 1977.

[199] Datta, B., "The Investment Criterion Problem," *Arthanitti*, Vol. 1, #2, May1958, pp. 81-92

[200] Davisson, W. "Public Investment Criteria," *Land Economics*, Vol. 40, #2, May, 1964, pp. 154-162.

[201] Dompere, K.K., *The Theory of Aggregate Investment in Closed Economic Systems*, Westport, Conn. Greenwood Press, 1999.

[202] Dompere, K.K., *The Theory of Aggregate Investment and Output Dynamics in Open Economic Systems*, Westport, Conn. Greenwood Press 1999.

[203a] Eckstein, O., "Investment Criteria for Economic Development and the Theory of Intertemporal Welfare Economics," *Quarterly Jour. of Econ.* Vol.71, Feb. 1957, pp.56-85.

[203b] Eckstein, O., *Water-resource Development*, Cambridge, Mass. Harvard Univ. Press, 1958.

[204] Feldstein, M.S. et al., "The Problem of Time-Stream Evaluation: Present Value Versus Internal Rate of Return Rules," *Bulletin of the Oxford University Institute of Economic Statistics*, Vol. 26, #1, 1964, pp. 79-85.

[205] Ferguson, C. E., *The Neoclassical Theory of Production and Distribution*, Cambridge, U.K., Cambridge University Press, 1971.

[206] Flemming, J.S. et al., "Present Value versus Internal Rate of Return: A Comment," *Economic Jour.*, Vol. 74, #294, 1964, pp. 490.

[207a] Grunfeld, Y., "The Determinants of Corporate Investment," in [207b], 1960, pp.211-266

[207b] Harberger,A. C. (ed), *Demand for Durable Goods*, Chicago, University of Chicago Press, 1960

[208] Haveman, R.H., *Water Resource Investment and Public Interest*, Nashville,Vanderbilt University Press, 1955

[209] Hirschleifer, J., "Investment Decision Under Uncertainty: Choice- Theoretic Approach," *Quarterly Jour. of Economics*, Nov. 1965, pp. 509-536.

[210] Hirschleifer, J., "Investment Decision under Uncertainty: Applications of the State-Preference Approach," *Quarterly Jour. of Economics,* Vol. 80, May 1966, pp. 252-277.

[211] Hirschleifer, J., "On the Theory of Optimal Investment Decision," *Jour. of Political Econ.* Vol. 66, #4, Aug 1958, pp. 329-352.

[212] Horvat, B., "The Optimum Rate of Investment Reconsidered," *Economic Jour.*, Vol. 75, #299, 1965, pp. 572-576.

[213] Jorgenson, D.W., “Theory of Investment Behavior,” in R. Ferber (ed.) *Determinants of Investment Behavior*, NBER Research, New York, Columbia University Press, 1967, pp. 129-155

[214] Lutz, F.A., *The Theory of Investment of the Firm*, Greenwood Press, New York, 1951

[215] Lutz, F.A. et al (eds.)., *The Theory of Capital*, Proceedings- International Economic Association, Macmillan. New York, 1963

[216a] Mainwaring, L., *Value and Distribution in Capitalist Economies*, Cambridge, U.K., Cambridge Univ. Press, 1984.

[216b] Marglin, S.A., *Public Investment Criteria: Benefit-Cost Analysis for Planned Economic Growth*, Cambridge, Mass, MIT Press, 1967.

[217] Marglin, S.A., *Approaches to Dynamic Investment Planning*, Amsterdam, North- Holland, 1963.

[218] Meek, R.L., “Ideal and Reality in the Choice Between Alternative Techniques,” *Oxford Economic Papers*, Vol. 16, #3, 1964, pp. 333-354.

[219] Merrett, A. et al., “Calculating the Rate of Return on Capital Projects,” *Jour. of Industrial Economics*, Vol. 9, #1, 1960, pp. 99-115.

[220] Mills, G., “ The Marginal Efficiency of Capital and Present Value Rule,” *Yorkshire Bulletin of Economic and Social Research*, Vol. 12, #1, 1960, pp. 28-31.

[221] Mishan, E.J., “Criteria for Public Investment: Some Simplifying Suggestions,” *Jour. of Political Economy*, Vol. 75, #2, April 1967, pp. 139-146

[222] Mishan, E.J., “A Proposed Normalization Procedure for Public Investment Criteria,” *Economic Jour.*, Vol. 77, Dec. 1967, pp. 777-796.

[223] Mishan, E.J., “Criteria for Public Investment: A Reply,” *Jour. of Political Econ.,* Vol.78., #1, Jan-Feb, 1970, pp. 178-180.

[224] Mishan, E.J., “Uncertainty and the Evaluation of Public Investment Decisions: Comment,” *Amer. Econ. Rev.,*Vol.62, March 1972, pp.161-164.

[225] Nickell, S.J., “On the Role of Expectations in the Pure Theory of Investment,” *Rev. Econ. Stud.*, Vol. 41, 1974, pp. 1-20.

[226] Nickell, S.J., *On Expectations, Government Policy and the Rate of investment*, Cambridge, Cambridge University Press, 1978.

[227] Nickell, S.J., *The Investment Decisions of Firms*, Cambridge, Cambridge University Press, 1978.

[228] Pegels, C. et al., " A Comparison of Decision Criteria for Capital Investment Decision," *Engineering Economist*, Vol. 13, #4, Summer 1968, pp. 211-220.

[229] Pitchford, J.D. et al., "A Note on the Marginal Efficiency of Capital," *Economic Jour.,* Vol. 68, #271, Sept. 1958, pp. 597-600.

[230] Reiter, S., "Choosing an Investment Program Among Interdependent Projects," *Rev. of Econ. Stud.* Vol. 30, Feb, 1963, pp. 32-36.

[231] Solow, R.M., *Capital Theory and the Rate of Return*, New York, North-Holland, 1963.

[232] Sundrum, R.M., "The Evaluation of Time Streams-An aspect of the Appraisal of Capital Projects," *Malayan Economic Rev.* Vol. 10, #2, Oct. 1965, pp. 7-34.

[233] Turvey, R., "Present Value Versus Internal of Return-An Essay in the Theory of the Third Best," *Economic Jour.*, Vol. 73, #289, March, 1963, pp. 93-98

On Economic Analysis, Cost-Benefit and Investment Criteria II

[234] Albegov, M.M. et al., "Optimization of the Location of a system of Industrial Complexes," *MATEKON,* Winter, 1971-72, pp. 132-143.

[235] Anchishkin, A.I., "On the Optimal Distribution of the Final Social Product," *MATEKON*, Summer 1971, pp. 331-369.

[236] Arzamastve, D.A., "Qualitative Analysis of the Relative Effectiveness of Capital Investment," *MATEKON*, Fall 1970, pp. 3-25.

[237] Chuev, I.V. et al., "The Generalized Equipment Replacement Problem, *MATEKON*, Fall 1969, pp. 75-91.

[238] Fedorovskii, Iu. P. "The Time Factor in the National Economic Optimality Criterion," *MATEKON,* Spring, 1971. pp. 281-299.

[239] Gladyshevskii, A.I. et al., "Microeconomic Calculations of Distributed Lags in Capital Construction," *MATEKON*, Spring, 1978, pp.58-79.

[240] Indin, D.B. et al., "Evaluation of the Information Complexity of Mathematical Program of the Mathematical Programming Programs," *MATEKON*, Winter 1976-77, pp. 3-25.

[241] Kantorovich, L.V., *The Best Use of Economic Resources*, Cambridge, Mass., Harvard University Press, Mass. 1965

[242] Katsenelinbogen, A.I. et al., "Capital Changes in Optimal Economic Models," *MATEKON*, Summer 1971, pp. 370-390.

[243] Khachturov, V.R., et al., "Dynamic Location Problem," *MATEKON*, Winter 1976-77 pp. 68-93.

[244] Liberman, Ia. G., "Intensity of Investment and the Modeling Economic Growth," *MATEKON*, Winter, 1970-71, pp. 91-127.

[245] Mikhalevskii, B.N., "Mechanical-Organic Economic Models and Models of an Open Multistage Dynamic System," *MATEKON,* Fall 1972, pp.3-32

[246] Nemeti, L., "On the Mathematical Treatment of Certain Problems, in Planning Socialist Macroeconomic Development," *Mathematical Studies*, Vol. IV, #2, pp. 28-45

[247] Presman, L.S., "A Model of Industrial Location," *MATEKON*, Spring, 1973, pp. 6-21.

[248] Pugacheve, V.F. et al., "An Interindustry Model in Multistage System of Optimal Long-Range Planning," *MATEKON*, Fall, 1975, pp. 78-99

[249] Temurdzhian, A.V., "On the Question of the Normative Coefficient of Effectiveness," *MATEKON*, Summer, 1972, pp. 355-363.

[250] Zauberman, A., *Mathematical Theory in Soviet Planning*, Oxford Univ. Press, New York, 1976.

[251] Zef'tsburg, L.M., "Precision and Reliability in Engineering Economic Calculations in the Selection of Capital Investment Alternatives," *MATEKON*, Spring, 1974, pp. 25-35.

[252] Zhdanko, A.V., "Modeling the Dynamics of Fixed Capital with Variable Service Life," *MATEKON*, Fall, 1972, pp. 75-100

On Economic Analysis, Shadow Prices and Efficient Pricing

[253] Alexandre, A., J. Barde, and D.W. Pearce, "Practical Determination of a Charge for Noise Pollution," *Jour. of Transport Economics and Policy*, Vol. 14, no. 2, 1980, pp. 205-220.

[254] Bacha, E et al., "Foreign Exchange Shadow Prices: A Critical Review of Currency Theories," *Quarterly Jour. of Econ.* Vol. LXXXV, May 1971, pp. 197-224

[255] W.J. Baumol, "The Role of Cost in the Minimum Pricing of Railroad Service," *Jour. of Business*, Vol. 35, 1961, pp. 357-366.

[256] Baumol, W.J et al., "Optimal Departures from Marginal Cost Pricing," *Amer. Econ. Rev.* Vol. 60, 1970, pp. 265-283.

[257a] Baumol, W.J. et al., "Optimal Depreciation Policy: Pricing the Products of Durable Assets," *Bell Jour. of Economics and Management Science,* Vol. 2, Autum 1971, pp. 365-376.

[257b] Baumol, W. J., "Scale Economics, Average Cost, and Profitability of Marginal-Cost Pricing," in R. E. Grieson 9ed.*), Essays in Urban Economics and Public Finance in Honor of William S. Vickrey,* Lexington, Mass, D. C. heath, 1975, pp. 43-57.

[258a] Beckwith, B. P., *Marginal-Cost Price-Output Control: A Critical History and restatement of the Theory*, New York, Columbia Univ. Press, 1955.

[258b] Bell, C. et al., "Shadow prices and semi-input: a critical note," *Oxford Bull. Econ. Stat.* XLII(42), 1980, pp. 251-256.

[259] Bell, C. et al., "Shadow Pricing for Project Evaluation under Alternative Macroeconomic Specifications", *Quart. J. Econ.* Vol. XCVIII(98) ,1983, pp. 457-477.

260] A. Bergson, "Optimal pricing for a public enterprise", *Quart. J. Econ.*, Vol. 86 , 1972, pp. 519-544.

[261] S.E. Berki, "The pricing of hospital services," in [317], 1972. pp. 351-370.

[262] Bhagwati, J. et al., "Value Subtracted, Negative Shadow Prices in Project Evaluation and Immiserising Growth: Three Paradoxes in the Presence of

Trade Distortions," *Econ. Jour.*, Vol. LXXXVIII(88), March, 1978, pp. 121-125.

[263] Brown, R.A., and C.H. Green, "Threats to Health and Safety: Perceived Risk and Willingness-to-Pay," *Social Science and Medicine*, Vol.15, no2, part D, 1981, pp. 67-75.

[264] Buchanan, J.M., "A Public Choice Approach to Public Utility Pricing," *Public Choice*, Vol. 5, Fall, 1968. pp. 1-17

[265] Bos, D., "Cost of Living Indices and Public Prices," *Economica*, Vol. 45, 1978, pp. 59-69.

[266] Carter, R.L., "Pricing and the Risk of Fire," in *Essays in the Theory and-Practice of Pricing*, London, 1967, The Inst. of Econ. Affairs, pp. 25-53.

[267] Cauley, S.D., "The Time Price of Medical Care," *Review of Economics and Statistics*, Vol. 69, no. 1, 1987, pp.59-66.

[268a] Chernickovsky, D., and I. Zmora, "A Hedonic Prices Approach to Hospitalization Costs: The Case of Israel," *Jour. of Health Economics*, Vol. 5, no. 2, 1986,, pp.179-191.

[268b] Cinnhetti, C.J., *The Marginal Cost and Pricing of Electricity: An Applied Approach*, Cambridge Mass. Ballinger Pub., 1977.

[269] Clarke, E.H., "Multipart Pricing of Public Goods," *Public Choice*, Vol. 2, 1971, pp. 17-33.

[270] Clarke, E.H., "Multipart Pricing of Public Goods: An Example," in [317], 1972, pp. 125-130.

[270b] Coase, R.H., "The Marginal Cost Controversy," *Economica,* (New Series), Vol. 13, #51, Aug. 1946, pp. 169-182.

[271] Csiko's-Nagy, B., *Towards a New Price Revolution*, Budapest, Akademiai Kiado, 1979.

[272] Debreu, G., *Theory of Value*, New York, Wiley, 1959.

[273] Diewert, W.E., "Cost-benefit analysis and project evaluation," *J. Public Econ.*, Vol. 22, 1983, pp. 265-302.

[274] W.E. Diewert, "The measurement of deadweight loss revisited," *Econometrica*, Vol. 49 ,1981, pp.1225-1244.

[275] Dompere, K.K., "The Inventory Control Problem and the Optimal Supply Price: Microtheoretic," *Eng. Costs Prod. Econ.*, Vol. 20, 1990, pp. 51-64.

[276] Dompere, K.K., "Technological Progress and Optimal Supply Price," *Inter. J. Prod. Econ.,* Vol. 32 1993 pp.365-381

[277] Dompere, K.K. and K.O. Nti, "Learning by Doing and Optimal Factor Demand," *Inter. J. Prod. Econ.*, Vol. 22, 1992, pp. 67-79.

[278] Dorfman, R. et al., *Linear Programming and Economic Analysis*, New York, McGraw-Hill, 1958

[279a] Dorward, N. *The Pricing Decision: Economic Theory and Business Practice.* New York, Harper and Row, 1987.

[279b] Duba, J.G., "Urban Transport..A Survey of Some Pricing Aspects,"in [317], 1972, pp. 267-282.

[280] Dunn, R.M., Jr., "Problems of Bias in Input Pricing for Benefit- Cost Analysis: A Reply," *Southern Econ. Jour.*, Vol. 34, #4, April 1968, pp.571.

[281] Farber, S., and A. Rambaldi, "Willingness to Pay for Air Quality: The Case of Outdoor Exercise," *Contemporary Policy Issues*, Vol. 11, no. 4 1993, pp. 19-30.

[282] Feldman, S. L. "Waste Collection Services: A Survey of Costs and Pricing," in [317] 1972, pp. 217-242.

[283] Feldstein, M.S., "Equity and Efficiency in Public Sector Pricing: The Optimal two-part Tariff," *Quarterly Jour. of Econ.*, Vol. 86, 1972, pp. 175-187.

[284] Findley, R., et al., "Project Evaluation, Shadow Prices and Trade Policy," *Jour. of Polit. Econ.* Vol. 84, 1972, pp. 175-187.

[285] Foster, E., et al., "Pricing Distortion and Economic Welfare," *Econometrica*, Vol. 38, 1970, pp. 281-297.

[286] Freixas, X. et al., "Average Cost Pricing versus Marginal Cost Pricing under Moral Hazard," *Jour. Pub. Econ.* Vol. 26 ,1985, pp.135-146.

[287] Gaffney, M., "The Valuation of Public Good-Discussion," in M.E. Garnsey et al., (Eds.),*Social Sciences and the Environment* Boulder, CO., University of Colorado Press, 1967, pp. 154-160

[288a] Gray, Clive. "On Measuring the Shadow Price of Uncommitted Fiscal Resources in Africa," *World Development*, Vol. 17 ,Feb. 1989, pp. 213-221

[288b] Griliches, Z. (eds.), *Price Index and Quality charge*, Cambridge Mass., Harvard Univ. Press, 1971.

[289] Gomory, R.E. et al., "Interger Programming and Pricing," *Econometrica*, Vol. 28, #3, July 1960, pp. 521-550.

[290] Grossman, S.J., et al., "Information and Competitive Price System," *Amer. Econ. Rev., Proceeding*, Vol. 66, 1976, pp. 246-253.

[291] Guesnerie, R., "Second best pricing rules in the Boiteux tradition," *J. Pub. Econ.*, Vol. 13, 1980, pp. 51-80.

[292] Hanke, S.H., "Some Behavioral Characteristics Associated with Residential Water Price Changes," *Water Resources Research*, Vol. 6, 1970, pp. 1383-1386.

[293] Hanke, S.H., "Pricing Urban Water," in [317], 1972, pp. 283-306.

[294] Harrison, D., Jr., and D.L. Rubinfeld, "Hedonic Housing Prices and the Demand for Clean Air," *Jour. of Environmental Management*, Vol. 5, no. 4, 1978, pp. 81-102

[295] Heckman, J.J., "Shadow Prices, Market Wages and Labor Supply," *Econometrica,* Vol. 42, 1974, pp. 679-694.

[296] Hirschleifer, J., "Peak Loads and Efficient Pricing: Comment," *Quarterly Jour. of Econ.*, Vol. 72, 1958, pp. 451-468.

[297] Ise, J., "The Theory of Value as Applied to Natural Resources," *Amer. Econ. Rev.*, Vol. 15, June 1925, pp. 284-291.

[298] Johnson, J.A., "The Distribution of the Burden of Sewer User Charges Under Various Formulas," *National Tax Jour.*, Vol. 22, Dec. 1969, pp. 472-485.

[299] Katzman, M.T., "Pricing Primary and Secondary Education," in [317], 1972, pp. 371-394.

[300] Kitchen, J.W. et al., "Land Values Adjacent to an Urban Neighborhood Park," *Land Economics*, Vol. 43, Aug. 1967, pp. 357-361.

[301] Knetsch, J.L., "The Influence of Reservoir Projects on Land Values," *Jour. of Farm Economics*, Vol. 46, Feb. 1964, pp. 231-243.

[302] Koopmans, T. C. ,*Three Essays on the State of Economic Science* , New York, McGraw-Hill, 1957.

[303] Koshmura, S., *Theory of Capital Reproduction and Accumulation*, DPG Pub. Co. Kitcherer, Ont., Canada, 1975.

[304] Langholm, O., *Full Cost and Optimal Price: A Study in the Dynamics of Multiple Production*, Oslo, Universities of Laget, 1969.

[305] Layard, R (ed.), *Cost-Benefit Analysis,* Penguin Press, Baltimore Md., 1972.

[306] Little, I.M.D.et al.,"the Use of World Price," in [305] 1972, pp. 60-172.

[307] Margolis, J., "Shadow Prices for Incorrect or Nonexistent Market Values," in *the Analysis and Evaluation of Public Expenditures: The PP.B. System*,Washington, D.C., Joint Econ. Committee, Vol. 1 (91st Congress 1st sess. 1969.) pp. 533-54

[307b] Margolis, J., 'Welfare Criteria, Pricing and Decentralization of Public Services," *Quarterly Jour. of Econ.*, Vol. 71, #3, Aug. 1957, pp.448-463.

[308] Marx, K., *Value, Price and Profit*, London, Allen and Unwin, 1950.

[309] Mashayekhi, Afsaneh. *Shadow Prices for Project Appraisal in Turkey.* Washington, D. C., World Bank Staff Working Paper No. 392, May 1980.

[310] McConnell, K.E., "Congestion and Willingness to Pay: A Case Study of Beach Use," *Land Economics*, Vol. 53, no. 2, 1977, pp. 185-195.

[311] McKean, R.N., "The Use of Shadow Prices," in [306], 1972, pp. 119-139

[312] Menz, F.C., and J.K. Mullin, "Expected Encounters and Willingness to Pay for Outdoor Recreation," *Land Economics*, Vol. 57, no.1, 1981, pp.33-40.

[313] Mishan, E.J., "Evaluation of Life and Limb: A Theoretical Approach," *The Jour. of Political Econ.,* Vol. 79, #4, 1971, pp. 687-705.

[314] Millman, J.W., "Beneficiary Charges-Toward a Unified Theory," in [317], 1972, pp. 27-52.

[315] Morishing, M. et al., *Value, Exploitation and Growth*, New York, McGraw-Hill, 1978.

[316] Moss, P.I., "Pricing Recreation Services," in [317], 1972, pp. 335-350

[317] Mushkin, S., (Eds.) *Public Prices for Public Products*, Washington, D.C., The Urban Institute, 1972.

[318] Mushkin, S. et al., "Public Prices: An Overview," in [317], 1972, pp. 3-26.

[319] Nelson, J.P., "Residential Choice, Hedonic Prices, and the Demand for Urban Air Quality," *Jour. of Urban Economics*, Vol. 5, no. 3, 1978, pp.357-369.

[320] Nelson, J.R., (Eds.) *Marginal Cost Pricing in Practice*, Englewood Cliffs, Prentice-Hall, 1964.

[321] Nordhaus, W.D. "Recent developments in price dynamics," in:[98], pp. 16-46.

[322] Paul, E.S., "Pricing Rules and Efficiency," in [317], pp. 73-95, 1972.

[323] Pollack, W., "Pricing Fire Protection Services," in [317], pp. 307-334

[324] Qayum, A., *Theory and Practice eof Accounting Prices*. New York, North- Holland, 1960

[325] Reese, R., "Second Best Rules for Public Enterprise Pricing," *Economica*, Vol. 35, pp. 260-273.

[326] Reese, R., *Public Enterprise Economics*, London, Weidenfeld and Nix-holson, 1976.

[327] Richer, J., "Willingness to Pay for Desert Protection," *Contemporary Economic Policy*, Vol.13, no. 4, 1995, pp. 93-104.

[328] Roberts, K.W.S., "Welfare Considerations of Nonlinear pricing," *Econ. Jour.*, Vol. 89, 1979, pp. 66-83.

[329] Roncaglia, A., *Sraffa and the Theory of Prices*, New York, Wiley and Sons, 1978.

[330] Rosenthal, D.H., and T.C. Brown, "Comparability of Market Prices and Consumer Surplus for Resource Allocation Decisions," *Jour. of Forestry*,Vol. 83, no. 2, 1985, pp.105-109.

[331] Ruggles, N., "The Welfare Basis of Marginal-Cost Pricing Principle," *Rev. of Econ. Studies*, Vol. 17, 1949, pp. 24-46.

[332] Ruggles, N., Recent Developments in the Theory of Marginal Cost-Pricing," *Rev. of Econ. Studies*, Vol. 17, 1949-56, pp. 107-126.

[333] Scarf, H.E., "The computation of equilibrium prices," in H.E. Scarf et al., (eds)., *Applied General Equilibrium Analysis*, New York, Cambridge Univ. Press, 1984, pp. 1-49.

[334] Sen, A.K., "Feasibility Constraint: Foreign Exchange and Shadow Wages," in [305], 1972, pp. 140-159.

[335] Sen, A.K., "Control Areas and Accounting Prices: An Approach to Economic Evaluation," *Econ. Jour.*, Vol. 82, 1972, pp. 486-501.

[336]Sherman, R. "Pricing Behavior of the Budget-Constrained Public Enterprise," *Jour. of Economic Behavior and Organization*, Vol. 4, 1983, pp. 381-393

[337] Shubik, M., "Voting or a Price System in a Competitive Market Structure," *Amer. Polit. Science Rev.* Vol. 64, 1970, pp. 179-181.

[338] Smolensky, E. et al.,"Waiting time as a congestion Charges," in [317], 1972, pp. 95-104.

[339] Squire, Lyn, I.M. D. Little, and M. Durdag, *Application of Shadow Pricing to Country Economic Analysis with an Illustration.* Washington, D.C., World Bank Staff Working Paper, no. 330, June 1979.

[340] Sraffa, P., *Production of Commodities by Means of Commodities. Prelude to Critique of Economic Theory*, Cambridge, England, Cambridge University Press, 1960.

[341] Steiner, P.O., "Peak Loads and Efficient Pricing," *Quarterly Jour. of Econ.*, Vol. 71, 1957, pp. 584-568.

[342] Steiner, P.O., "Peak Loads and Efficient Pricing: Reply," *Quarterly Jour. of Econ.*, vol. 72, Aug. 1958, pp. 465-468.

[343] Tandon, Pankaj, "A Note on Optimal Pricing of Publicly Produced Intermediate Inputs," *Atlantic Economic Jour.*, Vol. 17, Sept. 1989, pp. 43-46.

[344] The Board of Governors of Fed. Res. System and Social Science Research Council, *Econometrics of Price Determination Conference*; Oct. 30-31, 1970, Washington, D.C., The Board of Governors of Federal Reserve Bank System1972.

[345] Turvey, R. "How to Judge When Price Change Will Improve Resource Allocation," *Economic Jour.*, Vol. 84, 1974, pp. 825-832.

[346] Turvey, R., "Marginal Cost Pricing," *Economic Jour. Vol.* 79, June 1969, pp. 282-294.

[347] Vickery, W.S., "Economic Efficiency and Pricing," in [317], 1972, pp. 53-72

[348a] Vickrey, W.S., "Some Implications of Marginal Cost Pricing for Public Utilities," *Amer. Econ. Rev.*, vol. 45, May 1955, pp. 605-620.

[348b] Vickrey, W. S., "Some objections to Marginal Cost Pricing," *Jour. of Political Econ.*, Vol 56, #3, June 1948, pp218-238.

[349a] Willcox, M.C. et al., "Public Pricing and Family Income: Problems of Eligibility Standards," in [317], 1972 pp. 395-406.

[349b] Williams, T.S., *Pricing, Estimating, and Budgeting*, Cincinnati, Ohio, North Light Books, 1996

[350] Williamson, O., "Peak-Load Pricing and Optimal Capacity Under Indivisibility Constraint," *Amer. Econ. Rev.*, Vol. 56, Sept. 1966, pp. 810-827.

[351] Wohl, M., congestion Toll Pricing for Public Transport Facilities," in [317], 1972, pp. 243-266.

[352] Wohl, M., "The Short-Run Congestion Cost and Pricing Dilemma," *Traffic Quarterly*, Vol. 20, Jan. 1966.pp. 48-70.

[353] A. Zauberman, *Mathematical Theory in soviet Planning*, New York, Oxford Univ. Press, , 1976.

[354] Zettel, R. et al., *The Basic Theory of Efficiency Tolls: The Tolled, The Tolled-Off and Untolled*, Highway Research Board Record #47, 1964 .

On Economic Analysis and Social Rate of Discount

[355] Alchian, A.A., "The Rate of Interest, Fisher's Rate of Return Over Costs and Keynes Internal Rate of Returns," *Amer. Econ. Rev.*, Vol. 45, #5, 1995, pp. 938-943.

[356] Arrow, K.J., "Discounting and Public Investment Criteria," in [371], 1966, pp. 13-32.

[357a] Arrow, K.J., "The Social Discount Rate," in [167], 1969, pp. 56-75.

[357b] Arrow, K. J. et al., *Public Investment, and Optimal Fiscal Policy*, Baltimore, MD, The Johns Hopkins Press, 1970.

[358] Baumal, W.J., "On the Social Rate of Discount," *Amer. Econ. Review.* Vol.58, #4, 1968, pp. 788-802.

[359] Baumal, W.J., "On the Social Rate of Discount: Comment on the Comments," *Amer. Econ. Rev.*, Vol. 59, 1969, pp. 930.

[360] Bradford, D.F., "Constraints on Government Investment Opportunities and the Choice of Discount Rate," *Amer. Econ. Rev.*, Vol. 5, 1975, pp. 887-895

[361] Broussalian, V.L., *On Discounting and Risk in Military Decisions,* Washington, DC, Center for Naval Analysis, Aug. 1966.

[362] Denison, E.F., "Capital Theory and the Rate of Return," *Amer. Econ. Rev*Vol.54, #5, 1964, pp. 721-725.

[363] Dreze, J., "Discount Rates and Public Investment: A Post-Scriptum," *Economica,* 1974, pp. 52-61.

[364] Felstein, M.S., "The Social Time Preference Discount Rate in Cost-Benefit Analysis," *Economic Jour.*, Vol. 74, 1964. pp. 360-379.

[365] Feldstein, M.S., "The Derivation of Social Time Preference Rates," *Kyklos,* Vol. 18, 1965, pp. 277-287.

[366] Fisher, A.C. et al., "Resource Conservation Environmental Preservation and the Rate of Discount," *Quarterly Jour. of Econ.*, Vol. 89, Aug. 1975, pp. 358-369.

[367] Fisher, I., *The Rate of Interest*, New York, Macmillan Publishing Co. 1907.

[368] Hann, F.H. et al., (eds.) *The Theory of Interest Rates*, Proceedings of Conference Intern. Econ. Assoc., New York, Macmillan, 1965.

[369] Harberger, A.C., "Professor Arrow on the Social Discount Rate," in [167] 1969, pp. 76-88.

[370] Jean, W., "On Multiple Rates of Returns," *Jour. Of Finance*, Vol. 23, #1, 1968, pp. 187-191

[371] Kneese et al., (eds), *Water Resource Research*, Baltimore, John Hopkins, University Press, 1966.

[372] Knight, F.H. "Professor Fisher's Interest Theory: A Case in Point," *Jour. of Political Econ.* April 1931, pp. 176-212.

[373] Kubinski, F.M., et al., "Marginal Efficiency of Capital and Effect of Discounting Period," *Economic Records*, Vol. 36, Aug. 1960, pp. 424-425.

[374] Lee, D.R. et al., "The Private Discount Rate and Resource Conservation," *Canadian Jour. of Econ.,* Vol.8, #3, Aug. 1975, pp.351-363

[375] Lind, R.C., "The Social Rate of Discount and the Optimal Rate of Investment: Further Comment," *Quarterly Jour. of Economics*, Vol. 78, #2, May 1964, pp. 336-345

[376] Lutz, F.A., *The Theory of Interest*, Chicago, Aldine Pub. Co., 1968.

[377] Marglin, S.A., "The Social Rate of Discount and Optimal Rate of Investment," *Quarterly Jour. of Econ.* Feb. 1963, pp. 95-111.

[378] Merrett, A. et al., "Calculating the Rate of Return on Capital Projects," *Jour. of Industrial Econ.* Vol. 9, #1, 1960, pp. 98-115.

[379] Mikesell, R.E., *The Rate of Discount For Evaluating Public Projects*, Washington, D.C. American Enterprise Institute for Public Policy Research, 1977.

[380] Miller, W.L., "The Magnitude of the Discount Rate for Government Projects," *Southern Economic Jour.*, Vol. 28, #4, 1962, pp. 348-356

[381] Nichols, A., "On the Social Rate of Discount: Comment," *Amer. Econ. Rev.*, Dec. 1969, pp. 909-911.

[382] Reuber, G.L., "The Social Discount Rate: Discussion," in [167], 1969, pp. 88-94.

[383] Robichek, A. et al., "Conceptual Problems in the Use of Risk-Adjusted Discount Rates," *Jour. of Finance*, Vol. 21, #4, 1966, pp.727-730.

[384] Robinson, R., "The Rate of Interest, Fisher's Rate of Return Over Costs and Keynes Internal Rate of Return: Comment," *Amer. Econ. Rev.* Vol. 46, #5, Dec. 1956, pp. 972-973.

[385] Sandmo, A. et al., "Discount Rates for Public Investment in Closed and Open Economies," *Economica,* Vol. 39, Nov. 1971, pp. 395-412.

[386] Sandmo, A., "Discount Rates for Public Investment under Uncertainty," *Intern. Econ. Rev.*, Vol.13.,June 1972, pp. 287-302.

[387] Seagraves, J.A., "More on the Social Rate of Discount," *Quarterly Jour. of Econ.* Vol. 84, Aug. 1970, pp. 430-450

[388] Sen, A.K., "Optimizing the Rate of Saving," *Econ. Jour.*, Vol. 71, #283, Sept. 1961, pp. 479-496

[389] Sen, A.K., "Isolation, Assurance and the Social Rate of Discount," *Quarterly Jour. of Econ.* Vol. 81, #1, Feb. 1967, pp. 112-124.

[390] Soloman, E., "The Arithmetic of Capital-Budgeting Decisions," *Jour. of Business*, Vol. 29, #2, April 1956, pp. 124-129.

[391] Somers, H.M., "On the Demise of the Social Rate," *Jour. of Finance*, May, 1974, pp. 565-578.

[392] Sommers, G.G. et al. (eds), *Cost-benefit analysis of manpower policies*, Proc. of North Amer. Conf. Industrial Relations Center, Kingston, Ontario, Queens University, 1969.

[393] Tullock, G., "The Social Rate of Discount and the Optimal Rate of Investment: Comment," *Quarterly Jour. of Econ.* Vol. 78, #2, May 1964, pp. 331-336.

[394] U.S. Congress, Joint Economic Committee, Subcommittee on Economy's Government; *Hearing on Economic Analysis Public Investment Decision: Interest Rate Policy and Discounting Analysis*, 84th Congress, 2nd Session, Washington D.C.,1968.

[395] Usher, D., "The Social Rate of Discount and the Optimal Rate of Investment: Comment, *Quarterly Jour. of Econ.* Vol. 78, Nov. 1964. pp.641-644.

On Economics and Social Choice

[396] Aaron, H.J. et al., "Efficiency and Equity in the Optimal Supply of a Public Good," *Rev. of Econ. Stat.,* Vol.51, 1969, pp. 31-39.

[397] Aaron, H.J. et al., "Public Goods and Income Distribution," *Econometrica,* Vol. 38, 1970, pp. 907-920.

[398] Archibald, G.C., "Welfare Economics Ethics and Essentialism," Economica, (New Series), Vol. 26, 1959, pp. 316-327.

[399a] Arrow, K.J., *Social Choice and Individual Values*, Wiley, New York, 1951

[399b] Arrow, K.J., *Behavior under Uncertainty and Its Implication for* Policy, Technical Report #399, Center for Research on Organizational Efficiency , Stanford, Stanford University. 1983,

[400] Arrow. K. J., "A Difficulty in the Concept of Social Welfare," *Jour. of Political Econ.*, Vol. 58, 1950, pp. 328-246.

[401] Arrow, K.J., "Equality in Public Expenditure," *Quarterly Jour. of Econ.* Vol. 85, 1971, pp. 409-415.

[402] Arrow, K.J. et al. (eds.), *Readings in Welfare Economics*, Homewood, Irwin, 1969

[403] Atkinson, A.B. et al., *Lecture on Public Economics*, McGraw-Hill, New York, 1980.

[404] Atkinson, A.B., "Optimal Taxation and the Direct Versus Indirect Tax Controversy," *Canadian Jour. of Economics*, Vol. 10, 1977, pp. 590-606

[405] Atkinson, A. B. "Optimal Taxation and Direct versus Indirect Tax Controversy," *Canadian Jour. of Economics*, Vol. 10, #4, Nov. 1977, pp. 590-606.

[406] Atkinson, A.B. et al., "The Structure of Indirect Taxation and Economic Efficiency," *Jour. of Public Economics*, Vol. 1, 1972, pp. 97-119.

[407] Atkinson, A.B., "The Design of Tax Structure: Direct Versus Indirect Taxation," *Jour. of Public Econ.* Vol. 6, 1977, pp. 55-75.

[408] Axelrod, R., *Conflict of Interest*, Chicago, Markham, 1970.

[409] Balassa, B. et al., *Economic Progress, Private Values and Public Policy*, New York, North-Holland, 1977.

[410] Barr, N.A. et al., *Self-Assessment for Income Tax*, London, Heinemann, 1977.

[411] Baumol, W.J., *Welfare Economics and the Theory of the State,* Cambridge, Harvard University Press, 1967.

[412a] Becker, G.S., "A Theory of the Allocation of Time," *Economic Jour.*, Vol. 75, 1965, pp. 493-517.

[412b] Becker, G. S., "A Theory of Competition among pressure Groups for Political Influence," *Quarterly Jour. of Econ.*, Vol.97 (XCVII), 1983, pp. 371-400

[413] Becker, G.S., "A Theory of Social Interactions," *Jour. of Political Economy*, Vol. 82, 1974, pp. 1063-1093.

[414] Bentley, A.F., *The Process of Government*, Chicago, University of Chicago Press, 1907.

[415] Bergson, A., "A Reformation of Certain Aspects of Welfare Economics," *Quart. Jour. of Econ.* Vol. 52, Feb 1938, pp. 314-344.

[416] Bergson, A., "On the Concept of Social Welfare," *Quarterly Jour. of Econ.*, Vol. 68, May 1954, pp. 233-253

[417] Bernholz, P., "Economic Policies in a Democracy," *Kyklos*, Vol. 19, fasc. 1, 1966, pp. 48-80.

[418a] Black, D., "On the Rationale of Group Decision Making," *Jour. of Political Economy*, Vol. 56, Feb. 1948, pp. 23-34.

[418b] Black, D., *The Theory of Committees and Elections*, Cambridge, Cambridge Univ. Press, 1958.

[419] Blackorby, C. et al., "Utility vs. Equity," *Jour. Public Economics*, Vol. 7, 1977, pp. 365-381

[420] Blau, J.H. et al., "Social Decision Functions and the Veto," *Econometrica*, Vol. 45, May 1977, pp. 871-879.

[421] Bowen, H.R., "The Interpretation of Voting in the Allocation of Economic Resources," *Quart. Jour. of Econ.* Vol. 58, 1943, pp. 27-48.

[422] Breton, A., *The Economic Theory of Representative Government*, Chicago, Aldine Pub. Co. 1974.

[423] Brown, D.J., "Aggregation of Preferences," *Quart. Jour. of Econ.* Vol. 89, 1975, pp. 456-469.

[424] Buchanan, J.M., "Individual Choice in Voting and the Market," *Jour. of Political Econ.* Vol. 62, Aug. 1954, pp. 334-343.

[425] Buchanan, J.M., “An Economic Theory of Clubs,” *Economica*, Vol. 32, 1965, pp. 1-14.

[426] Buchanan, J.M., “Notes for an Economic Theory of Socialism,” *Public Choice*, Vol. 8, Spring 1970, pp. 29-43.

[427] Buchanan, J.M., *The Demand and Supply of Public Goods*, Rand Chicago, McNally, 1968.

[428] Buchanan, J.M., “Public Finance and Public Choice,” *National Tax Jour.*, Vol. 28, 1975, pp..383-394

[429a] Buchanan, J.M. et al. (eds.), *Theory of Public Choice*, Ann Arbor, The University of Michigan Press, 1972.

[429b] Buchanan, J. M. et al. (eds.), *Toward a Theory of the Rent Seeking Society*, College Station, Texas, Texas A and M Univ. Press, 1980

[430] Buchanan, J.M. et al., *The Calculus of Consen*t, The University of Michigan Press, Ann Arbor, 1962.

[431] Chipman, J.S., “The Welfare Ranking of Pareto Distributions,” *Jour. of Econ. Theory*, Vol. 9, 1974, pp. 275-282.

[432] Coleman, J.S., “Foundations for a Theory of Collective Decisions,” *Amer. Jour. of Sociology*, Vol. 71, May 1966, pp.615-627.

[433] Coleman, J.S, “The Possibility of a Social Welfare Function,” *Amer. Econ. Rev.*, Vol. 56, 1966, pp. 1105-1112.

[434] Coleman, J.S., “The Possibility of a Social Welfare Function: A Reply,” *Amer. Econ. Rev.*, Vol. 57, 1967, pp. 1311-1317.

[435] Comanor, W.S., “The Median Voter and the Theory of Political Choice,” *Jour. of Public Econ.* Vol. 5, Jan-Feb, 1976, pp. 169-177.

[436a] Corlette, W. et al., “Complementarity and Excess Burden of Taxation,” *Rev. Econ. Studies*, Vol. 21, 1953-1954, pp. 21-30.

[436b] D’Aspermont, C. et al., “Equity and the Information Basis of Collective Choice,” *Rev. of Econ. Stud.* Vol.44, June 1977, pp. 119-209.

[437] DeMeyer, F. et al., “A Welfare Function Using Relative Intensity of Preferences,” *Quart. Jour. of Econ.*, Vol. 85, 1971, pp. 179-186.

[438] Diamond, P., "Cardinal Welfare, Individualistic Ethics, and Interpersonal Comparisons of Utility: A Comment," *Jour. of Political Econ.* Vol. 75, Oct 1967, pp. 765-766

[439] Dobb, M.A., *Welfare Economics and the Economics of Socialism*, Cambridge, Cambridge Univ. Press 1969.

[440] Dompere, Kofi K., "On Epistemology and Decision-Choice Rationality," in R., Trappl. (ed.) *Cybernetics and Systems Research*, North-Holland, New York, 1982, pp. 219-228

[441] Dompere, K.K. et al., *Epistemic of Development Economics: Toward a Methodological Critique and Unity*, Westport. Conn. Greenwood Press, 1995

[442] Downs, A., A., *Economic Theory of Democracy*, Harper and Row, New York, 1957.

[443a] Dubins, L.E. et al., "How to Cut a Cake Fairly," *Amer. Math Monthly* 1961, pp. 1-17; Reprinted in P.Newman (ed.), *Readings in Mathematical Economics, Vol. 1: Value Theory*, Baltimore, MD, John Hopkins Univ. Press, 1968.

[443b] Ellickson, B., "A Generalization of the Pure Theory of Public Goods," *American Econ. Rev.* Vol. 63, 1973, pp. 417-432.

[444a] Farquharson, R., *Theory of Voting*, New Haven, Yale University Press, 1969.

[444b] Fontaine, E., *Economic Principles for Project Evaluation*, Washington, DC, Organization of American States, 1975

[445] Geetz, C.J., "Earmarked Taxes and Majority Rule Budgetary Process," *Amer. Econ. Rev.,* Vol. 58, March 1968, pp. 128-136.

[446] Gigerenzer, G. et al. (Eds.), *Bounded Rationality*: The Adative Toolbox, Cambridge, Mass. MIT Press, 1999.

[447] Green, J., and J-J Laffont, "Characterization of Satisfactory Mechanisms for Revelation of Preferences for Public Goods," *Econometrica,* Vol. 45, March 1977, pp. 427-438.

[448] Green, J. et al., "Imperfect Personal Information and the Demand Revealing Process: A Sampling Approach," *Public Choice*, Vol. 29, Spring 1977, pp. 79-94.

[449] Green, J.R. et al. (eds.), *Incentatives in Public Decision-Making*, North-Holland, New York, 1979

[450a] Grieson, R.E. (ed.), *Public and Urban Economics*, Toronto, Heath and Co. 1976.

[450b] Griffin K.B., "Foreign Assistance: Objectives and Consequences," *Economic Development and Cultural Change*, Vol. 18, 1970, pp. 313-332.

[451] Haefele, E.T., "A Utility Theory of Representative Government," *Amer. Econ. Rev.,* Vol. 61, June 1971, pp. 350-367.

[452] Hammond, P.J., "Why Ethical Measures of Inequality Need Interpersonal Comparisons," *Theory and Decision*, Vol. 7, Oct. 1976, pp. 263-274.

[453] Harberger, A.C., "The Basic Postulates for Applied Welfare Economics; An Interpretive Essay," *Jour. of Econ. Lit.* Vol. 9, #3, 1971, pp.785-793

[454] Hardin, R., "Collective Action as an Agreeable n-Prisoner's Dilemma," *Behav. Science,* Vol. 16, Sept. 1971, pp. 471-481.

[455] Harsanyi, J.C., "Cardinal Utility in Welfare Economics and in the Theory of Risk-Taking," *Jour. Polit. Econ.* Vol. 61, Oct. 1953, pp. 434-435.

[456] Hause, J.C., "The Theory of Welfare Cost Measurement," *Jour. of Polit. Econ.*, Vol 83, 1975, pp.1145-1182..

[457a] Head, J.G. et al., "Public Goods, Private Goods and Ambiguous Goods," *Econ. Jour.*, Vol. 79, 1969, pp. 567-572.

[457b] Heineke, J. M., *Economic Model of Criminal Behavior*, New York, North-Holland, 1978.

[458] Hicks, J.R., "The Four Consumer's Surpluses," *Rev. of Econ. Stud.*Vol. 13, 1944., pp. 68-73.

[459] Hicks, J.R., "The Foundations of Welfare Economics," *Economic Jour.,* Vol. 59,1949, pp. 696-712.

[460] Intrilligator, M.D., "A Probabilistic Model of Social Choice," *Rev. Econ. Stud.* Vol. 40, Oct. 1973, pp. 553-560.

[461] Jakobsson, U., "On the Measurement of the Degree of Progression," *Jour. of Public Economics,* Vol. 14, 1976, pp. 161-168

[462a] Kaldor, N., “Welfare Propositions of Economics and Interpersonal Comparison,” *Economic Jour.* Vol, 49. 1939, pp.549-552.

[462b] Kaldor, N. (ed.), *Conflicts in Policy objectives*, Oxford, Oxford Univ. Press, 1971.

[463] Kaufmann, A. *The Science of Decision-Making: Introduction to Praxeology*, New York, McGraw-Hill, 1968.

[464a] Kemp, M.C. et al., “More on Social Welfare Functions: The Incompatibility of Individualism and Ordinalism,” *Economica*, Vol. 44, Feb. 1977, pp. 89-90

[464b] Kemp, M.C. et al., “On the Existence of Social Welfare Functions: Social Orderings and Social Decision Function,” *Economica*, Vol. 43, Feb. 1976, pp. 59-66.

[465a] Kickert, W.J.M., *Organization of Decision-Making: A Systems- Theoretical Approach,* New York, North-Holland, 1980.

[465b] Klappholz, K., “Value Judgments of Economics,” *British Jour. of Philosophy*, Vol. 15, 1964, pp. 97-114.

[466a] Kramer, G.H., “On a Class of Equilibrium Conditions for Majority Rule,” *Econometrica,* Vol.41, 1973, pp. 285-297.

[466b] Krueger,A. “The Political Economy of Rent Seeking Society,” *American Econ. Rev.*, Vol.64, 1974, pp.291-302.

[467] Laffont, J. -J. (ed.), *Aggregation and Revelation of Preferences*, New York, North-Holland, 1979.

[468] Lancaster, K.J., “A New Approach to Consumer Theory,” *Jour. of Political Economy*, Vol. 74, 1966, pp. 132-157.

[469] Lancaster, K.J., *Variety, Equity, and Efficiency,* New York, Columbia University Press, 1979.

[470] Lin, S.A.Y. (ed.), *Theory and Measurement of Economic Externalities*, New York, Academic Press, 1976.

[471] Lindsay, C.M., “A Theory of Government Enterprise,” *Jour. of Political Econ.* Vol. 84, Oct. 1976, pp. 1061-1077.

[472] Lipsey, R.G. et al., “The General Theory of Second Best,” *Rev. of Econ. Stud.* 1957, Vol. 24, pp. 11-32.

[473] Little, I.M.D., "Social Choice and Individual Values," *Jour. of Political Econ.* Vol. 60, Oct. 1952, pp. 422-432.

[474] Little, I.M.D., *A Critique of Welfare Economics*, Oxford, Clarendon Press, 1957.

[475a] Marschak, J., *Economic Information, Decision, and Prediction: Selected Essays, Vol. II*, Boston, Mass, D. Reidel Pub. Co, 1974.

[475b] McFadden, D., "The Revealed Preferences of a Public Bureaucracy : Theory," *Bell Jour. of Econ.* Vol. 6, #2, (Autumn),1975, pp. 55-72.

[476] McFadden, D., "The Revealed Preferences of a Government Bureaucracy: Empirical Evidence," *Bell Jour. of Econ.* Vol. 7, # 1, (Spring)1976, pp. 55-72.

[477] McGuire, M., "Private Good Clubs and Public Good Clubs: Economic Model of Group Formation," *Swedish Jour. of Econ.*, Vol. 74, 1972, pp. 84-99

[478] McGuire, M., "Group Segregation and Optimal Jurisdiction," *Jour. of Political Econ.* Vol. 82, Jan/Feb 1974, pp. 112-132.

[479] Mirkin, B., *Group Choice*, New York, John Wiley, 1979.

[480] Millerson, J.C., "Theory of Value with Public Goods: A Survey Article," *Jour. Econ. Theory*, Vol.5, Dec. 1972, pp. 419-477.

[481] Mishan, E.J., "Survey of Welfare Economics: 1939-1959," in *Surveys of Economic Theory*, *Vol.1*, New York, Macmillan,1968, pp. 156-222..

[482a] Mishan, E.J., "Welfare Criteria :Resolution of a Paradox," *Economic Jour.* Vol. 83, 1973, pp747-767.

[482b] Mishan, E. J., "Flexibility and Consistency in Cost Benefit Analysis," *Economica,* Vol. 41, 1974, pp81-96.

[483] Mishan, E.J., "The Use of Compensating and Equivalent Variation in Costs-Benefit Analysis," *Economica,* Vol.43, 1976, pp.185-197.

[484] Mishan, E.J., *Introduction to Normative Economics*, New York, Oxford Univ. Press, 1981.

[485] Mueller, D. C., "Voting Paradox," in C. K. Rowley (ed.), *Democracy and Public Choice*, New York, Basil Blackwell, 1987, pp. 77-102.

[486] Mueller, D.C., "Public Choice: A Survey," *Jour. of Econ. Lit.* Vol. 14, 1976, pp. 396-433.

[487] Mueller, D.C., "The Possibility of a Social Welfare Function: Comment," *Amer. Econ. Rev.*, Vol. 57, 1967, pp. 1304-1311.

[488] Mueller, D.C., "Constitutional Democracy and Social Welfare," *Quart. Jour. of Econ.* Vol. 87, Feb. 1973, pp. 60-80.

[489] Mueller, D.C., Allocation, Redistribution and Collective Choice," *Public Finance*, Vol. 32, 1977,pp. 225-244.

[490] Mueller, D.C., "Voting by Veto," *Jour. of Public Econ.,* Vol. 10, 1978, pp. 57-75

[491] Mueller, D.C., *Public Choice*, New York, Cambridge Univ. Press, 1979.

[492] Mueller, D.C. et al., "Solving the Intensity Problem in a Representative Democracy," in R.D. Leiter et al.(eds.), *Economics of Public Choice,* Cyro Press, New York, 1975, pp. 54-94.

[493a] Musgrave, R.A. et al. (eds.), *Classics in the Theory of Public Finance*, New York, St. Martin's Press, 1994

[493b] Musgrave, R. A., *Public Finance in a Democratic Society: Collected Papers*, New York, New York Univ. Press, 1986

[494a] Newbery, D. et al.(ed), *The Theory of Taxation for Developing Countries*, New York, Oxford University Press, 1987.

[494b] Niskanen, W. A., *Bureaucracy and Representative Government*, Chicago, Adine-Atherton, 1971.

[495a] Nti, K. O., "Rent-Seeking with Asymmetric Valuation," *Public Choice,* Vol. 98, 1999, pp415-430.

[495b] Olson, M., *The Logic of Collective Action*, Cambridge, Mass, Harvard Univ. Press, 1965.

[496] Ordover, J.A., "Distributive Justice and Optimal Taxation of Wages and Interest in a growing Economy," *Jour. of Public Economics*, Vol. 5, 1976, pp. 139-160

[497] Park, R.E., "The Possibility of a Social Welfare Function: Comment," *Amer. Econ. Rev.*, Vol. 57, Dec. 1967, pp. 1300-1304.

[498] Pattanaik, P.K., "Risk Impersonality and the Social Welfare Function," *Jour. of Political Econ.* Vol. 76, 1968, pp. 1152-1169.

[499] Pattanaik, P.K., *Voting and Collective Choice*, New York, Cambridge Univ. Press, 1971.

[500] Pauly, M.V., "Clubs Commonality, and the Core: An Integration of Game Theory and the Theory of Public Goods," *Economica,* Vol. 35, 1967, pp. 314-324.

[501] Pauly, M.V., "Cores and Clubs," *Public Choice*, Vol. 9, 1970, pp. 53-65.

[502] Plott, C.R., "A Notion of Equilibrium and Its Possibility Under Majority Rule," *Amer. Econ. Rev.* Vol. 57, 1967, pp. 787-806.

[503] Plott, C.R., "Ethics, Social Choice Theory and the Theory of Economic Policy," *Jour. of Math. Soc.*, Vol. 2, 1972, pp. 181-208.

[504] Plott, C.R., "Path Independence, Rationality and Social Choice," E*conometrica*, Vol. 41, Nov. 1973, pp. 1075-1091.

[505] Plott, C.R., "Axiomatic Social Choice Theory: An Overview and Interpretation," *Amer. Jour. Polit. Science*, Vol. 20, Aug. 1976, pp. 511-596.

[506] Prest, A.R., "The Budget and Interpersonal Distribution," *Public Finance*, vol. 23, 1968, p.p 80-98.

[507] Rae, D.W., "Decision-Rules and Individual Values in Constitutional Choice," *Amer. Polit. Science Rev.*, Vol. 63, March 1969, pp. 40-56.

[508a] Rae, D.W., "The Limit of Consensual Decision," *Amer. Polit. Science Rev.* Vol. 69, Dec. 1975, pp. 1270-1294.

[508b] Ramsey, F.P., "A Contribution to the Theory of Taxation," *Econ. Jour.,* Vol.37, March 1927, pp. 47-61.

[509] Rapport, A. et al., *Prisoner's Dilemma*, Ann Arbor, Michigan University Press, 1965.

[510] Rawls, J.A., *A Theory of Justice*, Cambridge, Mass., Harvard Univ. Press, 1971

[511] Rawls, J.A., "Concepts of Distributional Equity: Some Reasons for the maximum Criterion," *Amer. Econ. Rev.,* Vol. 64, May 1974, pp. 141-146.

[512] Ray, P., "Independence of Irrelevant Alternatives," *Econometrica,* Vol. 41, Sept. 1973, pp. 987-891.

[513] Reimer, M., "The Case for Bare Majority Rule," *Ethics*, Vol. 62, Oct. 1951, pp. 16-32.

[514] Reimer, M., The Theory of Political Coalition, Yale Univ. Press, New Haven, 1962.

[515] Roberts, F.S., *Measurement Theory: with Applications to Decision Making, Utility and the Social Science,* Reading, Addison-Wesley, 1979.

[516] Roberts, K.W.S., "Voting Over Income Tax Schedules," *Jour. of Public Econ.*,Vol. 8, 1977, pp. 329-340.

[517] Romer, T., "Individual Welfare, Majority Voting and the Properties of a Linear Income Tax," *Jour. of Public Econ.,* Vol. 4, 1975, pp. 163- 185.

[518a] Rothenberg, J., *The Measurement of Social Welfare*, Englewood, Cliff, N.J., Prentice-Hall, 1961

[518b] Rowley, C. K.(ed.), Democracy and Public Choice, New York, Basil Blackwell, 1987

[519a] Rubinstein, A., *Modeling Bounded Rationality*, Cambridge, Mass, MIT Press, 1998

[519b] Sandmo, S. "Optimal Taxation," *Jour. of Public Econ.* Vol. 15, July/Aug. 1972, pp. 37-54.

[519c] Saposnik, R., "On the Transitivity of the Social Preference Relation Under Simple Majority Rule," *Jour. Econ. Theory*, Vol. 10, Feb. 1975, pp. 1-7.

[520a] Satterthwaite, M.A., "Strategy- Proofness and Arrows Conditions: Existence and Correspondence Theorem for Voting Procedures and Social Welfare Functions," *Jour. of Econ. Theory*, Vol. 10, April 1975, pp.187-217.

[520b] Schneider, H., *National Objectives and Project Appraisal in Developing Countries*, Paris, Development Centre of OECD, 1975.

[521] Sen, A.K., "A Possibility Theorem on Majority Decisions," *Econometrica*, Vol. 34, April 1966, pp. 491-499.

[522] Sen, A.K., "Quasi-transitivity, Rational Choice and Collective Decisions," *Rev. Econ. Stud.* Vol. 36, July 1969, pp. 381-394.

[523] Sen, A.K., "Rawls versus Ben than: An Axiomatic Examination of the Pure Distribution Problem," *Theory and Decision*, vol. 4 Feb/April, 1974, pp. 301-310.

[524] Sen, A.K., "Informational Basis of Alternative Welfare Approaches Aggregation and Income Distribution," *Jour. Public Econ.* Vol. 3, Nov 1974, pp. 387-403.

[525] Sen, A.K., "Liberty, Unanimity and Rights," *Economica, Vol.* 43, Aug. 1976, pp. 217-245.

[526] Sen, A.K., "Social Choice Theory: A Re-examination," *Econometrica,* Vol. 45, Jan 1977, pp. 43-89.

[527] Sen, A.K., "On Weight and Measures Informational Constraints in Social Welfare Analysis," *Econometrica*, Vol. 45, Oct. 1977, pp.1539-1572

[528] Sen, A.K., *Collective Choice and Social Welfare*, Holden-Day, San Francisco, 1970.

[529] Sen, A.K. et al., "Necessary and Sufficient Condition for Rational Choice under Majority Decision," *Jour. of Econ. Theory*, Vol. 1, Aug. 1969, pp. 178-202.

[530a] Siegan, B. H., *Economic Liberties and the Constitution*, Chicago, Univ. of Chicago Press, 1980.

[530b] Spann, R.M., "Collective Consumption of Private Goods," *Public Choice,* Vol. 20, Winter, 1974, pp. 63-81.

[530c] Steinhouse, H., "Sur la Division Pragmatique," *Econometrica,* (Supplement),Vol. 17, 1949, pp. 315-319.

[531a] Stigler, G.J., "Director's Law of Public Income Redistribution," *Jour. Law and Econ.* Vol. 13, 1970, pp. 1-10.

[531b] Stigum, B.P et al., *Foundation of Utility and Risk Theory with Application*, Boston, Mass, D. Reidel Pub. 1983.

[532a] Stone, A.H. et al., "Generalized Sandwich Theorems," Duke Math. Jour., Vol. 9, 1942, pp. 356-359.

[532b] Taylor, M.J., "Graph Theoretical Approach to the Theory of Social Choice," *Public Choice*, Vol. 4, Spring 1968, pp. 35-48.

[533] Taylor, M.J., "Proof of a Theorem on Majority Rule," *Behavioral Science,* Vol. 14, May 1969, pp.228-231.

[534] Tideman, J.N., "Ethical Foundation of the Demand-Revealing Process," *Public Choice*, Vol. 29, Spring 1977, pp. 71-77.

[535] Tideman, J.N. et al., "A New and Superior Process for Making Social Choice," *Jour. of Polit. Economy*, Vol. 84, Dec. 1976, pp. 1145-1159.

[536a] Tollison, R. D., "Rent-Seeking: A Survey," *Kyklos*, Vol.35, 1982, pp. 575-602.

[536b] Tollison, R. M. et al., "Information and Voting: An Empirical Note," *Public Choice,* Vol. 24, 1975, pp. 43-49.

[536c] Tresch, R.W., *Public Finance: Normative Theory*, Plano. TX. Business Publications Inc., 1981.

[537a] Tullock, G., "Some Problems of Majority Voting," *Jour. of Polit. Econ.* Vol. 67, Dec. 1959, pp. 571-579.

[537b] Tullock, G., *The Politics of Bureaucracy*, Washington, D. C., Public Affairs Press, 1965.

[538] Varian, H.R., "Equity, Envy and Efficiency," *Jour. Econ. Theory*, Vol. 9, 1974, pp. 63-91.

[539] Varian, H.R., "Two Problems in the Theory of Fairness," *Jour. of Public Econ.* Vol. 5, April-May 1976, pp. 249-260.

[540] Williamson, O.E. et al., "Social Choice: A Probabilistic Approach," *Econ. Jour.* Vol. 77, Dec. 1967, pp.797-813.

[541] Wilson, R., "A Game-Theoretic Analysis of Social Choice" in B. Liebermann (ed.) *Social Choice*, Gordon and Breach, New York, 1971, pp.393-407

[542] Wilson, R.A., "Stable Coalition Proposals in Majority-Rule Voting," *Jour. of Econ. Theory*, Vol. 3, 1971, pp. 254-271.

[543a] Wingo, L. et al. (eds.), *Public Economics and Quality of Life*, John Baltimore, Hopkins Univ. Press, 1977.

[543b] Yaari, M.E. et al., "On Dividing Justly," Social Choice and Welfare, Vol. 1, 1984, pp. 1-24.

On Economic and General Cost-Benefit Analysis: Contingent Valuation Method (CVM)

[544a] Adamowicz, W., J. Louviere, and M. Williams, "Combining Revealed and Stated Preference Methods for Valuing Environmental Amenities," *Jour. of Environmental Economics and Management,* Vol.26, 1994, pp. 271-292.

[544b] Ajzen, I., and M. Fishbein, *Understanding Attitudes and Predicting Social Behavior*. Englewood Cliffs, New Jersey: Prentice-Hall, Inc., 1980.

[545] Arrow, K. et al., Repeat of NOAA Panel on Contingent Valuation," *Federal Register,* Vol. 58, 1993, pp. 4601-4614.

[546] Bateman, Ian, and Ken Willis (eds.), *Valuing Environmental Preference: Theory and Practice of the Contingent Valuation Method in the US, EC and Developing Countries*. Oxford , UK, Oxford University Press, 2002

[547] Batie, S.S. et al., "Valuing Non-Market Goods-Conceptual and Empirical Issues: Discussion," *Amer. Jour. Of Agricultural Economics*, Vol. 61, #5, 1979, pp. 931-932.

[548] Bentkover J. D. et al. (eds.), *Benefit Assessment: The State of the Art*, Boston, Mass. D.Reidel, 1986.

[549] Bishop, R.C. et al., "Measuring Values of Extra-Market Goods: Are Indirect Measures Biased?," *Amer. Jour. of Agricultural Economics,* Vol. 61, 1979, pp. 926-930.

[550] Bishop, Richard C., and Thomas A. Heberlein, "The Contingent Valuation Method.," In Johnson, Rebecca L., and Gary V. Johnson, (eds.), *Economic Valuation of Natural Resources: Issues, Theory, and Applications,* Boulder: Westview Press, 1990, pp. 81-104

[551] Bishop, R.C. et al., Contingent Valuation of Environmental Assets: Comparison with a Simulated Market," *National Resources Jour.*, Vol.23, #3, 1983, pp. 619-634.

[552] Boyle, Kevin J., Michael P. Welsh, and Richard C. Bishop, "The Roleof Question Order and Respondent Experience in Contingent-Valuation Studies," *Jour. of Environmental Economics and Management*, 1993, Vol. 25, pp. S-80-S-99

[553] Brookshire, D.S. et al., "Estimating Option Prices and Existence value for Wildlife Resources," *Land Economics*, Vol. 59, 1983, pp. 1-15

[554] Brookshire, D.S. et al., "The Advantage of Contingent Valuation Methods for Benefit Cost Analysis," *Public Choice*, Vol. 36, #2, 1981, pp. 235-252.

[555] Brookshire, D.S. et al., "Valuing Public Goods: A Compromise of Survey and Hedonic Approaches," *Amer. Economic Review*, Vol. 72, #1, 1982, pp. 165-177.

[556] Burness, H.S. et al., Valuing Policies which Reduce Environmental Risk," *Natural Resources, Jour.*, Vol. 23, #3, 1983, pp. 675-682.

[557] Carson, R.T. et al., *Temporal Reliability of Estimates from Contingent Valuations , Discussion Paper 95-37*, Washington, D.C. Resource for the Future, Aug. 1995.

[558] Carson , R.T. et al., *Referendum Design and Contingent Valuation: the NOAA Panels no-vote Recommendation, Discussion Paper 96-05,* Washington, D.C., Resources for the Future, 1996.

[559] Carson, Richard T., and Robert Cameron Mitchell, "The Value of Clean Water: The Public's Willingness to Pay for Boatable, Fishable, and Swimmable Quality Water," *Water Resources Research*, 1993, Vol. 29, pp. 2445-2454

[560] Carson,R., et al., *A Contingent Valuation Study of Lost Passive Use Values Resulting from the Exxon Valdex Oil Spill*, Report to the Attorney General of Alaska, Natural Resource Damage Assessment, Inc. La Jolla, CA., November 1992.

[561] Carson, R., et al., *A Bibliography of Contingent Valuation Studies and Papers*, La Jolla, CA Natural Resource Damage Assessment, Inc., ,March1994.

[562] Cummings, Ronald G., David S. Brookshire, and William D. Schulze, et al., (eds.), *Valuing Environmental Goods: An Assessment of the Contingent Valuation Method.* Totowa, New Jersey: Rowman and Allanheld, 1996.

[563] Desvouges, William H., et al., *Measuring Nonuse Damages Using Contingent Valuation: An Experimental Evaluation of Accuracy.* North Carolina: Research Triangle Institute Monograph, 1992.

[564] Diamond, Peter A., Jerry Hausman, Gregory K. Leonard, and Mike A Denning, "Does Contingent Valuation Measure Preferences? Experimental Evidence," In Hausman, J.A.,(eds.), *Contingent Valuation: A Critical Assessment.* New York: North-Holland, 1993, pp. 41-89.

[565] Diamond, P.A., and J.A. Hausman, "On Contingent Valuation Measurement of Nonuse Values," In Hausman, J., (ed.), *Contingent Valuation: A Critical Assessment*. Amsterdam: North-Holland Press, 1993, pp. 3-38.

[566] Diamond, P.A. et al., "Contingent Valuation: Is Some Number Better then no Number?," *Jour. Economics Perspectives*, Vol. 8, 1994, pp. 45-64.

[567a] DiBona, Charles J., "Assessing Environmental Damage," *Issues in Science and Technology*, Fall 1992, Vol.8, pp. 50-54.

[567b] Dickie, M. A. Fisher, and S. Gerking, "Marketing Transactions and Hypothetical Demand Data: A Comparative Study," *Jour. of American Statistical Association*, March 1987, 82, 69-75

[568] Fishkin, J.S., *Democracy and Deliberation: New Direction for Democratic Reform*. New Haven: Yale University Press, 1991.

[569] Frosythe, R. et al., "Demand-Revealing Mechanisms for Private Goods Actions," in [595], 1982 pp. 21-33.

[570] Freeman, A. Myrick, *The Measurement of Environment and Resource Values: theory and Method*. Washington D.C.: Resources for the Future, 1993.

[571] Freeman, A.M. III., "Approaches to Measuring Public Goods Demands," *Amer. Jour. of Agricultural Economics Proceedings*, Vol. 61, #5, pp. 915-920.

[572] Frisch, Deborah, "Reasons for Framing Effects," *Organizational Behavior and Human Decision Process*, Vol. 54, 1993, pp. 399-429.

[573] Gregory, R., "Interpreting Measures of Economic Loss: Evidence from Contingent Valuation and Experimental Studies," *Jour. Of Environmental Economics and Management*, Vol. 13, 1986, pp. 325-337.

[574] Groves, Robert M., *Survey Errors and Survey Costs*. New York: John Wiley and Sons, 1989.

[575] Hanemann, W. Michael, "Welfare Evaluations in Contingent Valuation Experiments with Discrete Responses," *American Jour. of Agricultural Economics*, August 1984, Vol. 66, pp.332-341.

[576] Hanemann, W. Michael, "Preface: Notes on the History of Environmental Valuation in the USA." In Navrud, Stale, (ed.), *Pricing the Environment: The European Experience*. Oxford, UK: Oxford University Press, 1992.

[577] Hanemann, W.M., "Valuing the Environment through Contingent Valuation," *Jour. of Economic Perspectives*, Vol. 8, 1994, pp. 19-44.

[578] Harrison, Glenn W., "Valuing Public Goods with the Contingent Valuation Method: A Critique of Kahneman and Knetsch," *Jour. of Environmental Economics and Management*, 1992, Vol. 23, pp. 248-57.

[579] Hausman, J.A., (ed.), *Contingent Valuation: A Critical Assessment*. New York: North-Holland, 1993.

[580] Hoehn, J.P., and A. Randall, "Too Many Proposals Pass the Benefit Cost Test," *American Economic Review*, June 1989, Vol. 79, pp. 544-51

[581] Kahneman, Daniel, "Comments on the Contingent Valuation Method," In Cummings, Ronald G., David S. Brookshire, and William D. Schulze, (eds), *Valuing Environmental Goods: A State of the Arts Assessment of the Contingent Valuation Method.* Totowa: Rowman and Allanheld1986, pp. 185-94.

[582] McNeal, B.J., "On the elicitation of Preferences for alternative Therapies," *New England Jour. of Medicine,* Vol. 306, 1982, pp. 1259-1262.

[583] Mitchell, R.C. et al., *An Experiment in Determining Willingness to Pay for National Water Quality Improvements*, Report Prepared for U.S. Environmental Protection Agency, Washington, D.C., Resources for the Future, 1981.

[584] Mitchell, Robert Cameron, and Richard T. Carson, *Using Surveys to Value Public Goods: the Contingent Valuation Method.* Washington, D.C.: Resources for the Future, 1989.

[585] Navrud, Stale, *Pricing the European Environment*, New York: Oxford University Press, 1992.

[586] Neill, Helen R., et al., "Hypothetical Surveys and Real Economic Commitments," *Land Economics*, May 1994, Vol.70, pp. 145-54.

[587] O'Hanlon, P. et al., "Scope for Valuation of Environmental goods: Comment," *Land Economics*, Vol. 54, #3, pp. 381-387.

[588] Portney, P.R., "The Contingent Valuation Debate: Why Economists Should Care," *Jour. of Economic Perspectives,* Vol. 8, 1994, pp. 3-17.

[589] Randall, A. et al., Contingent Valuation Survey for Evaluating Environmental Assets," *Natural Resources Jour.* Vol. 23, 1983, pp. 635-648.

[590] Seip, K., and J. Strand, "Willingness to Pay for Environmental Goods in Norway: A Contingent Valuation Study With Real Payment," *Environmental and Resource Economics*, Vol. 2, 1992, pp. 91-106.

[591] Smith, V.L., "Experimental Economics: Induced Value Theory," *American Economic Review*, Proceeding, Vol. 66, 1976, pp.274-279.

[592] Smith, V.L., " Experiments with a Decentralized Mechanism for Public Good Decision," *American Economic Review*, Vol. 70, 1980, pp. 584-599.

[593] Smith, V.L., " Microeconomic System as an Experimental Science," *American Economic Review*, Vol. 72, #5, 1982, pp. 923-955.

[594] Smith, V.K.(ed.) *Advances in Applied Microeconomics*, Vol. 1, New York, JAI Press, 1981.

[595] Smith, V.K.(ed.) *Research in Experimental Economics, Vol. II*, New York, JAI Press, 1982.

[596] Smith, V.K., "An Experimental Comparison of Three Public Good Decision Mechanisms," *Scandinavian Jour. of Economics*, Vol. 81, 1979, pp.198-215.

[596b] Smith, V.K., "Indirect Revelation of demand for Public Goods: An Overview and Critique," *Scot. Jour. of Political Econ.*, Vol. 26, June, 1979, pp183-189

[597] Strauss, R.P. et al., "A New Approach to the Demand for Public Goods," *Jour. of Public Economics*, Vol. 6, 1976, pp. 191-204.

[598] Thaler, R., "Toward a Positive Theory of Consumer Choice," *Jour. of Economic behavior and Organization*, Vol. 1, 1980, pp. 39-60.

[599] Thayer, M.A., "Contingent Valuation Technique for Assessing Environmental Impacts: Further Evidence," *Jour. of Environmental Economics and Management,* Vol. 8, 1981, pp. 27-44.

[600] Tideman, N. et al., " A New and Superior Process for Making Social Choices," *Jour. of Political Economy*, Vol. 84, 1976, pp. 1145-1159.

[601] Tversky, A. et al., "The Framing of Decisions and the Psychology of Choice," *Science,* Vol. 211, 1981, pp. 453-458.

[602] B.A. Weisbrod, Costs and Benefits of Medical Research: A Case Study of Poliomyelitis, in: *Benefit-Cost and Policy Analysis 1971*, Chicago, Aldine, 1972, pp.142-160.

[603] B.A. Weisbrod, *A Guide to Benefit-Cost Analysis, as seen Through a Controlled Experiment in Treating the Mentally Ill*, Institute for Research on Poverty, Discussion Papers, DP #559-798, University of Wisconsin- Madison, 1979

On Economic and General Cost-Benefit Analysis: Revealed Preference Approach (RPA)

[604] Arrow, K.J., "Political and Economic Evaluation of Social Externalities," in[156], pp. 1-23.

[605] Bain, J.S., "Criteria for Undertaking Water-Resource Development," *Amer. Econ. Rev. Papers and Proceedings,* Vol. 50, #2, 1960, pp. 310-320.

[606] Barsb, S.L., *Cost-Benefit Analysis and Manpower Programs*, Toronto, D.C. Heath& Co, 1972.

[607] Bator, F.M., "The Anatomy of Market Failure," *Quarterly Jour. of Econ* Vol. 72, 1958, pp. 351-379.

[608] *Benefit-Cost and Policy Analysis1973*, An Aldine Annual, Chicago, Aldine Pub. Co.1974.

[609] *Benefit-Cost Analysis and Policy Analysis1974*, An Aldine Annual, Chicago, Aldine Pub. Co., 1995

[610a] Benefit-*Cost Analysis and Policy Analysis 1971*, An Aldine Annual, Chicago, Aldine Pub. Co., 1972.

[610b] Brookings Institution, *Applying Benefit-Cost Analysis to Public Programs*, Brookings Research Report #79, Washington, D.C., 1968

[611] Borus, M.E., "A Benefit-Cost Analysis of the Economic Effectiveness of Retraining the Unemployed," *Yale Economic Essays*, Vol. 4, #2, 1964, pp. 371-430.

[612] Buchanan, J.M., "Politics, Policy and the Pigovian Margins," *Economica,* Vol. 29, #113, 1962, pp. 17-28.

[613] Devine, E.T., "The Treatment of Incommensurables in Cost-Benefit Analysis," *Land Economics*, Vol. 42, #3, 1966, pp.383-387.

[614] Freeman, H. et al., *Evaluating Social Projects in Developing Countries,* Paris, Organization for Econ. Co-operation and Dev., 1979

[615] Gramlich, E.M., *Benefit-Cost Analysis of Government Programs*, Englewood Cliffs, N.J. Prentice-Hall, Inc. 1981.

[616] Hansen, W.L. et al., "the Distribution of Costs and Direct Benefits of Public Higher Education. The Case of California," *Jour. of Human Resources*, Vol. 4, #2, 1969, pp.176-191.

[617] Haveman, R.H., "Benefit-Cost Analysis: Its Relevance to Public Investment Decisions: Comment," *Quarterly, Jour. of Econ.,*Vol. 81, #4, 1967, pp. 695-702.

618] Haveman, R.H., *Water Resource Investment and Public Interest*, Vanderbilt Univ. Press, Nashville, 1965

[619] Haveman, R.H., et al., "Unemployment Excess Capacity and Benefit-Cost Investment Criteria," *Rev. of Econ. Stat.*,Vol. 49, 1967, pp. 382-392

[620] Hochman, H. M., et al., "Pareto Optimal Redistribution," *Amer. Econ.* Review, Vol.59, 1969, pp542-557.

[621] Intrilligator, M.D. (ed.), *Frontiers of Quantitative Economics*, North-Holland, New York, 1971.

[622] Knesse A.V., "Research Goals and Progress Toward them," in H. Jarrett, (ed.), *Environmental Quality in a Growing Economy*, Washington, DC, John Hopkins Press, 1966, pp. 69-87.

[623] Knetsch, J.L., "Economics of Including Recreation as a Purpose of Eastern Water Projects," Jour. of Farm Econ. Vol. 46, #5, 1964, pp. 1148-1157.

[624] Krutilla, J.V., "Conservation Reconsidered," *Amer. Econ. Rev.*, Vol. 57, 1967, pp.777-786.

[625] Krutilla, J.V., (ed.), *Natural Environments: Studies in Theoretical and Applied Analysis*, Baltimore, John Hopkins Univ. Press 1972.

[626] Krutilla, J.V. et al., *Multiple Purpose River Development Studies* in *Applied Economic Analysis*, Baltimore, John Hopkins Univ. Press, 1958.

[627] Layard, R. (ed.), Cost-Benefit Analysis, Baltimore, Penguin, 1972.

[628] Lesourne, J., *Cost-Benefit Analysis and Economic theory,* New York, North-Holland, 1975.

[629] Little, I.M.D. et al., *Manual of Industrial Project Analysis in Developing Countries*, Vol. II: Social Cost Benefit Analysis, Paris, OECD Pub., 1968.

[630] Mass, A., "Benefit-Cost Analysis: Its Relevance to Public Investment Decisions," *Quarterly Jour. of Economics*, Vol. 80, #2, 1966, pp. 208-226.

[631] MacDonald, J.S., "Benefits and Costs: Theoretical and Methodological Issues: Discussion" in [647], 1969, pp.30-37.

[632] Marciariello, J.A., *Dynamic Benefit-Cost Analysis*, Toronto, Heath and Co., 1975.

[633] May, K. O., "A Set of Independent, Necessary and Sufficient Conditions for Simple Majority Decision," *Econometrica*, Vol. 20, #4, Oct. 1952, pp.680-684.

[634] Mishan, E.J., *Cost-Benefit Analysis*, New York, Praeger, 1976.

[635] Mishan, E.J., "Reflections on Recent Developments in the Concept of External Effects," *Canadian Jour. of Econ. and Political Science*, Vol. 31, #1, 1965, pp. 3-34.

[636] Musgrave, R.A., "Cost-Benefit Analysis and the Theory of Public Finance," *Jour. of Econ. Literature*, Vol. 7, 1969, pp797-806

[637] Prest, A.R. et al., "Cost-Benefit Analysis: A Survey" *Survey in Economic Theory*, Vol. III, New York, St. Martin Press, 1966 (Also in Economic Jour., Vol. 75, Dec. 1965, pp.685-705)

[638] Raynauld, A., "Benefits and Costs: Theoretical and Methodological Issue: Discussion" in [647] pp. 37-41.

[639] Samuleson, P.A., "Aspects of Public Expenditures Theories," *Rev. of Econ. and Stat.*, Vol.40, 1958, pp. 332-338.

[640] Samuelson, P.A., The Pure Theory of Public Expenditure," *Rev. of Econ. and Stat.*, Vol. 36, 1954, pp. 387-387.

[641] Schwartz, H. et al., (eds.) *Social and Economic Dimensions of Project Evaluation, Symposium on the Use of Socioeconomic Investment Criteria*, Washington, D.C., Inter-American Dev. Bank, 1973 .

[642] Scitovsky, T., "Two Concepts of External Economies," *Jour. of Political Econ.*, Vol.17, 1954, pp.143-151.

[643] Sen, A. and S. Marlgin., *Guidelines for project Evaluation*, New York, U.N. publication Series #2, 1972

[644] Solo, R.A., "Benefit-Cost Analysis and Externalities in Private Choice: Comment," *Southern Econ. Jour.,* Vol. 34, #4, 1968, pp. 569-570.

[645] Solow, R.M., "The Economics of Resources or the Resources of Economics," *Amer. Econ. Rev.*, Vol. 64, #2, 1974, pp. 1-14

[646] Solow, R.M. et al., "Extraction Costs in the Theory of Exhaustible Resources," *Bell Journal of Economics*, Vol. 7, 1977, pp 359-370.

[647] Somers, G.G. et al. (eds.), *Cost-Benefit Analysis of Manpower Policies*, Proceedings of a North American Conference., Kingston. Ontario,1969

[648] Stern, N., "Welfare Weights and the Elasticity of Marginal Valuation of Income" in M. Artis et al(eds.), *Studies in Modern Economic Analysis,* Oxford, Blackwell, 1977.

[649] Sugden, R. et al., *The Principles of Practical Cost-Benefit Analysis*, Oxford, Oxford Univ. Press, 1978.

[660] U.S. Congress, Subcommittee on *Benefit and Costs of Federal Inter-Agency River Basin Committee* ,82nd Congress, 1st Session, 1950.

[651] United States of America Committee on Interstate and Foreign Commerce House of Representatives, *Use of Cost-Benefit Analysis By Regulatory Agencies,* Joint and Subcommittee on Consumer Protection and Finance, Serial #96-157 (1979)

[652] United States of America Department of Transportation, *Guidance for Regulatory Evaluation: Handbook for DOT Benefit-Cost Analysis,* Washington, DC1984.

[653] United States of America Environmental Protection Agency, *Guidelines for Performing Regulatory Impact Analysis,* EPA-230-01-84-00-3 Washington, DC (1983).

[654] *United States of America Federal Highway Administration, Costs and benefits of requiring conspicuity retrofits, Draft Evaluation Paper* Washington, D. C., October 1994.

[655] Water Resource Council of USA, *Procedures for evaluation of national economic development (NED) benefits and costs in water resources planning (level C) final rule*, Fed. Register44 (1979) 72892-72977

On Fuzzy Decisions

[656] Adamo, J.M., "Fuzzy Decision Trees", *Fuzzy Sets and Systems*, Vol. 4, #3, 1980, pp. 207-219.

[657] Asai, K. and H. Tanaka, "Applications of Fuzzy Sets Theory To Decision Making and Control," *Jour. JAACE*, Vol. 19, 1975, pp. 235-242.

[658] Asai, K., et. al., "Decision Making and Its Goal in Fuzzy Environment," in [806], pp. 257-277.

[659] Baldwing, N.C.F., et. al., "Comparison of Fuzzy Sets on the Decision Spaces," *Fuzzy Sets and Systems*, Vol. 2, #3, 1979, pp. 213-231.

[660a] Banerjee, A., "Fuzzy Choice Functions, Revealed Preference and Rationality,". *Fuzzy Sets and Systems*, Vol. 70, 1995, pp. 31-44.

[660b] Basu, A. and A. Dutta, "Reasoning with Imprecise Knowledge to enhance Intelligent Decision Support," *IEEE Tran. On System, man, and Cybernetics*, Vol. 19, #4,1989, pp.756-770

[661] Bellman, R.E. and L.A. Zadeh, "Decision-making in Fuzzy Environment," *Management Science*, Vol. 17, 1970, pp. 141-164.

[662] Bezdek, J., et. al., "Fuzzy Measures of Preference and Consensus in Group Decision-Making," in *Proceeding of the 1977 IEE Conference on Decision and and Control,* 1977, pp1303 - 1308

[663] Bezdek, J., "A Fuzzy Relation Space for Group Decision Theory," *Fuzzy Sets and Systems*, Vol. 1, 1978, pp. 255-268.

[664] Bezdek, J., "Fuzzy Relation Spaces for Group Decision Theory: An Application," *Fuzzy Sets and Systems*, Vol. 2, #1, 1979, pp. 5-14.

[665] Bezdek, J.C. et al., "Convex Decomposition of Fuzzy Partitions," *Jour. of Math. Analysis and Applications*" Vol. 67 #2 1979, pp. 490-512.

[666] Bezdek, J.C., *Pattern Recognition with Fuzzy Objective Function Algorithms*. New York, Plenum Press, 1981.

[667] Bezdek, J.C., (ed.) *Analysis of Fuzzy Information*. Boca Raton, FL CRC Press, (3 volumes). 1987.

[668] Bezdek, J.C. and S.K. Pal, (eds.), *Fuzzy Models for Pattern Recognition: Methods that Search for Patterns in Data,* New York IEEE Press, 1992

[669] Bhakat, S.K. and P. Das, "On the definition of a fuzzy subgroup," *Fuzzy Sets and Systems*, Vol. 51, #2, 1992, pp. 235-241.

[670] Bit, A.K. et al., "Fuzzy Programming Approach to Multicriteria decision-Making Transportation Problem," *Fuzzy Sets and Systems*, vol. 50 #2, 1992, pp. 135-141.

[671] Bliss, J.M., "Fuzzy Sets in Multiple Criteria Decision Making," *TIMS Studies in Management Science*, Vol. 6, 1977, pp. 21-30.

[672] Blin, J.M., "Fuzzy Relations in Group Decision Theory," *Jour. Cybernetics*, Vol. 4, 1974, pp. 17-22.

[673] Blin, J.M., et. al., "Fuzzy Sets and Social Choice", *Jour. Cybernetics*, Vol. 3, 1973, pp. 28-33.

[674] Bocklish, S. et al; *Fuzzy Sets Applications, Methodological Approaches and Results*, Berlin, Verlag, 1986.

[675] Borisov, A.N., "A Theory of Possibility for Decision-Making", *Fuzzy Sets and Systems*, Vol. 9, #1, 1983, pp. 13-23.

[676] Bouchon, B., "Fuzzy Questionnaires," *Fuzzy Sets and Systems*, Vol. 6 #1, 1981, pp. 1-9.

[677] Bouchon, B. et. al. (eds.), *Uncertainty and Intelligent Systems*, New York, Springer-Verlag, 1988.

[678] Bouchon, B., et. al. (eds.), *Uncertainty and Knowledge-Based Systems,* New York Springer-Verlag, 1987.

[679] Bouchon-Meunien, B. et. al. "Towards General Measures of Comparison of Objects," *Fuzzy Sets and Systems*, Vol. 84, #2, 1996, pp. 143-154.

[680] Buckley, J.J., "Fuzzy Decision Making with Data: Applications to Statistics," *Fuzzy Sets and Systems*, Vol. 16, #2, 1985, pp. 139-147.

[681] Capocelli, R.M., "Fuzzy Sets and Decision Theory," *Information and Control*, Vol. 23, 1973, pp. 446-473.

[682] Carlsson, C., *Fuzzy Set Theory for Management Decision*, Koln, Verlag TUV Rheinland, 1990.

[683] Chen, S.J. et al., *Fuzzy Multiple Attribute Decision Making: Methods and Applications*, New York, Springer-Verlag 1992.

[684] Chen, S.J. et al., *Fuzzy Multiple Decision Making, Lecture Notes in Economics and Math. Systems*, #375, New York Springer-Verlag 1992.

[685] Chen, S.M., "A New Method for Handling Multicriteria Fuzzy Decision-Making Problems," *Cybernetics and Systems*, Vol. 25, #5, 1994,pp. 409-420.

[686] Cox, E., *The Fuzzy Systems Handbook*, Cambridge, MA, Academic Press, 1994.

[687] Capocelli, R.M. et al., "Fuzzy Sets and Decision Theory," *Information and Control*, Vol. 23, 1973, pp. 446-473.

[688] Czogala, E., et. al., "Aspects for the Evaluation of Decision Situations," in [709], 1982, pp. 41-49..

[689] Dimitrov, V., "Group Choice Under Fuzzy Information," *Fuzzy Sets and Systems*, Vol. 9, #1, 1983, pp. 25-40.

[690] Dimitrov, V., "Creative Decision-Making Through Fuzzy Catastrophes," in [710]. 1985, pp. 391-399.

[691] Di Nola, A., et. al., "Hyperpatial Representation of a Particular Set of Fuzzy Preference Relations", *Fuzzy Sets and Systems*, Vol. 7, #1, 1982, pp. 79-88.

[692] Di Nola, A., et. al., "Fuzzy Relation Equations Theory as a Basis of Fuzzy Modelling: An Overview," *Fuzzy Sets and Systems* Vol. 40 #3, 1991, pp. 415-429.

[693] Di Nola, A., et. al., "Fuzzy Relational Equation Under a Class of Triangular Norms: A Survey and New Results," *Stochastica*, Vol. 8, #2, 1984.pp. 99-146.

[694] Dompere, K.K., "The Theory of Fuzzy Decisions," in [709] 1982, pp. 365-379.

[695] Dubois, D. et. al., *Fuzzy Sets and Systems: Theory and Applications*, New York, Academic Press, 1980.

[696] Dubois, D. et. al., "Measuring and Updating information," *Information Science*, Vol. 57-58, 1991, pp. 181-195.

[697] Dubois, D. et. al., *Possibility Theory*, New York, Plenum Press, 1988.

[698] Dubois, D., et. al., "The Use of Fuzzy Numbers in Decision Analysis," in [709], 1982, pp. 309-322.

[699] Efstathiou, J., et. al., "Multi-attribute Decision-making Using a Fuzzy Heuristic Approach," IEEE *Trans On System, Man and Cybernetics*, Vol. 9, 1979, pp. 326-333.

[700] Eick, C.F. et. al., "Decision Making Involving Imperfect Knowledge," *IEEE Transactions on Systems, Man and Cybernetics*, Vol. 23, #3, 1993, pp. 840-851.

[701] Ento, Y., "Fuzzy Decision Theory," in [1003], 1980, pp. 2980-2990.

[702] Francioni, J.M., et. al., "Imprecise Decision Tables and Their Optimization," in [710], 1985, pp. 89-96.

[703] Freire, E., et. al., "A Method of Multi-criteria Analysis", in [710], 1982, pp. 289-300.

[704] Fung, L.W., "An Axiomatic Approach to Rational Decision-Making in a Fuzzy Environment," in [806], 1976, pp. 227-256.

[705] Fung, L.W., et. al., "The K'th Optimal Policy Algorithm for Decision-Making in Fuzzy Environments," in P. Eykhoff, *Identification and System Parameter Estimation*, New York, North-Holland, pp. 1025-1059, 1974

[706] Gluss, B., "Fuzzy Multistage Decision-Making," *Intern. Jour. of Control*, Vol. 17, 1973, pp. 177-192.

[707] Gupta, M.M., et. al., "Decision and Control in a Fuzzy Environment: A Rationale," *Proc. 3rd IFAC Symp. Identification and System Parameter Estimation*, Hague, June, 1973, pp. 1048-1049.

[708] Gupta, M.M., et. al. (eds.), *Fuzzy Automata and Decision Processes*, New York, North-Holland, 1977.

[709] Gupta, M.M. et. al. (Eds), *Fuzzy Information and Decision Processes*, New York North-Holland, 1982.

[710] Gupta, M.M. et. al. (Eds), Approximate Reasoning in Expect Systems, New York, North-Holland, 1982.

[711] Gupta, M.M. et. al. (eds), *Fuzzy Computing: Theory, Hardware and Applications*, New York, North-Holland, 1988.

[712] Gupta, M.M. et. al. (eds), *Fuzzy Logic In Knowledge-Based Systems, Decision and Control*, New York, North-Holland, 1988.

[713] Gupta, M.M. et. al., (eds.), *Advances in Fuzzy Set Theory and Applications*, New York, North-Holland, 1978.

[714] Gupta, M.M. et. al., "On the Principles of Fuzzy Neural Networks," Fuzzy Sets and Systems, Vol. 61, #1, 1994 pp. 1-18.

[715] Hagg, C., "Possibility and Cost in Decision Analysis," *Fuzzy Sets and Systems,* Vol. 1, #2, 1978., pp. 81-86

[716] Herrera, F. et. al., "A Rationale Consensus Model in Group Decision Making Using Linguistic Assessments," *Fuzzy Sets and Systems*, Vol. 88 #1, 1997, pp.31-49.

[717] Hipel, K.W., "Fuzzy Set Methodologies in Multicriteria Modelling," in [709], 1982, pp. 279-288.

[718] Hwang, C.L. et. al., *Multiple Attribute Decision Making: Methods and Applications*, New York, Springer-Verlag, 1981.

[719] Jocobson, D.H., "On Fuzzy Goals and Maximizing Decisions in Stochastic Optimal Control," *Jour. Math., Anal. and Appl.*, Vol 55, 1976, pp434-440

[720] Jain, R., "Decison-making in the Presence of Fuzzy Variable," *IEEE Trans. System, Man and Cybernetics,* 1976, pp. 698-703.

[721] Jain, R., Procedure for Multi-aspect Decision Making Using Fuzzy Sets," *Intern. Jour. of Systems Science*, Vol. 8, 1977, pp. 1-7.

[722] Jain, R., "Decision-Making in the Presence of Fuzziness and Uncertainty," *Proceedings of the IEEE Conference on Decision and control, New Orleans,* 1978, pp. 1318-1323.

[723] Jamshidi, M., et al. (eds), *Fuzzy Logic and Control*, Englewood Cliffs, N.J., Prentice Hall, 1993.

[724] Janko, W.H., et. al. (eds*), Progress in Fuzzy Sets and Systems*, Boston, Mass, Kluwer 1990.

[725] Johnston, D.M. et. al., "Application of Fuzzy Decision Making: An Evaluation," *Civil Engineering Systems*; Vol. 5; #2, 1988 pp. 87-92.

[726] Jones, A. et. al. (Eds.); *Fuzzy Sets Theory and Applications*, Boston, Mass., D. Reidal, 1986.

[727] Kacprzyk, J. et. al., *Combining Fuzzy Imprecision with Probabilistic Uncertainty in Decision Making*, New York, Springer-Verlag, 1988.

[728] Kacprzyk, J., "Multistage Decision Processes in a Fuzzy Environment: A Surey," in [709], 1982, pp. 251-266.

[729] Kacprzyk, J., *Multistage Decision-Making Under Fuzziness*, Koln, Verleg TUV Rheinland, 1983.

[730] Kacprzyk, J., "Group Decision-Making with a Fuzzy Majority via Linguistic Quantifiers, Part I: A Consensory-Like Pooling," *Cybernetics and Systems*, Vol. 16 #(2-3), 1985, pp. 119-129.

[731] Kacprzyk, J., "Group Decision-Making with a Fuzzy Majority via Linguistic Quantifiers. Part II: A Competitive-Like Pooling," *Cybernetics and Systems*, Vol. 16 #(2-3) pp. 131-144.

[732] Kacprzyk, J. et. al. (eds.), *Multiperson Decision Making Using Fuzzy Sets and Possibility Theory*, Dordrecht, Kluwer, 1990.

[733] Kacprzyk, J., et. al., (eds), *Non-Conventional Preference Relations in Decision Making*, New York, Springer-Verlag 1988.

[734] Kaymak, U. et. al., "A Sensitivity Analysis Approach to Introducing Weights Factors into Decision Functions in Fuzzy Multicriteria Decision Making," *Fuzzy Sets and Systems*, Vol. 97, 1988, pp. 169-182.

[735] Kickert, W.J.M., *Fuzzy Theories on Decision-Making*, Maritinus Nighoff, Boston, 1978.

[736] Kim, J.B., "Fuzzy Rational Choice Functions," *Fuzzy Sets and Systems*, Vol. 10, #1, 1983, pp. 37-44.

[737] Kitainik, L., *Fuzzy Decision Procedures with Binary Relations: Toward a Unified Theory*, Boston, Mass. Kluwer, 1993.

[738] Kling, R., "Fuzzy Planner: Reasoning with Inexact Concepts in a Procedural Problem-solving Language," *Jour. Cybernetics*, Vol. 3, 1973, pp. 1-16.

[739] Klir, G.J. et al., *Fuzzy Sets and Fuzzy Logic: Theory and Applications*, Englewood Cliffs, N.J., Prentice-Hall, 1995.

[740] Kuz'min, V.B. and S.V. Ovchinnikov, "Grope Decision I: In Arbitrary Spaces of Fuzzy Binary Relations," *Fuzzy Sets and Systems*, Vol. 4, #1, 198, pp. 53-62.

[741] Kuz'min,V.B. and S. V. Ovchinnikov, "Design of Group Decisions II: In Spaces of Partial Order Fuzzy Relations," *Fuzzy Sets and Systems*, Vol. 4, #2, 1980, pp. 153-166.

[742] Leung, Y., "A Concept of a Fuzzy Ideal for Multicriteria Conflict Resolution" in P.P. Wang (ed.), *Advances in Fuzzy Set Theory and Application*, Plenum, New York, 1982.

[743] Leung, Y., "Dynamic Conflict Resolution Through A Theory of A Displaced Fuzzy Ideal," in [710], 1982, pp. 381-390.

[744] Liou, T.S. et. al., "Subjective Assessment of Mental Workload: A Fuzzy Multicriteria Approach," *Fuzzy Sets and Systems*, Vol. 62, 1994, pp. 155-165.

[745] Loo, S.G., "Fuzzy Relations in the Social and Behavoral Sciences", *Jour. of Cybernetics*; Vol. 8, #1, 1978, pp. 1-16

[746] Lowen, R., "Comments on Zeleny's Paper Concerning Cognitive Equilibria," *Intern. Jour. of General Systems*, Vol. 19, #4, 1991, pp. 387-393.

[747] Mamdani, E.H. et. al. (eds.), *Fuzzy Reasoning and its Applications*, London, Academic Press, 1981.

[748] Manes, E.G., "A Class of Fuzzy Theories," *Jour. of Math. Analysis and Applications*, Vol. 85 #2, 1982, pp. 409-451.

[749] Mares, M.,*Computation Over Fuzzy Quantities*, Boca Raton, Fl., CRC Press, 1994.

[750] Marin, R.S. et. al., "Modeling the representation of Time from a Fuzzy Perspective," *Cybernetics and Systems*, Vol. 25, #2, 1994, pp. 217-231.

[751] Marks II, R.J. (Ed.), *Fuzzy Logic Technology and Applications*, New York, IEEE Press, 1994.

[752] Menges, G. (ed.), *Information, Inference and Decision*, D. Reidel Pub., Dordrecht, Holland, 1974.

[753] Menges, G., et. al., "Linear Partial Information as Fuzziness," in H. Bossel et. al. (Eds.), *Systems Theory in the Social Sciences,* Birkhauser, Verlag, Vasel, 1976, pp. 307-322.

[754] Menges, G., et. al., "On the Problem of Vagueness in the Social Sciences," in [752], 1974, pp. 51-61.

[755] Nakamura, K., Preference Relation on a Set of Fuzzy Utilities as a Basis for Decision Making," *Fuzzy Sets and Systems*, Vol. 20, 1986, pp. 147-162.

[756] Negoita, C.V. et. al., *Applications of Fuzzy Sets to Systems Analysis*, New York, Birkhauser, Basel an Stuttgart and Halster Press 1975.

[757] Nie, J. et. al., *Fuzzy Neural Control: Principles, Algorithms and Applications,* Englewood Cliffs, N.J., Prentice Hall, 1994.

[758] Nojiri, H., "On the Fuzzy Team Decision in a Changing Environment", *Fuzzy Sets and Systems*, Vol. 3, #2, 1980, pp. 137-150.

[759] Nojiri, H., "A Model of Fuzzy Team Decision", *Fuzzy Sets and Systems*, Vol. 2, #3, 1973, pp. 201-212.

[760] Novak, V.J. et. al., *Fuzzy Approach to Reasoning and Decision-Making*, Boston, Mass. Kluwer, 1992.

[761] Novak, V., Fuzzy Sets and Their Applications, Bristol, Adam Hilger 1987.

[762a] Nowakowska, M., "A New Theory of Time: Generation of Time from Fuzzy Temporal Relation," in G.E. Lasker (Ed.), *Applied Systems and Cybernetics*, Vol. VI, New York, Pergamon Press, 1981, pp. 2742-2747.

[762b] Nurmi, H., "A Fuzzy Solution to a Majority Voting Game," *Fuzzy Sets and Systems*, Vol. 5, #2, 1981, pp. 187-198.

[763] Nurmi, H., "Approaches to Collective Decision-making with Fuzzy Preference Relations," *Fuzzy Sets and Systems*, Vol. 6, #3, 1981, pp. 249-260.

[764] Orlovsky, S.A., "Decision-making With a Fuzzy Preference Relation," *Fuzzy Sets and Systems*, Vol. 1, 1978, pp. 155-167.

[765] Orlowska, E., "Representation of Vague Information," *Information Systems*, Vol. 13, #2, 1988, pp. 167-174.

[766] Ostermark, R., "Temporal Interdependence in Fuzzy MCDM Problems," *Fuzzy Sets and Systems*, Vol. 88 #1, 1997, pp. 69-80.

[767] Otto, K.N. et. al., "Approximating α -cuts with the Vertex Method," *Fuzzy Sets and Systems*, Vol. 55 #1 1993, pp. 43-50.

[768] Ovchinnikov, S.V., "The Duality Principle in Fuzzy Set Theory," *Fuzzy Sets and Systems*, Vol. 42, #1, 1991 pp. 133-144.

[769] Ovchinnikov, S.V., "Structure of Fuzzy Binary Relations," *Fuzzy Sets and Systems*, Vol. 6, #2, 1981, pp. 169-196.

[770] Ovchinnikov, S.V., "Choice Theory for Cardinal Scales," in [709], 1982, p. 323-338.

[771] Pedrycz, W., "Why triangular Membership Functions?" *Fuzzy Sets and Systems*, Vol. 64 #1, 1994, pp. 21-30.

[772] Predrycz, W., "Neural Structures of Fuzzy Decision-making," *Jour. of Intelligent and Fuzzy Systems*, Vol. 2, #2, 1994, pp. 161-178.

[773] Prade, H. et. al., *Fuzzy Logic in Knowledge Engineering*, Koln, Verlag TUV Rheinland, 1986.

[774] Prade, H. et. al., "The Treatment of Uncertainty in Knowledge-based Systems Using Fuzzy Sets and Possibility Theory," *Intern. Jour. of Intelligent Systems*, Vol. 3, #2, 1988, pp. 141-165.

[775] Ragade, R.K. and Gupta, M.M. (1979). "Fuzzy sets theory and applications - a synthesis," in [713] 1978, pp. 19-25.

[776] Ragade, R.K., "Profile Transformation Algebra and Group Consensus Formation Through Fuzzy Sets," in [708], 1977, pp. 331-356.

[777] Roubens, M., "Choice Procedures in Fuzzy Multicriteria Decision Analysis Based on Pairwise Comparisons," *Fuzzy Sets and Systems*, Vol. 84, #2, 1996, pp. 135-142.

[778] Saaty, T.L., "Exploring the Interface Between Hierarchies, Multiple Objectives and Fuzzy Sets," *Fuzzy Sets and Systems*, Vol. 1, #1, 1978. pp. 57-68.

[779] Sakawa, M. et. al., "Interactive Decision Making for Multiobjective linear Fractional Programming Problem with Fuzzy Parameters," *Cybernetics and Systems*, Vol. 16 #4, 1985, pp. 377-394.

[780a] Sanchez, E. (Ed.), *Fuzzy Information, Knowledge Representation and Decision Analysis*, Oxford, Pergamon Press 1984.

[780b] Sanchez, E. et. al. (Eds), *Approximate Reasoning in Intelligent Systems, Decision and Control*, Oxford U.K., Pergamon Press, 1987.

[781] Sekita, Y., et. al., "A Fuzzified Relevance Tree Approach for Solving the Complex Planning," in [710],1982, pp. 401-410.

[782] Seo, F., et. al., "Fuzzy Multiatribute Utility Analysis for Collective Choice," *IEEE Transactions on Systems, Man and Cybernetics,* Vol. 15, #1, 1985, pp. 45-53.

[783] Shimura, M., "Fuzzy Sets Concept in Rank-ordering Objects", *Jour. Math. Anal. and Appl.*, Vol. 43, 1973. pp. 717-733.

[784] Spillman, B., et. al., "A Dynamic Perspective On Leadership: Development of a Fuzzy Measurement Procedure," *Fuzzy Sets and Systems*, Vol. 7, 1982, pp. 19-33.

[785] Sugeno, M., "Fuzzy Decision-making Problems," *Trans. SICE*, Vol. 11, 1975, pp. 709-714.

[786] Takeda, E., "Interactive Identification of Fuzzy Outranking Relation in Multi-criteria Decision Problem," in [709], 1982, pp. 301-322.

[787] Takeda, E., et. al., "Multiple Criteria Decision Problems with Fuzzy Domination Structures," *Fuzzy Sets and Systems*, Vol. 3, #2, 1980, pp. 123-136.

[788] Tanaka, H., et. al., "A Formulation of Fuzzy Decision Problems and Its Application to an Investment Problem," *Kybernetics*, Vol. 5, 1976, pp. 25-30.

[789] Tanino, T., "Fuzzy Preference Ordering in Group Decision Making," *Fuzzy Sets and Systems*, Vol. 12, 1984, pp. 117-131.

[790] Tanino, T., "On Group Decision Making, Under Fuzzy Preferences," in [732], 1990, pp. 172-185.

[791] Tseng, T.Y. et. al., "A New Algorithm for Fuzzy Multicriteria Decision Making," *Intern. Jour. of Approximate Reasoning*, Vol. 6, 1992, pp. 45-66.

[792] Uemura, Y., "A Simple Decision Rule on Fuzzy Events," *Cybernetics and Systems*, vol. 24 #5, 1993, pp. 509-521.

[793] Uemura, Y., "A Decision Rule on Fuzzy Events Under an Observation," *Jour. of Fuzzy Mathematics*, Vol. 1 #1, 1993, pp. 39-52.

[794] Uhr, L., "Toward Integrated Cognitive Systems Which must Make Fuzzy Decisions About Fuzzy Problems," in [806], 1975, pp. 353-393.

[795] Wang, P.P., et. al., (Eds.), *Fuzzy Sets*, New York, Plenum Press, 1980.

[796] Yager, R.R., "A Model of Participatory Learning," *IEEE Transactions on Systems, Man and Cybernetics,* Vol. 20, #5, 1990, pp. 1229-1234.

[797] Yager, R.R., et. al., "Decision-making with Fuzzy Sets," *Decision Science*, Vol. 6, 1975, pp. 590-500.

[798] Yager, R.R., "Fuzzy Decision Making Including Unequal Objectives," *Fuzzy Sets and Systems*, Vol. 1, #2, 1978, pp. 87-95.

[799] Yager, R.R., "Decisions with Ordinal Preferences and Importances," *Kybernetics,* Vol. 9, 1980, pp. 109-114.

[800] Yager, R.R., "Satisfaction and Fuzzy Decision Functions," in [795], 1980, pp. 171-194.

[801] Yager, R.R., "On Considerations of Credibility of Evidence," *Intern. Jour. of Approximate Reasoning*, Vol. 71, # (1-2), 1992, pp. 45-72.

[802] Yao, Y.Y. et. al., "A Decision Theoretic Framework for Approximating Concepts," *Intern. Jour. of Man-Machine Studies*, Vol. 37 #6, 1992, pp. 793-809.

[803] Yu, P. L., "Cone Convexity, Cone Extreme Points and nondominated Solutions in Decision Problem with Multiobjectives," *Jour. of Optim., Theory and Appl.*, Vol. 14, 1974, pp. 319-377.

[803] Yu, P.L., "Cone Convexity, Cone Extreme points and Nondominated Solutions in Decision Problems with Multiobjectives," in G. Leitman (ed) . *Multicriteria Decision Making and Differential Games*, New York, Plenum press, 1979, pp.1-59.

[804] Zadeh, L.A., "Similarity Relations and Fuzzy Ordering," *Information Science,* Vol. 3, 1971, pp. 117-200.

[805] Zadeh, L.A., "Outline of a New Approach to the Analysis of Complex System and Decision Processes," *IEEE Trans. on System, Man., and Cybernetics*, Vol. 3, #1, 1973, pp. 28-44.

[806] Zadeh, L.A., et. al., (eds.), *Fuzzy Sets and Their Applications to Cognitive and Decision Processes*, New York, Academic Press, 1975.

[807] Zhang, D. et. al., "State-dependent Weights in Multicriteria Value Functions," *Jour. of Optimization Theory and Applications*, Vol. 74, #1, 1992, pp. 1-21.

[808] Zhang, H. et. al., "A Technique for Handling Fuzzy Decision-Making Problems Concerning two kinds of Uncertainty," *Cybernetics and Systems*, vol. 22, #6, 1991 pp. 681-698.

[809] Zimmerman, H.J., *Fuzzy Sets, Decision Making and Expert Systems*, Boston, Kluwer, 1987.

[810] Zimmerman, H.J., *Fuzzy Set Theory and its Applications*, Boston, Mass, Kluwer, 1991.

[811] Zimmerman, H.J. et. al., (eds), *Fuzzy Sets and Decision Analysis,* TIMS Studies in the Management Science, Vol. 20, Amsterdam, North-Holland Elsevier, 1984.

[812] Zimmerman, H.J., et. al., Latent Connectives in Human Decisions Making," *Fuzzy Sets and Systems*, Vol. 4, #1, 1980.

[813] Zimmerman, H-J., et. al., "Decisions and Evaluation by Hierarchical Aggregation of Information," *Fuzzy Sets and Systems*, Vol. 10, #3, 1983, pp. 243-260.

On Fuzzy Decisions and Some Applications

[814] Adlassning, K.P., "A Survey on Medical Diagnosis and Fuzzy Subsets," in [709], pp. 203-218.

[815] Adlassning, K.P., et. al., "Computer-Assisted Medical Diagnosis Using Fuzzy Subsets," in [709], 1982, pp. 219-248.

[816] Albin, M., *Fuzzy Sets and Their Application to Medical Diagnosis*, A Ph.D. Dissertation, Dept. of Math., University of California, Berkeley, California, USA, 1975.

[817] Allen, A.D., "A Method of Evaluating Technical Journals on the Basis of Published Comments Through Fuzzy Implication: A Survey of the Major IEEE Transactions," *IEEE Trans. Syst. Man and Cybernetics*, SMC-3, 1973, pp. 422-425.

[818] Allen, A.D., "Measuring the Empirical Properties of Sets," *IEEE Trans. System, Man and Cybernetics*, SMC-A 1974, pp. 66-73.

[819] Bardossy, A. et. al., "Fuzzy Fatigue Life Prediction." *Structural Safety*, Vol. 6, 1989, pp. 25-38.

[820] Bardossy, A., et. el., "Kriging with Imprecise (Fuzzy) Variogram," Part I: Theory Part II: Application," *Mathematical Geology*, vol. 22 #1, 1990, pp. 63-94.

[821] Bardossy, A. et. al., *Fuzzy Rule-Based Modeling with Applications to Geophysical, Biological and Engineering Systems*, New York, CRS Press, 1995.

[822] Bardossy, A. et. al., "Fuzzy rule-based Models for Infiltration," *Water Resources Research,* Vol. 29 1993, pp. 373-382.

[823] Bardossy, A., "Analysis of a Karstic aquifer Management Problem by Fuzzy Composite Programming," *Water Resource Bulletin*, vol. 28, #1, 1992, pp. 63-74.

[824] Bezded, J.C., "A Physical Interpretation of Fuzzy ISODATA," *IEEE Transactions, Syst. Man and Cybern.,* vol. 6, 1976, pp. 387-389.

[825] Bezded, J.C. et. al., "Optimal Fuzzy Partitions: A Heuristic For Estimating the Parameters in Mixture of Normal Distributions," *IEEE Trans. Comp.* C-24, 1975 pp. 835-838.

[826] Cao, H. et. al.: "Some Applications of Fuzzy Sets to Meterological Forecasting", *Fuzzy Sets and Systems*, vol. 9, #1, 1983, pp. 1-12.

[827] Carlsson, C. et. al., (eds.), *Multiple Criteria Decision Making: Selected Case Studies*, New York, McGraw-Hill, 1981.

[828] Carlsson, C., "Tackling an MCDM - problem with the Help of Some Results from Fuzzy Sets Theory," *European Jour. of Operational Research*, vol. 10, #3, July, 1982, pp. 270-281.

[829a] Cavallo, R.E. (ed.), *Systems Methodology in Social Science Research*, Boston, Kluwer, 1982.

[829b] Cervin, V.B. et. al., "Toward Fuzzy Algorithms in Behavioral Modification", in [709], 1982, pp. 291-296.

[830] Chanas, S. et. al., "Maximum Flow in a Network with Fuzzy Arc Capacities," *Fuzzy Sets and Systems*, vol. 8, #2, 1982, pp. 165-173.

[831] Chang, R.L.P. et. al., "Application of Fuzzy Sets in Curve Fitting," *Fuzzy Sets* and *Systems*, Vol. 2 #1, 1979, pp. 67-74.

[832] Cheng, Y.Y.M. et. al., "An Algorithm for Multiple Attribute, Multiple Alternative Decision Problems Based on Fuzzy Sets with Application to Medical Diagnosis," *IEEE Transactions on Systems, Man and Cybernetics*, vol. 10, #10, 1980, pp. 645-650.

[833] Davidson, J. et. al., "A Fuzzy Decision Model for the Design of Rural Natural Gas Networks," *Fuzzy Sets and Systems*, vol. 53, #3, 1993, pp. 241-252.

[834] Degani, R. et. al., "The Problem of Linguistic Approximation in Clinical Decision Making," *Intern. Jour. of Approximate Reason*, vol. 2, #2, 1988, pp. 143-162.

[835] Dockery, J.T., "Fuzzy Design of Military Information Systems," *Intern Jour. of Man-Machine Studies*, vol. 16, #1, 1982, pp. 1-38.

[836] Esogbue, A.O. et. al., "Fuzzy Sets and the Modelling of Physician Decision Processes: Part I: The Initial Interview-Information Gathering Process," *Fuzzy Sets and Systems*, vol. 2, 1979, pp. 279-291.

[837] Esogbue, A.O. et. al., "Fuzzy Sets and the Modelling of Physicial Decision process Part II: Fuzzy Diagnosis Decision Models," *Fuzzy Sets and Systems*, vol. 3, #1, 1980, pp. 1-9.

[838] Esogbue, A.O. et. al., "Measurement and Valuation of a Fuzzy Mathematical Model for Medical Diagnosis," *Fuzzy Sets and Systems*, vol. 10, 1983, pp. 223-242.

[839] Esogbue, A.O. et. al., "On the Application of Fuzzy Set Theory to the Optimal Flood Control Problem Arising in Water Resources Systems," *Fuzzy Sets and Systems*, vol. 48, #2, 1992, pp. 155-172.

[840] Esogbue, A.O., "A Fuzzy Set Model For the Evaluation of Best Management Practices in Nonpoint Source Water Pollution Policy Formulation," in [709] pp. 431-438.

[841] Feng, D.Y. et. al., *Fuzzy Mathematics in Earthquake Research*, vol. 1 and 2. Beijing, Seismological Press, 1986.

[842] Feng, D.Y. et. al., "Application of Fuzzy Decision-making in Earthquake Research," *Fuzzy Sets and Systems,* vol. 36, #1, 1990, pp. 15-26.

[843] Fordon, W.A. et. al., "The Application of Fuzzy Set Theory to Medical Diagnosis," in [713], 1979, pp. 445-461.

[844] Frank, P.M. et. al., "Fuzzy Supervision and Application to Lean Production", *Intern. Jour. of Systems Science*, vol. 24, #10, 1993, pp. 1935-1944.

[845] Fieschi, M. et. al., "SPHINX: An Interactive System for Medical Diagnosis Aids," in [709], 1982, pp. 269-276.

[846] Gale, S., "Inexactness, Fuzzy Sets and the Foundations of Behavioral Geography," *Geographical Analysis,* vol. 4, 1972, pp. 337-349.

[847b] Gale, S., "Inexactness, Fuzzy Sets and the Foundation of Behavioral Geography," *Geographical Analysis*, Vol.4, #4, 1972, pp.337-349.

[848] Giering III, E.W. et. al., "The Application of Fuzzy Set Theory t the Modeling of Competition in Ecological Systems," *Fuzzy Sets and Systems*, vol. 9, #2, 1983, pp. 103-128.

[849] Groguen, J., "On Fuzzy Robot Planning" in [806], 1975, pp. 429-447.

[850] Gunaratne, M. et. al., "Fuzzy Multi-attribute Decision Making in Pavement Management," *Civil Engineering Systems*, vol. 2, #3, 1985, pp. 166-170.

[851] Hammerbarch, I.M. et. al., "The Personalization of Security Selection: An Application of Fuzzy Sets Theory," *Fuzzy Sets and Systems*, Vol. 5, #1, 1981, pp. 1-10.

[852] Hora, F., "Fuzzy Simulation Model of Civil Evacuation From Large - Scale Earthquake - Generated Fires," in [710], 1982, pp. 313-328.

[853] Karwowski, W. et. al., (Eds), *Applications of Fuzzy Set Theory in Human Factors*, New York, Elsevier, 1986.

[854] Kerre, E.E., "The Use of Fuzzy Set Theory in Electrocardiological Diagnostics," in [709] pp. 277-282.

[855] Kochen, M., "Applications of Fuzzy Sets in Psychology," in [806], 1975, pp. 395-408.

[856] Kokawa, M. et. al., "Experimental Approach to Fuzzy Simulation of Memorizing Forgetting and Interference Process," in [806], 1975, pp. 409-428.

[857] Kokawa, M. et. al., "Fuzzy-theoretic and Concept - Formational Approach to Hint-effect Experiments in Human Decision Processes," *Fuzzy Sets and Systems*, vol. 2, #1, 1979, pp. 25-36.

[858] Lee, S.C. et. al., "Fuzzy Sets and Neural Networks," *Jour. of Cybernetics*, Vol. 4, 1974, pp. 83-103.

[859] LeFaivre, R., *Fuzzy Problem Solving*, Ph.D. Dissertation, Dept. of Computer Science, Univ. of Wisconsin, Madison, Wisconsin, 1974.

[860] Leung, Y., *Spacial Analysis and Planning Under Imprecision*, New York, North-Holland, 1988.

[861] Li, Zy. et. al., A Model of Weather Forecast by Fuzzy Grade Statistics," *Fuzzy Sets and Systems*, Vol. 26, #3, 1988, pp. 275-281.

[862] Lusk, E.J., "Evaluating Performance Statistics Used to Monitor Performance: A Fuzzy Approach," *Fuzzy Set and Systems*, vol. 5, #1, 1981, pp. 149-158.

[863] Mou-chao, M. et. al., "The Multistage Evaluation Method in Psychological Measurement: An Application of Fuzzy Sets Theory to Psychology," in [709],1982, pp. 307-312.

[864] Narasimhan, R., "A Fuzzy Subset Characterization of a Site-selection Problem," *Decision Sciences*, vol. 10, #10, 1979, pp. 618-628.

[865] Negoita, C.V., "Fuzzy Models for Social Processes," in H. Bossel et. al. (eds), *Systems Theory in the Social Science*, Birkhauser Verlag Basel, 1976, pp. 283-291.

[866] Negoita, C.V., et. al., "Fuzzy Systems and Artificial Intelligence," *Kybernetics,* vol. 3, 1974, pp. 173-178.

[867] Negoita, C.V., *Management Applications of Systems Theory*, Basel and Stuttgart, Birkhauser Velag, 1979.

[868] Oguntade, O.O. et. al., Ophthalmological Prognosis via Fuzzy Subsets", *Fuzzy Sets and Systems*, vol. 7, #2, 1982, pp. 123-138.

[869] Oguntade, O.O. et. al., "Evaluation of Architectural Design Profiles Using Fuzzy Sets," *Fuzzy Sets and Systems*, vol. 5, #3, 1981, pp. 221-234.

[870] Oheigeartaigh, M., "A Fuzzy Transportation Algorithm," *Fuzzy Sets and Systems,* vol. 8, #3, 1982. pp. 225-234.

[871] Olbero, A. et. al., "The Structure of Relations in Personnel Management," *Fuzzy Sets and System*, vol. 5, #2 1981, pp. 115-126.

[872] Onisawa, T. et. al., "Fuzzy Human Reliability Analysis on the Chernobyl Accident," *Fuzzy Sets and Systems*, vol. 28, #2, 1988, pp. 115-127.

[873] Roberts, D.W., "*Analysis of Forest Succession with Fuzzy Graph Theory,*" *Ecological Modelling*, vol. 45, #4, 1989, pp. 261-274.

[874] Saaty, T.L., "Exploring the Interface Between Hierarchies, Multiple Objectives and Fuzzy Sets," *Fuzzy Sets and Systems,* vol. 1, #1, 1978, pp. 57-68.

[875] Saitto, P. et. al., "Fuzzy Characterization of Coronary Disease," *Fuzzy Sets and Systems.* Vol. 5, #3, 1981, pp. 245-258.

[876] Sakawa, M. et. al., "An Interactive Fuzzy Satisfying Method Using Augmented Minimax Problems and its Application to Environmental Systems," *IEEE Transactions on Systems, Man and Cybernetics*, vol. 15, #6, 1985 pp. 720-729.

[877] Sanchez, E., "Medical Applications with Fuzzy Sets in [726], 1986, pp. 331-347.

[878] Sanchez, E., "Medical Diagnosis and Composite Fuzzy Relations," in [713] 1979, pp. 437-444.

[879] Sanchez, E., "Inverse of Fuzzy Relations: Application to Possibility Distributions and Medical Diagnosis," *Fuzzy Sets and Systems*, vol. 2, #1, 1979, pp. 75-86.

[880] Sanchez, E., "Solutions in Composite Fuzzy Relation Equations: Application to Medical Diagnosis in Brouwerian Logic," in [708], 1977, pp. 387-402.

[881] Saridis, G.N., et. al., "Fuzzy Decision-making in Prosthetic Devices," in [708], 1977, pp. 387-402.

[882] Seif, A. et. al., "Multi-group Classification Using Fuzzy-Correlation," *Fuzzy Sets, and Systems*, Vol. 3, #1, 1980, pp. 109-122.

[883] Sekita, Y. et. al., "A Fuzzified Relevance Tree Approach for Solving the Complex Planning," in [806], 1972, pp. 401-410.

[884] Shortliffe, E.H. et. al., "A Model of Inexact Reasoning in Medicine," *Math. Biosciences*, Vol. 23, 1975, pp. 351-379.

[885] Smets, P., "Medical Diagnosis: Fuzzy Sets and Degrees of Belief," *Fuzzy Sets and Systems*, Vol. 5, #3, 1981, pp. 259-266.

[886] Song Q. et. al., "Forecasting Enrollments with Fuzzy Time Series - Part I," *Fuzzy Sets and Systems*, Vol. 54, #1, 1993, pp. 1-9.

[887] Song, Q., et. al., "Forecasting Enrollments with Fuzzy Time Series - Part II," *Fuzzy Sets and Systems*, Vol. 62, #1, 1994, pp. 1-8.

[888] Soula, G. et. al., "Soft Deduction Rules in Medical Diagnosis Processes," in [109], 1982 pp. 89-98.

[889] Spillman, B. et. al., "A Dynamic Perspective on Leadership: Development of a Fuzzy Measurement Procedure," *Fuzzy Sets and Systems*, Vol. 1, #1, 1982, pp. 19-34.

[890] Tazaki, E. et. al., "Structural Modeling in a Class of Systems Using Fuzzy Sets Theory," *Fuzzy Sets and Systems*, Vol. 2, #1, 1979, pp. 87-104.

[891] Van de Walle, B. et. al., "Fuzzy Multi-criteria Analysis of Cutting Techniques in Nuclear Reactor Dismantling Project," *Fuzzy Sets and Systems*, Vol. 74, #1, 1995, pp. 115-126.

[892] Vila, M.A. et. al., "Problems of Classification in a Fuzzy Environment," *Fuzzy Sets and Systems*, Vol. 9, #3, 1983, pp. 229-241.

[893] Vila, M.A. et. al., "On Medical Diagnosis Using Possibility Measures," *Fuzzy Sets and Systems*, Vol. 10, #3, 1983, pp. 211-222.

[894] Wierzechon, S.T., "Application of Fuzzy Decision-making Theory to Coping with Ill-dedined Problems," *Fuzzy Sets and Systems*, Vol. 7, #1, 1982, pp. 1-18.

[895] Wierzechon, S. T., " An Algorhythm for Identification of Fuzzy Measure," *Fuzzy Sets and Systems*, Vol. 9, #1, 1983, pp 69-78.

[896] Wierzechon, S.T., "Ill-Defined Problems - How to Cope with them," in G.E. Lasker (ed.) *Applied Systems and Cybernetics,* Vol. VI, Pergamon Press, New York, 1981, pp. 2748-2752.

[897] Woodbury, M.A. et. al., "Clinical Pure Types as a Fuzzy Partition," *Jour. of Cybernetics*, Vol. 4, #3, 1974, pp. 111-121.

[898] Yen, J. et. al., *Industrial Applications of Fuzzy Logic and Intelligent Systems*, New York, IEEE Press 1994.

[899] Zadeh, L.A., "Biological Applications of the Theory of Fuzzy Sets and Systems," in L.D. Proctor (ed.) *Biocybernetics of Central Nervous System*, Co. Boston, Mass., Little Brown and 1969, pp. 199-212.

[900] Zetenyi, T. (ed.), *Fuzzy Sets in Psychology*, New York, North-Holland, 1988.

[901] Zhang, W. et. al., "Mathematical Models of Multifactorial Decision and Weather Forecast," in [780a] 1984, pp. 265-269.

On Fuzzy Decisions, Risks, Games and Economic Analysis

[902] Aluja, G.J., "Towards a New Paradigm of Investment Selection in Uncertainty," *Fuzzy Sets and Systems*, Vol. 84, #2, 1996, pp. 187-198.

[903a] Aubin, J.P., "Cooperative Fuzzy Games," Mathematics of Operations Research, Vol. 6, 1981, pp. 1-13.

[903b] Aubin, J. P., Mathematical Methods of Game and Economic Theory, New York, North-holland, 1979.

[904] Ayyub, B.M., et. al., (eds), *Analysis and Management of Uncertainty*, New York, Elsevier 1992.

[905a] Basu, K., "Fuzzy Revealed Preference Theory", *Jour. of Econ. Theory*, vol. 32, 1984, pp. 212-227.

[905b] Bhattacharya, U. et. al., "Fuzzy Multi-criteria Facility Location Problem," *Fuzzy Sets and Systems*, Vol. 51, #3, 1997, pp. 277-287.

[906] Billot, A., *Economic Theory of Fuzzy Equilibria: An Axiomatic Analysis*, New York, Springer-Verlag, 1992.

[907] Blin, J.M. et. al., "Fuzzy Sets and Social Choice," *Jour. of Cybernetics*, Vol. 3, #4, 1973, pp. 28-36.

[908] Buckles, B.P. et. al., "A Fuzzy Representation of Data for Relational Databases," *Fuzzy Sets and Systems*, Vol. 7, #3, 1982, pp. 213-226.

[909] Buckley, J.J., "The Fuzzy Mathematics of Finance," *Fuzzy Sets and Systems*, Vol. 21, #3, 1987, pp. 257-273.

[910] Buckley, J.J., "Solving Fuzzy Equations in Economics and Finance," *Fuzzy Sets and Systems*, Vol. 48, #3, 1992, pp. 289-296.

[911] Buckley, J.J. et al., "The Utility of Information and Risk-taking Fuzzy Expert Systems," *Intern. Jour. of Intelligent Systems*, Vol. 3, #2, 1988, pp. 17-197.

[912] Buckley, J.J., "Portfolio Analysis Using Possibility Distributions" in [780b], 1987, pp. 69-76.

[913] Buckley, J.J., "Fuzzy Input-Output Analysis," *European Jour. of Operations Research*, Vol. 30, 1989, pp. 56-60.

[914] Buckley, J.J., "Fuzzy PERT" in [936] 1989, pp. 103-114.

[915] Butnariu, D. D., "Fuzzy Games: A Description of the Concept," *Fuzzy Sets and Systems*, Vol. 1, 1978, pp. 181-192.

[916] Butnariu, D., "Stability and Shapely Value for a n-Persons Fuzzy Game," *Fuzzy Sets and Systems*, Vol. 4, #1, 1980, pp. 63-72.

[917] Calzi, M.L., Towards a General Setting for the Fuzzy Mathematics of Finance," *Fuzzy Sets and Systems*, Vol. 35, 1990, pp. 265-280.

[918] Cao, B., "Input-Output Mathematical Model with T-Fuzzy Data," *Fuzzy Sets and Systems*, Vol. 59, #1, 1993, pp. 15-23.

[919] Carlucci, D. et. al., "Fuzzy Cluster of Demand With a Regional Service System," in [708], pp. 379-386.

[920] Cavallo, R.E. (ed.), *Systems Methodology in Social Science Research*, Boston, Mass. Kluwer, 1982.

[921] Chang, P.L. et. al., "A Fuzzy Multi-criteria Decision Making Method for Technology Transfer Strategy Selection in Biotechnology," *Fuzzy Sets and Systems*, Vol. 63, #2, pp. 131-139.

[922] Chanas, S. et. al., "The Use of Fuzzy Variables in PERT," *Fuzzy Sets and Systems*, Vol. 5, 1981, pp. 11-20.

[923] Chang, S.S.L., "On Risk and Decision-making in a Fuzzy Environment," in [806], 1975pp. 219-226, ,

[924a] Chen, G.Q. et. al., "Application of Fuzzy Set Theory to Economics," in [1202] 1983, pp. 275-305.

[924b] Cobb, L. et. al. (eds.) *Mathematical Frontiers of the Social and Policy Science*, Boulder, Colorado, Westview Press, 1981.

[925] Czogola, E., *Probabilistic Sets: Fuzzy and Stochastic Approach to Decision, Control and Recognition Processes*, Koln, Verlag TUV ,Rheinland, 1986.

[926] Dompere, K.K., "The Theory of Fuzzy Decisions, Cost Distribution Principle in Social Choice and Optimal Tax Distribution," *Fuzzy Sets and Systems*, Vol. 53, 1993, pp. 253-274.

[927] Dompere, K.K., "A Fuzzy-Decision Theory of Optimal Social Discount Rate: Collective-Choice-Theoretic," *Fuzzy Sets and Systems*, Vol. 58, 1993, pp. 269-301.

[928] Dompere, K.K., "The Theory of Social Costs and Costing for Cost-Benefit Analysis in a Fuzzy Decision Space," *Fuzzy Sets and Systems*, Vol. 76, 1995, pp. 1-24

[929] Dompere, K.K., "The Theory of Approximate Prices: Analytical Foundations of Experimental Cost-Benefit Analysis in a Fuzzy-Decision Space," *Fuzzy Sets and Systems*, Vol. 87, 1997, pp. 1-26.

[930a] Dompere, K.K., "Cost-Benefit Analysis, Benefit Accounting and Fuzzy Decisions I: Theory," *Fuzzy Sets and Systems,* Vol. 92, 1987, pp. 275-287.

[930b] Dompere, K.K., "Cost-Benefit Analysis, Benefit Accounting and Fuzzy Decisions: Part II, Mental Illness in Hypothetical Community," *Fuzzy Sets and Systems*, Vol. 100, 1998, pp. 101-116.

[931a] Dompere, K.K., "Cost-Benefit Analysis of Information Technology," in Allen Kent et. al. (eds.), *Encyclopedia of Computer Science and Technology*, vol. 41, Supp. 26, New York, Marcal Dekker, 1999, pp.27-44.

[931b] Dompere, K. K., *Cost-Benefit Analysis and the Theory of Fuzzy Decision: Vol. II: Fuzzy Value Theory*, Springer-Verlag, New York, Forthcoming.

[932] Dufresne, D., "The Distribution of Perpetuity, with Application to Risk Theory and Pension Funding," *Scandinavian Actuarial Jour.*, 1990, pp. 39-79.

[933] Dumitru, V., et. al., "Membership Functions, Some Mathematical Programming Models and Production Scheduling," *Fuzzy Sets and Systems*, Vol. 8, #1, 1982, pp. 19-33.

[934] Ellsberg, D., "Risk, Ambiguity, and the Savage Axioms," *Quart. Jour. of Economics*, Vol. 75, 1961, pp. 643-669.

[935] Esogbue, A.O., "Dynamic Programming, Fuzzy Sets, and the Modeling of R&D Management Constol Systems," *IEEE Transactions, on System, Man and Cybernetics*, Vol. 13, #1, 1983, pp. 18-29.

[936] Evans, G.W. et. al. (eds), *Applications of Fuzzy Sets Methodologies in Engineering*, Amsterdam, New York, Elsevier, 1989.

[937] Fodor, J.C., "An Axiomatic Approach to Fuzzy Preference Modelling," *Fuzzy Sets and Systems*, Vol. 52, #1, 1992, pp. 47 52.

[938] Giles, R., "A Utility-valued Logic for Decision Making", *International Jour. of Approximate Reasoning,* Vol. 2, #2, 1988, pp. 113-142.

[939] Gupta, C.P., "A Note on the Transformation of Possibilistic Information into Probabilistic Information for Investment Decisions," *Fuzzy Sets and Systems*, Vol. 56, #2, 1993, pp. 175-182

[940] Ikejima, K., et. al., "Risk assessment for gas Pipelines Using Fuzzy Sets," *Civil Engineering Systems*, Vol. 4, #3, pp. 147-152.

[941] Kacprzyk, J. et. al., "Long-Term Inventory Policy-making Through Fuzzy Decision - making Models," *Fuzzy Sets and Systems*, Vol. 8, #2, 1982, pp. 117-132.

[942] Kacprzyk, J. et. al., "A Fuzzy Approach to the Stability of Integrated Regional Development," in GE Laskers (ed.), *Applied Systems and Cybernetics*, Vol. VI, New York, Pergamon Press, 1981, pp. 2997-3004

[943] Kandel, A., "Application of Fuzzy Logic to the Detection of Static Hazards in Combinational Switching Systems," *Intern. Jour. Comp. Inf. Science,* Vol. 3, 1974, pp. 129-139

[944] Karwowski, W. et. al., "Fuzzy Concepts in Production Management Research: A Review," *International Journal of Production Research*, Vol. 24, #1, 1986, pp. 129-147.

[945] Lasek, M., "Hierarchical Structures of Fuzzy Rating in the Analysis of Strategic Goals of Enterprises," *Fuzzy Sets and Systems*, Vol. 50, #2, 1992, pp. 127-134.

[946] McCain, R.A., "Fuzzy Confidence Intervals in a Theory of Economic Rationality," *Fuzzy Sets and Systems*, Vol. 23, #2, 1987, pp. 205-218.

[947] Nakamura, K., "Preference Relation on a Set of Fuzzy Utilities as a Basis for Decision Making," *Fuzzy Sets and Systems*, Vol. 20, 1986, pp. 147-162.

[948] Nishizaki, I. et. al., "A Solution Concept Based on Fuzzy Decisions in n-Person Cooperative Games," *Japanese Jour. of Fuzzy Theory and Systems*, Vol. 4, #2, 1992, pp. 183-194.

[949] Nojiri, H., "A Model of the Executive's Decision Processes in New Product Development," *Fuzzy Sets and Systems*, Vol. 7, #3, 1982, pp. 227-242.

[950] Nowakowska, M., "Methodological Problems of Measurement of Fuzzy Concepts in the Social Sciences," *Behavioral Science*, Vol. 22, #2, 1977, pp. 107-115.

[951a] Nurmi, H., "A Fuzzy Solution to a Majority Voting Game," fuzzy Sets and Systems, Vol. 5, 1981, pp. 187 - 198

[951b] Ostaszewski, K., *An Investigation into Possible Application of Fuzzy Sets Method in Acturial Science*, Schaumburg, Illinois, Society of Actuaries, 1993.

[952] Ostermark, R., "Profit Apportionment in Concerns with Mutual Ownership - An Application of Fuzzy Quantities," *Fuzzy Sets and Systems*, Vol. 26, #3, 1988, pp. 283-297.

[953] Park, K.S., "Fuzzy Set Theoretic Interpretation of Economic Order Quantity," *IEEE Transaction on Systems, Man and Cybernetics*, Vol. 17, #6, 1987, pp. 1082-1084.

[954] Ponsard, C., "An Application of Fuzzy Sets Theory to the Analysis of Consumer's Spatial Preferences", *Fuzzy Sets and Systems*, Vol. 5, #3, 1981, pp. 235-244.

[955] Ponsard, C., "Fuzzy Mathematical Models in Economics," *Fuzzy Sets and Systems*, Vol. 28, #3, 1988, pp. 273-283.

[956] Ponsard, C. et. al., *Fuzzy Economics and Spatial Analysis*, Dijon, France, Librairie de L'Universite, 1986.

[957] Prade, H., "Using Fuzzy Set Theory in a Scheduling Problem. A Case Study," *Fuzzy Sets and Systems*, Vol. 2, #2, 1979, pp. 153-166.

[958] Regade, R.K., "Fuzzy Games in the Analysis of Options", *Jour. Cybernetics*, Vol.6, 1976, pp. 213-221.

[959] Schmucker, K.J., *Fuzzy Sets, Natural Language Computation and Risk Analysis*, Rockville, MD., Computer Science Press, 1984.

[960] Schott, B. et. al., "Fuzzy Uncertainty in Imperfect Competition," Information *Science: Intelligent Systems*, Vol. 76, #(3-4) 1994, pp. 339-354

[961] Seo, F., et. al., "On Construction of a Cooperative Fuzzy Game in International Fuzzy Decision Environment: A Possibilitic Approach," *Control and Cybernetics*, Vol. 21, #1, 1992, pp. 277-294.

[962] Shnaider, E., et. al., "The Use of Fuzzy Set Theory for Forecasting Corporate Tax Revenues," *Fuzzy Sets and Systems*, Vol. 31, #2, 1989, pp. 187-204.

[963] Smithson, M., "Applications of Fuzzy Set Concepts to Behavioral Sciences," *Jour. of Mathematical Social Science*, Vol. 2, 1982, pp. 257-274.

[964] Smithson, M., *Fuzzy Set Analysis for Behavioral and Social Sciences*, New York, Springer-Verlag, 1987.

[965] Smithson, M., "Fuzzy Set Theory and the Social Sciences: The Scope for Applications," *Fuzzy Sets and Systems*, Vol. 26, #1, 1988, pp. 1-21.

[966] Spillman, B., et. al., "Coalition Analysis with Fuzzy Sets", *Kybernetes*, Vol. 8, 1979, pp. 203-211.

[967] Tanaka, H., et. al., "A Formulation of Fuzzy Decision Problems and Its Applications to an Investment Problem," *Kybernetics*, Vol. 5, 1976, pp. 25-30.

[968] Teng, J.-Y. et. al., "Transportation Investment Project Selection Under Fuzzy Multi-objective Programming," *Fuzzy Sets and Systems*, Vol. 96, #3, 1998, pp. 259-280.

[969] Turksen, I.B., "Approximate Reasoning for Production Planning," *Fuzzy Sets and Systems*, Vol. 26, #1, 1988, pp. 23-37.

[970] Turksen, I.B. et. al., "An Expert System Prototype for Inventory Capacity Planning: An Approximate Reasoning Approach," *Intern. Jour. of Approximate Reasoning*, Vol. 5, #3, 19981, pp. 223-252.

[971a] Wang, P.P. et. al. (eds.), *Fuzzy Sets: Theory and Applications to Policy Analysis and Information Systems*, New York, Plenum Press 1980.

[971b] Wernerfelt, B., "Semifuzzy Games," *Fuzzy Sets and Systems*, Vol. 19, 1986, pp.21-28

[972] Zadeh, L.A., "Possibility Theory and Soft Data Analysis, in [924b], 1981, pp. 69-129.

[973] Zadeh, L.A., "Fuzzy Systems Theory: A Framework for the Analysis of Humanistic System" in [829a], pp. 25-41.

On Fuzzy Logic

[974] Baldwin, J.F., "A New Approach to Approximate Reasoning Using a Fuzzy Logic," *Fuzzy Sets and Systems*, Vol. 2, #4, 1979, pp. 309-325.

[975] Baldwin, J.F., "Fuzzy Logic and Fuzzy Reasoning," *Intern. J. Man-Machine Stud.*, Vol. 11, 1979, pp. 465-480.

[976] Baldwin, J.F., "Fuzzy Logic and Its Application to Fuzzy Reasoning," In. [996], 1979, pp.96-115.

[977] Baldwin, J.F. et al., "Fuzzy Relational Inference Language," *Fuzzy Sets and Systems*, Vol. 14, #2, 1984, pp. 155-174.

[978] Baldsin, J. and B.W. Pilsworth, "Axiomatic Approach to Implication For Approximate Reasoning With Fuzzy Logic," *Fuzzy Sets and Systems*, Vol. 3, #2, 1980, pp. 193-219.

[979] Baldwin, J.F., et. al., "The Resolution of Two Paradoxes by Approximate Reasoning Using A Fuzzy Logic," *Synthese*, Vol. 44, 1980, pp. 397-420.

[980] Chatterji, B.N., "Character Recognition Using Fuzzy Similarity Relations," in [996], 1982, pp. 131-137.

[981] Fine, K., "Vagueness, Truth and Logic," *Synthese*, Vol. 30, 1975, pp. 265-300.

[982] Fukami, S., et. al., "Some Considerations On Fuzzy Conditional Inference," *Fuzzy Sets and Systems*, Vol. 4, #3, 1980.pp. 243-273.

[983] Gaines, B.R., "Fuzzy Reasoning and the Logic of Uncertainty," *Proc. 6th International Symp. of Multiple-Valued Logic, IEEE 76CH* 1111-4C, 1976, pp. 179-188.

[984] Gaines, B.R., "Foundations of Fuzzy Reasoning," *Inter. Jour. of Man-Machine Studies*, Vol. 8, 1976, pp. 623-668.

[985] Gaines, B.R., "Foundations of Fuzzy Reasoning," in [994], 1977, pp. 19-75.

[986a] Gaines, B. R., "Precise Past, Fuzzy Future," intern Jour. of Man-Machine Stud., Vol. 19, #1,1983, pp.117-134.

[986b] Gains, B.R., "Fuzzy and Probability Uncertainty Logics," Information and Control, Vol.38, 1978, pp.154-169.

[986c] Gains, B. R., "Modeling Practical Reasoning," *Intern. Jour. of Intelligent Systems*, Vol. 8, #1, 1993, pp.51-70.

[987] Giles, R., "Lukasiewics Logic and Fuzzy Set Theory," *Intern. J. Man-Machine Stud.*, Vol. 8, 1976, pp. 313-327.

[988] Giles, R., "Formal System for Fuzzy Reasoning," *Fuzzy Sets and Systems*, Vol. 2, #3, 1979, pp. 233-257.

[989] Goguen, J.A., "The Logic of Inexact Concepts," *Synthese*, Vol. 19, 1969, pp. 325-373.

[990] Gottinger, H.W., "Towards a Fuzzy Reasoning in the Behavioral Science," *Cybernetica*, Vol. 16, #2, 1973, pp. 113-135.

[991] Gottinger, H.W., "Some Basic Issues Connected With Fuzzy Analysis", in H. Klaczro and N. Muller (eds.), *Systems Theory in Social Sciences*, Birkhauser Verlag, Basel, 1976, pp. 323-325.

[992] Gottwald, S., "Fuzzy Propositional Logics," *Fuzzy Sets and Systems*, Vol. 3, #2, 1980, pp. 181-192.

[993] Gupta, M.M. et. al., (Eds.), *Fuzzy automata and Decision Processes*, North Holland, New York, 1977.

[994] Gupta, M.M. and E. Sanchez (Eds.), *Fuzzy Information and Decision Processes,* North Holland, New York, 1982.

[995] Gupta, M.M. et. al., (eds.) *Approximate Reasoning In Decision Analysis,* North Holland, New York, 1982.

[996] Gupta, M.M. et. al., (Eds.), Advances in Fuzzy Sets Theory and Applications, New York North Holland, , 1979.

[997] Hersh, H.M. et. al., "A Fuzzy Set Approach to Modifiers and Vagueness in Natural Language," *J. Experimental*, Vol. 105, 1976, pp. 254-276.

[998] Hohle, U. and E.P. Klement , *Non-classical Logics and Their Applications - A Handbook*, Boston, Kluwer,. 1994.

[999] Kaipov, V., Kh. et. al., "Classification in Fuzzy Environments," in [996], 1982,pp. 119-124,

[1000] Kandel, A., "Codes Over Languages," *IEEE Transactions on Systems Man and Cybernetics*, Vol. 4, 1975, pp. 135-138.

[1001] Kaufman, A., "Progress in Modeling of Human Reasoning of Fuzzy Logic" in[994], 1982, pp. 11-17.

[1002] Lakoff, G., "Hedges: A Study in Meaning Criteria and the Logic of Fuzzy Concepts," *Jour. Philos. Logic,* Vol. 2, 1973, pp. 458-508.

[1003] Lasker, G.E. (Eds.), *Applied Systems and Cybernetics,* Vol. 6, Pergamon Press, New York, 1980.

[1004] Lee, E.T. et. al., "Some Properties of Fuzzy Logic," *Information and Control,* Vol. 19, 1971, pp. 417-431.

[1005] Lee, E.T., et. al., "Notes On Fuzzy Languages," *Information Science*, Vol. 1, 1969, pp. 421-434.

[1006] Lee, R.C.T., "Fuzzy Logic and the Resolution Principle," *Jour. of Assoc. Comput.* Mach., Vol. 19, 1972, pp. 109-119.

[1007] Lee, R.C.T., "Some Properties of Fuzzy Logic," *Information and Control*, Vol. 19, 1971, pp. 417-431.

[1008] LeFaivre, R.A., "The Representation of Fuzzy Knowledge", *Jour. of Cybernetics*, Vol. 4, 1974, pp. 57-66.

[1009] Machina, K.F., "Truth, Belief and Vagueness," *Jour. Philos. Logic*, Vol. 5, 1976, pp. 47-77.

[1010] Mitra, S. and S.K. Pal, " Logical Operation based Fuzzy MLP for Classification and Rule Generation," *Neural Networks*, Vol.7, #2 1994, pp.353-373.

[1011] Mizumoto, M.., "Fuzzy Conditional Inference Under Max-$\odot$ Composition," *Information Sciences*, Vol.27, #3, 1982, pp. 183-207.

[1012] Mizumoto, M. et. al., "Several Methods For Fuzzy Conditional Inference," *Proc. of IEEE Conf. on Decision and Control*, Dec. 12-14, Florida, 1979, pp. 777-782.

[1013] Montero, F.J., "Measuring the Rationality of a Fuzzy Preference Relation," *Busefal.* Vol. 26, 1986, pp. 75-83.

[1014] Morgan, C.G., "Methods for Automated Theorem Proving in Non-Classical Logics," *IEEE Trans. Compt.* C-25, 1976, pp. 852-862.

[1015] Nowakowska, M., "Methodological Problems of Measurements of Fuzzy Concepts in Social Sciences", *Behavioral Sciences*, Vol. 22, #2, 1977, pp. 107-115.

[1016] Negoita, C.V., "Representation Theorems For Fuzzy Concepts," *Kybernetes*, Vol. 4, 1975, pp. 169-174.

[1017] Pinkava, V., "'Fuzzification of Binary and Finite Multivalued Logical Calculi," *Intern. Jour. Man-Machine Stud.*, Vol. 8, 1976, pp. 171-730.

[1018] Skala, H.J., *Non-Archimedean Utility Theory*, D. Reidel Dordrecht, 1975.

[1019] Skala, H.J., "On Many-Valued Logics, Fuzzy Sets, Fuzzy Logics and Their Applications," *Fuzzy Sets and Systems*, Vol. 1, #2, 1978, pp. 129-149.

[1020] Sugeno, M. and T. Takagi, "Multi-Dimensional Fuzzy Reasoning," *Fuzzy Sets and Systems*, Vol. 9, #3, 1983, pp. 313-325.

[1021] Tamburrini, G. et. al., "Some Foundational Problems in the Formalization of Vagueness", in [994],1982, pp. 161-166.

[1022] Tamura, S. et. al., "learning of Formal Language," *IEEE Trans. Syst. Man. Cybernetics*, SMC-3, 1973, pp. 98-102.

[1023] Tan, S K.et al., "Fuzzy Inference Relation Based on the Theory of Falling Shadows," Fuzzy Sets and Systems, Vol. 53, #2, 1993, pp. 179-188.

[1024] Tong, R.M. et. al., "A Critical Assessment of Truth Functional Modification and its Use in Approximate Reasoning", *Fuzzy Sets and Systems*, Vol. 7, #1, 1982.pp. 103-108.

[1025] Van Fraassen, B.C., "Comments: Lakoff's Fuzzy Propositional Logic," in D. Hockney et. al., (Eds.), *Contemporary Research in Philosophical Logic and Linguistic Semantics* Holland, Reild, 1975, pp. 273-277.

[1026] Verma, R.R., "Vagueness and the Principle of the Excluded Middle," *Mind*, Vol. 79, 1970, pp. 66-77.

[1027] Wiredu, J.E., "Truth as a Logical Constant With an Application to the Principle of the Excluded Millde," *Philos. Quart.*,Vol. 25, 1975, pp. 305-317.

[1028] Zadeh, L.A., "Quantitative Fuzzy Semantics," *Inform. Science*, Vol. 3, 1971, pp. 159-176.

[1029] Zadeh, L.A., "A Fuzzy Set Interpretation of Linguistic Hedges," *Jour. Cybernetics*, Vol. 2, 1972, pp. 4-34.

[1030] Zadeh, L.A., "Fuzzy Logic and Its Application to Approximate Reasoning," *Information Processing 74, Proc. IFIP Congress 74*, #3, North Holland, New York, pp. 591-594, 1974.

[1031] Zadeh, L.A, "The Concept of a Linguistic Variable and Its Application to Approximate Reasoning," in K.S. Fu et. al. (eds.), *Learning Systems and Intelligent Robots*, Plenum Press, New York, 1974, pp. 1-10.

[1032] Zadeh, L.A., "Fuzzy Logic and Approximate Reasoning," *Syntheses*, Vol. 30, 1975, pp. 407-428.

[1033] Zadeh, L. A., et al.(eds.), *Fuzzy Logic for the Management of Uncertainty*, New York, Wily and Sons, 1992.

[1034] Zadeh, L.A., et. al., (eds.), *Fuzzy Sets and Their Applications to Cognitive and Decision Processes*, New York, Academic Press, 1974.

On Fuzzy Mathematical Programming and the Theory of Fuzzy Optimization

[1036] Bose, R.K. and Sahani D, "Fuzzy Mappings and Fixed Point Theorems," *Fuzzy Sets and Systems*, Vol. 21, 1987, pp. 53-58.

[1037] Buckley, J.J, "Fuzzy Programming And the Pareto Optimal Set," *Fuzzy Set and Systems,* Vol. 10, #1, 1983, pp. 57-63.

[1038] Butnariu D. "Fixed Points for Fuzzy Mappings," *Fuzzy Sets and Systems*, Vol. 7, 1982, pp.191-207.

[1039] Carlsson, G., "Solving Ill-Structured Problems Through Well Structured Fuzzy Programming," in J.P. Brans (ed.), *Operation Research 81*, Amsterdam, North-Holland, 1981.pp. 467-477.

[1040] Carlsson, C., "Tackling an AMDM - Problem with the Help of Some Results From Fuzzy Set Theory," *European Journal of Operational Research*, Vol. 10, #3, 1982, pp. 270-281.

[1041] Cerny, M., "Fuzzy Approach to Vector Optimization," *Intern. Jour. of General Systems*, Vol. 20, #1, pp. 23-29.

[1042] Chang, C.L., "Interpretation and Execution of Fuzzy Programs," in [806], 1975, pp. 191-218.

[1043] Chang, S.K., "On the Execution of Fuzzy Programs Using Finite State Machines," *IEEE, Trans. Comp.* C-12, 1982, pp. 214-253.

[1044] Chang, S.S. [1985]. "Fixed Point Theorems for Fuzzy Mappings," *Fuzzy Sets and Systems*, 17, 181-187.

[1045] Chang, S.S.L., "Fuzzy Dynamic Programming and the Decision Making Process," in *Proc. 3rd Princeton Conference on Information Science and Systems*, Princeton, 1969, pp. 200-203.

[1046] Chang, S.Y., et. al., "Modeling To Generate Alternatives: A Fuzzy Approach," *Fuzzy Sets and Systems*, Vol. 9 #2, 1983, pp. 137-151.

[1047] Dubois, D. et. al., "Systems of Linear Fuzzy Constraints," *Fuzzy Sets and Systems*. Vol. 3, #1, 1980, pp.37-48.

[1048] Dubois, D., "An Application of Fuzzy Arithmetic to the Optimization of Industrial Machining Processes," *Mathematical Modelling*, Vol. 9, #6, 1987, pp. 461-475.

[1049] Eaves, B.C., "Computing Kakutani Fixed Points," *Journal of Applied Mathematics*, Vol. 21, 1971,pp. 236-244.

[1050] Feng, Y. J., "A Method Using Fuzzy Mathematics to Solve the Vector Maxim Problem," *Fuzzy Set and Systems.* , 1983, Vol. 9, #2, pp.129-136.

[1051] Hamacher, H., et. al., "Sensitivity Analysis in Fuzzy Linear Programming," *Fuzzy Sets and Systems*, Vol. 1, 1978, pp.269-281.

[1052] Hannan, E.L., "On the Efficiency of the Product Operator in Fuzzy Programming with Multiple Objectives," *Fuzzy Sets and Systems*, Vol. 2, #3, 1979, pp. 259-262.

[1053] Hannan, E.L., "Linear Programming with Multiple Fuzzy Goals," *Fuzzy Sets and Systems*, Vol. 6, #3, 1981.pp.235-248.

[1054] Heilpern, S. "Fuzzy Mappings and Fixed Point Theorem," *Journal of Mathematical Analysis and Applications*, Vol. 83, 1981, pp.566-569.

[1055] Ignizio, J.P., et. al., "Fuzzy Multicriteria Integer Programming via Fuzzy Generalized Networks," *Fuzzy Sets and Systems*, Vol. 10, #3, 1983, pp. 261-270.

[1056] Jarvis, R.A., "Optimization Strategies in Adaptive Control: A Selective Survey," *IEEE Trans. Syst. Man. Cybernetics*, SMC-5, 1975, pp. 83-94.

[1057] Jakubowski, R., et. al., "Application of Fuzzy Programs to the Design of Machining Technology," *Bulleting of the Polish Academy of Science*, Vol. 21, 1973, pp. 17-22.

[1058] Kabbara, G., "New Utilization of Fuzzy Optimization Method," in [995], 1982, pp. 239-246.

[1059] Kacprzyk, J. et. al., (Eds), *Optimization Models Using Fuzzy Sets and Possibility Theory*, Boston, Mass., D. Reidel, 1987.

[1060] Kakutani, S. [1941]. "A Generalization of Brouwer's Fixed Point Theorem," *Duke Mathematical Journal*, Vol. 8, pp. 416-427.

[1061] Kaleva, O. [1985]. "A Note on Fixed Points for Fuzzy Mappings", *Fuzzy Sets and Systems,* Vol. 15, 1985, pp. 99-100

[1062] Kandel, A., "On Minimization of Fuzzy Functions," *IEEE Trans. Comp. C-22*, 1973, pp. 826-832.

[1063] Kandel, A., "Comments on Minimization of Fuzzy Functions", *IEEE Trans. Comp.*, C-22, 1973, pp. 217.

[1064] Kandel, A., "On the Minimization of Incompletely Specified Fuzzy Functions," *Information, and Control*, Vol. 26, 1974, pp. 141-153.

[1065] Lai, Y., et. al., *Fuzzy Mathematical Programming*, New York, Springer-Verlag, 1992.

[1066] Leberling, H., "On Finding Compromise Solution in Multcriteria Problems, Using the Fuzzy Min-Operator," *Fuzzy Set and Systems*, Vol. 6, #2, 1981, pp. 105-118.

[1067] Lee, E.S. et. al., "Fuzzy Multiple Objective Programming and Compromise Programming with Pareto Optimum," *Fuzzy Sets and Systems*, Vol. 53, #3, 1993, pp. 275-288.

[1068] Lowen, R., "Connex Fuzzy Sets," *Fuzzy Sets and Systems*, Vol. 3, 1980, pp. 291-310

[1069] Luhandjula, M.K., "Compensatory Operators in Fuzzy Linear Programming with Multiple Objectives," *Fuzzy Sets and Systems*, Vol. 8, #3, 1982, pp. 245-252.

[1070] Luhandjula,M.K., "Linear Programming Under Randomness and Fuzziness," *Fuzzy Sets and Systems*, Vol. 10, #1, 1983. pp. 45-54.

[1071] Negoita, C.V. et. al., "Fuzzy Linear Programming and Tolerances in Planning," *Econ. Group Cybernetic Studies*, Vol. 1, 1976, pp. 3-15.

[1072] Negoita, C.V., and A.C. Stefanescu, "On Fuzzy Optimization," in [995],1982, pp. 247-250.

[1073] Negoita, C.V., "The Current Interest in Fuzzy Optimization," *Fuzzy Sets and Systems*, Vol. 6, #3, 1981, pp. 261-270.

[1074] Negoita, C.V., et. al., "On Fuzzy Environment in Optimization Problems," in J. Rose et. al., (Eds), Modern Trends in Cybernetics and Systems, Springer, Berlin, 1977, pp.13-24.

[1075] Orlovsky, S.A., "On Programming with Fuzzy Constraint Sets," *Kybernetes*, Vol. 6, 1977, pp. 197-201.

[1076] Orlovsky, S.A., "On Formulation of General Fuzzy Mathematical Problem," *Fuzzy Sets and Systems*, Vol. 3, 1980, pp. 311-321.

[1077] Ostasiewicz, W., "A New Approach to Fuzzy Programming," *Fuzzy Sets and Systems*, Vol. 7, #2, 1982. 139-152.

[1078] Pollatschek, M.A., "Hieranchical Systems and Fuzzy-Set Theory," *Kybernetes*, Vol. 6, 1977, pp. 147-151.

[1079] Ponsard, G., "Partial Spatial Equilibra With Fuzzy Constraints," *Journal of Regional Science*, Vol. 22, #2, 1982, pp. 159-175.

[1080] Prade, M., "Operations Research with Fuzzy Data," in P.P. Want et. al., (eds.), *Fuzzy Sets* , New York, Plenum, 1980, pp. 155-170.

[1081] Ralescu, D., "0ptimization in a Fuzzy Environment," in [996], 1979, pp.77-91.

[1082] Ralescu, D.A., "Orderings, Preferences and Fuzzy Optimization" in J. Rose (ed.) *Current Topics in Cybernetics and Systems*, Springer, Berlin, 1978.

[1083] Sakawa, M., *Fuzzy Sets and Interactive Multiobjective Optimization*, New York, Plenum Press, 1993.

[1084] Sakawa, M. et. al., "Feasibility and Pareto Optimality for Multi-objective Nonlinear Programming Problems with Fuzzy Parameters," *Fuzzy Sets and Systems*, Vol. 43, #1, pp. 1-15.

[1085] Tanaka, K., et. al., "Fuzzy Programs and Their Execution," in [1034], 1974, pp. 41-76.

[1086] Tanaka, K., et. al., "Fuzzy Mathematical Programming," *Transactions of SICE*, 1973, pp. 109-115.

[1087] Tanaka, H., et. al., "On Fuzzy-Mathematical Programming," *Journal of Cybernetics*, Vol. 3, #4, 1974, pp. 37-46.

[1088] Tanaka, H., et. al., Fuzzy Linear Programming, Based on Fuzzy Functions," *Bulletin of Univ. of Osaka Prefecture*, Series A, Vol. 29, #2, 1980, pp. 113-125.

[1089] Tsichritzis, D., "A Model for Iterative Computation," *Information Science*, Vol. 5, 1973, pp. 187-197.

[1090] Vira, J., "Fuzzy Expectation Values in Multistage Optimization Problems," *Fuzzy Sets and Systems*, Vol. 6, #2, 1981, pp. 161-168.

[1091] Verdegay, J.L., "Fuzzy Mathematical Programming," in [709], 1982, pp. 231-238.

[1092] Warren, R.H., "Optimality in Fuzzy Topological Polysystems," *Jour. Math. Anal.*, Vol. 54, 1976, pp. 309-315.

[1093] Weiss, M.D., "Fixed Points, Separation and Induced Topologies for Fuzzy Sets," *Jour. Math. Anal. and Appln.*, Vol. 50, 1975, pp. 142-150.

[1094] Wiedey, G., and H.J. Zimmermann, "Media Selection and Fuzzy Linear Programming," *Journal Oper. Res. Society* Vol. 29, 1978, pp. 1071-1084.

[1095] Wilkinson, J., "Archetypes, Language, Dynamic Programming and Fuzzy Sets," in J. Wilkinson et. al., (eds.), *The Dynamic Programming of Human Systems*, MSS New York ,Information Corp., 1973, pp. 44-53.

[1096] Yager, R.R., "Mathematical Programming with Fuzzy Constraints and Preference on the Objective," *Kybernetes*, Vol. 8, 1979, pp. 285-291.

[1097] Zadeh, L.A., "Outline of a New Approach to the Analysis of Complex Systems and Decision Process," in J.L. Cochrane, et. al., (eds), *Multiple Criteria Decision Making*, Columbia, Univ. of South Carolina Press, 1973. Also in *IEEE Transactions on System, Man and Cybernetics* Vol. 1, pp. 28-44

[1098] Zadeh, L. A., "The Role of Fuzzy Logic in the Management of Ucertainty in expert Systems," *Fuzzy Sets and Systems*, Vol. 11, 1983, pp199-227.

[1099] Zimmerman, H.-J., "Description and Optimization of Fuzzy Systems," *Intern. Jour. Gen. Syst.* Vol. 2, #4, 1975, pp. 209-215.

[1100] Zimmerman, H-J, "Fuzzy Programming and Linear Programming with Several Objective Functions," *Fuzzy Sets and Systems*, Vol. 1, #1, 1978, pp. 45-56.

[1101] Zimmerman, H.J., "Applications of Fuzzy Set Theory to Mathematical Programming," *Information Science*, Vol. 36, #1, 1985, pp. 29-58.

On Fuzzy Mathematics

[1102] Bandler, W., et. al., "Fuzzy Power Sets and Fuzzy Implication Operators," *Fuzzy Sets and Systems*, Vol. 4 #1, 1980, pp. 13-30.

[1103] Banon, G., "Distinction between Several Subsets of Fuzzy Measures," *Fuzzy Sets and Systems*, Vol. 5, #3, 1981, pp. 291-305.

[1104] Bellman, R.E., "Mathematics and Human Sciences," in J. Wilkinson et. al. (eds.), *The Dynamic Programming of Human Systems*, New York, MSS Information Corp., 1973, pp. 11-18.

[1105] Bellman, R.E and Glertz, M., "On the Analytic Formalism of the Theory of Fuzzy Sets," *Information Science*, Vol. 5, 1973, pp. 149-156.

[1106] Brown, J.G., "A Note On Fuzzy Sets," *Information and Control*, Vol. 18, pp. 32-39, 1971.

[1107] Butnariu, D., Fixed Points For Fuzzy Mapping," *Fuzzy Sets and Systems*, Vol. 7, #2, pp. 191-207, 1982.

[1108] Butnariu, D., "Decompositions and Range For Additive Fuzzy Measures", *Fuzzy Sets and Syst ems*, Vol. 10, #2, pp. 135-155, 1983

[1109] Cerruti, U., "Graphs and Fuzzy Graphs" in [994], 1982, pp.123-131.

[1110] Chakraborty, M.K. et. al., "Studies in Fuzzy Relations Over Fuzzy Subsets," *Fuzzy Sets and Systems*, Vol. 9, #1, 1983,pp. 79-89.

[1111] Chang, C.L., "Fuzzy Topological Spaces," *J. Math. Anal. and Applications*, Vol. 24, 1968,pp. 182-190.

[1111] Chang, S.S.L., "Fuzzy Mathematics, Man and His Environment", *IEEE Transactions on Systems, Man and Cybernetics*, SMC-2 1972, pp. 92-93,

[1113] Chang, S.S.L. et. al., "On Fuzzy Mathematics and Control," *IEEE Transactions, System, Man and Cybernetics*, SMC-2, 1972, pp. 30-34.

[1114] Chang, S.S., "Fixed Point Theorems for Fuzzy Mappings," *Fuzzy Sets and Systems*, Vol. 17, 1985, pp. 181-187.

[1115] Chapin, E.W., "An Axiomatization of the Set Theory of Zadeh", *Notices, American Math. Society*, 687-02-4 754, 1971.

[1116] Cheng-Zhong, L., "Generalized Inverses of Fuzzy Matrix" in [995], 1982, pp.57-60.

[1117] Chitra, H., and P.V. Subrahmanyam, "Fuzzy Sets and Fixed Points," *Jour. of Mathematical Analysis and Application*, Vol. 124, 1987, pp. 584-590.

[1118] Czogala, J. et. al., Fuzzy Relation Equations On a Finite Set," *Fuzzy Sets and Systems,* Vol. 7, #1,1982. pp. 89-101.

[1119] DiNola, A. et. al., (Eds.), *The Mathematics of Fuzzy Systems*, Koln, Verlag TUV Rheinland, 1986.

[1120] DiNola, A. et. al., "On Some Chains of Fuzzy Sets," *Fuzzy Sets and Systems*, Vol. 1980.4, #2, pp. 185-191,

[1121] Dombi, J., "A General Class of Fuzzy Operators, the DeMorgan Class of Fuzzy Operators and Fuzzy Measures Induced by Fuzzy Operators," *Fuzzy Sets and Systems*, Vol. 8, #2, 1982,.pp. 149-163.

[1122] Dubois, D. and H. Prade, "Towards Fuzzy Differential Calculus, Part I: Integration of Fuzzy Mappings," *Fuzzy Sets and Systems*, Vol. 8, #1,1982 , pp. 1-17.

[1123] Dubois,D. and H. Prade, "Towards Fuzzy Differential Calculus, Part 2: Integration On Fuzzy Intervals," *Fuzzy Sets and Systems*, Vol. 8, #2, pp. 105-116, 1982.

[1124] Dubois.D and H. Prade, "Towards Fuzzy Differential Calculus, Part 3: Differentiation", *Fuzzy Sets and Systems*, Vol. 8, #3, 1982, pp. 225-233.

[1125] Dubois, D. and H. Prade, *Fuzzy Sets and Systems*, New York, Academic Press, 1980.

[1126] Dubois, "Fuzzy Real Algebra: Some Results," *Fuzzy Sets and Systems*, Vol. 2, #4, pp. 327-348, 1979.

[1127] Dubois, D. and H. Prade , "Gradual Inference rules in approximate reasoning," *Information Sciences*, Vol. 61(1-2), 1992, pp. 103-122.

[1128] Dubois, D. and H. Prade, "On the combination of evidence in various mathematical frameworks,." In: Flamm. J. and T. Luisi, (eds.), *Reliability Data Collection and Analysis*. Kluwer, Boston, 1992,pp. 213-241.

[1129] Dubois, D. and H. Prade, (eds.) , *Readings in Fuzzy Sets for Intelligent Systems*. San Mateo, CA, Morgan Kaufmann,. 1993

[1130] Dubois, D. and H. Prade, "Ffuzzy sets and probability: Misunderstanding, bridges and gaps." *Proc. Second IEEE Intern. Conf. on Fuzzy Systems*, San Francisco, 1993, pp. 1059-1068.

[1131] Dubois, D. and H. Prade [1994], "A survey of belief revision and updating rules in various uncertainty models," *Intern. J. of Intelligent Systems*, Vol. 9, #1, pp. 61-100.

[1132] Erceg, M.A., "Functions, Equivalence Relations, Quotient Spaces and Subsets in Fuzzy Set Theory," *Fuzzy Sets and Systems*, Vol. 3, 1980, #1, pp. 79-92.

[1133] Feng, Y.-J., "A Method Using Fuzzy Mathematics to Solve the Vectormaximum Problem," *Fuzzy Sets and Systems*, Vol. 9, #2, 1983, pp. 129-136.

[1134] Filev, D.P. et. al., "A Generalized Defuzzification Method via Bag Distributions," *Intern. Jour. of Intelligent Systems*, Vol. 6, #7, 1991, pp. 687-697.

[1135] Goetschel, R. Jr., et. al., "Topological Properties of Fuzzy Number," *Fuzzy Sets and Systems*, Vol. 10, #1, pp. 87-99, 1983.

[1136] Goguen, J.A., "Mathematical Representation of Hierarchically Organized System," in E.O. Attinger (ed.) *Global System Dynamics*, pp. 111-129, S. Karger, Berlin, 1970.

[1137] Goodman, I.R., "Fuzzy Sets As Random Level Sets: Implications and Extensions of the Basic Results," in [1159], 1981, pp. 2756-2766.

[1138] Goodman, I.R., "Fuzzy Sets As Equivalence Classes of Random Sets" in [1216],1992, pp. 327-343.

[1139] Gupta, M.M. et. al., (Eds), *Fuzzy Antomata and Decision Processes*, New York , North-Holland, 1977.

[1140] Gupta, M.M. and E. Sanchez (Eds.), *Fuzzy Information and Decision Processes*, New York, North-Holland, 1982.

[1141] Gupta,M. M. and E. Sanchez , (Eds.), *Approximate Reasoning In Decision Analysis*, New York, North-Holland, 1982.

[1142] Gupta, M.M et. al., (eds.), *Advances in Fuzzy Set Theory and Applications*, North-Holland, New York, 1979.

[1143] Higashi, M. and G.J. Klir, "On measure of fuzziness and fuzzy complements," *Intern. J. of General Systems*, Vol. 8 #3, 1982, pp. 169-180.

[1144] Higashi, M. and G.J. Klir, "Measures of uncertainty and information based on possibility distributions," *Intern. J. of General Systems*, Vol. 9 #1, 1983, pp. 43-58.

[1145] Higashi, M. and G.J. Klir, "On the notion of distance representing information closeness: Possibility and probability distributions," *Intern. J. of General Systems,* Vol. 9 #2, 1983, pp. 103-115.

[1146] Higashi, M. and G.J. Klir , "Resolution of finite fuzzy relation equations," *Fuzzy Sets and Systems*, Vol. 13, #1,1984, pp. 65-82.

[1147] Higashi, M. and G.J. Klir , "Identification of fuzzy relation systems," *IEEE Trans. on Systems, Man, and Cybernetics*, Vol.14 #2, 1984, pp. 349-355.

[1148] Jin-wen, Z., "A Unified Treatment of Fuzzy Set Theory and Boolean Valued Set theory: Fuzzy Set Structures and Normal Fuzzy Set Structures," *Jour. Math. Anal. and Applications*, Vol. 76, #1, 1980, pp. 197-301.

[1149] Kandel, A., *Fuzzy Mathematical Techniques with Applications*, Reading, MA, Addison-Wesley 1986.

[1150] Kandel, A. and W.J. Byatt, "Fuzzy Processes," *Fuzzy Sets and Systems*, Vol. 4, #2, 1980, pp. 117-152.

[1151] Kaufmann, A. and M.M. Gupta , *Introduction to fuzzy arithmetic: Theory and applications*, New York Van Nostrand Rheinhold,. 1991.

[1152] Kaufmann, A., *Introduction to the Theory of Fuzzy Subsets*, Vol. 1, New York, Academic Press, 1975.

[1153a] Kaufmann,A., *Theory of Fuzzy Sets*, Paris, Merson Press, 1972.

[1153b] Kaufmann, A. et. al., *Fuzzy Mathematical Models in Engineering and Management Science*, New York, North-Holland,1988

[1154] Kim, K.H. et. al., "Generalized Fuzzy Matrices," *Fuzzy Sets and Systems*, Vol. 4, #3, 1980, pp. 293-315.

[1155] Klement, E.P. and W. Schwyhla, "Correspondence Between Fuzzy Measures and Classical Measures," *Fuzzy Sets and Systems,* Vol. 7, #1, 1982.pp. 57-70.

[1156] Kokawa, M. et. al. "Fuzzy-Theoretical Dimensionality Reduction Method of Multi-Dimensional Quality," in [994], 1982, pp. 235-250.

[1157] Kramosil, I. et. al., "Fuzzy Metrics and Statistical Metric Spaces," *Kybernetika,* Vol. 11, 1975, pp. 336-344.

[1158] Kruse, R., "On the Construction of Fuzzy Measures," *Fuzzy Sets and Systems,* Vol. 8, #3, 1982, pp. 323-327.

[1159] Lasker, G.E. (ed.), *Applied Systems and Cybernetics, Vol. VI: Fuzzy Sets and Systems*, Pergamon Press, New York, 1981.

[1160] Lake, L., "Fuzzy Sets, Multisets and Functions", I. *London Math. Soc.* Vol. 12 , #2, 1976, pp. 323-326.

[1161] Lientz, B.P., "On Time Dependent Fuzzy Sets", *Inform, Science*, Vol. 4, 1972, pp. 367-376.

[1162] Lowen, R., "Fuzzy Uniform Spaces," *Jour. Math. Anal. Appl.*, Vol. 82, #2 1981, pp. 367-376.

[1163] Martin, H.W., "Weakly Induced Fuzzy Topological Spaces," *Jour. Math. Anal. and Application*, Vol. 78, 1980, pp. 634-639.

[1164] Michalek, J., "Fuzzy Topologies," *Kybernetika*, Vol. 11, 1975, pp. 345-354.

[1165] Negoita, C.V. et. al., *Applications of Fuzzy Sets to Systems Analysis*, Wiley and Sons, New York, 1975.

[1166] Negoita, C.V., "Representation Theorems For Fuzzy Concepts," *Kybernetes*, Vol. 4, 1975, pp. 169-174.

[1167] Negoita, C.V. et. al., "On the State Equation of Fuzzy Systems," *Kybernetes*, Vol. 4, 1975, pp. 231-214.

[1168] Negoita, C.V., "Fuzzy Sets in Topoi," *Fuzzy Sets and Systems*, Vol. 8, #1, 1982, pp. 93-99

[1169] Nguyen, H. T., "Possibility Measures and Related Topics," in [994], 1982, pp 197-202

[1170] Nowakowska, M., "Some Problems in the Foundations of Fuzzy Set Theory," in [994], 1982, pp. 349-360.

[1171] Ovchinnikov, S.V., "Structure of Fuzzy Binary Relations," *Fuzzy Sets and Systems*, Vol. 6, #2, 1981, pp. 169-195.

[1172] Pedrycz, W., "Fuzzy Relational Equations with Generalized Connectives and Their Applications," *Fuzzy Sets and Systems*, Vol. 10, #2, 1983, pp. 185-201.

[1173] Raha, S. et. al., "Analogy Between Approximate Reasoning and the Method of Interpolation," *Fuzzy Sets and Systems*, Vol. 51, #3, 1992, pp. 259-266.

[1174] Ralescu, D., "Toward a General Theory of Fuzzy Variables," *Jour. of Math. Analysis and Applications*, Vol. 86, #1, 1982, pp. 176-193.

[1175] Rao, M.B. et. al., "Some Comments On Fuzzy Variables," *Fuzzy Sets and Systems*, Vol. 6 #2, 1981, pp. 285-292.

[1176] Rodabaugh, S.E., "Fuzzy Arithmetic and Fuzzy Topology," in [1159], 1981, pp. 2803-2807.

[1177] Rodabaugh, S. et. al., (eds.), *Application of Category Theory to Fuzzy Subsets*, Boston, Mass., Kluwer 1992.

[1178] Roubens, M. et. al., "Linear Fuzzy Graphs," *Fuzzy Sets and Systems*, Vol. 10, #1, 1983, pp. 798-86.

[1179] Rosenfeld, A., "Fuzzy Groups," *Jour. Math. Anal. Appln.*, Vol. 35, 1971, pp. 512-517.

[1180] Rosenfeld, A., "Fuzzy Graphs," in [1034], 1974, pp. 77-95.

[1181] Rubin, P.A., "A Note on the Geometry of Reciprocal Fuzzy Relations," *Fuzzy Sets and Systems*, Vol. 7, #3, 1982, pp. 307-309.

[1182] Ruspini, E.H., "Recent Developments In Mathematical Classification Using Fuzzy Sets," in [1159], 1981, pp. 2785-2790.

[1183] Sanchez, E., "Resolution of Composite Fuzzy Relation Equations," *Information and Control*, Vol. 3, 1976, pp. 39-47.

[1184] Santos, E.S., "Maximin, Minimax and Composite Sequential Machines," *Jour. Math. Anal. and Appln.* Vol. 24, 1968, pp. 246-259.

[1185] Santos, E.S., "Maximin Sequential Chains,"*Jour. of Math. Anal. and Appln.*, Vol. 26, 1969, pp. 28-38.

[1186] Santos, E.S., "Fuzzy Algorithms," *Inform. and Control*, Vol. 17, 1970, pp. 326-339.

[1187] Sarkar, M., "On Fuzzy Topological Spaces", *Jour. Math. Anal. Appln.*, Vol. 79, 1981, pp. 384-394.

[1188] Skala, H. et. al. (eds.), *Aspects of Vagueness*, Boston, D. Reidel, 1984.

[1189] Smithson, M., *Ignorance and Uncertainty, Emerging Paradigms*, New York, Springer-Verlag, 1989.

[1190] Sugeno, M., et. al., "Multidimensional Fuzzy Reasoning," *Fuzzy Sets and Systems*, Vol. 9, #3, 1983, pp. 313-325..

[1191] Sugeno, M., "Inverse Operation of Fuzzy Integrals and Conditional Fuzzy Measures," *Transactions SICE*, Vol. 11, 1975, pp. 709-714.

[1192] Terano, T. et. al., *Fuzzy Systems Theory and its Applications*, New York, Academic Press, 1992.

[1193] Terano, T. et. al., *Applied Fuzzy Systems*, New York, AP Professional, 9994.

[1194] Thornber, K.K., "A Key to Fuzzy Logic Inference," *Intern. Jour. of Approximate Reasoning*, Vol. 8, #2, 1993, pp. 105-129.

[1195] Triantaphyllon, E. et. al., "The Problem of Determining Membership Values in Fuzzy Sets in Real World Situations," in D.E. Brown et. al. (eds), *Operations Research and Artificial Intelligence: The Integration of Problem-solving Strategies*, Boston, Mass., Kluwer, 1990, pp. 197-214.

[1196] Tsichritzis, D., "Participation Measures," *Jour. Math. Anal. and Appln.*, Vol. 36, 1971, pp. 60-72.

[1197] Tsichritzis, D., "Approximation and Complexity of Functions on the Integers," *Inform. Science*, Vol. 4, 1971, pp. 70-86.

[1198] Turksens, I.B., "Four Methods of Approximate Reasoning with Interval-Valued Fuzzy Sets," *Intern. Journ. of Approximate Reasoning*, Vol. 3, #2, 1989, pp. 121-142.

[1199] Turksen, I.B., "Measurement of Membership Functions and Their Acquisition," *Fuzzy Sets and Systems*, Vol. 40, #1, 1991, pp. 5-38.

[1200] Verdegay, J. et. al., *The Interface Between Artificial Intelligence and Operations Research in Fuzzy Environment*, Koln, Verlag TUV Rheinland, 1989.

[1201] Wang, L.X., *Adaptive Fuzzy Sets and Control: Design and Stability Analysis*, Englewood Cliffs, N.J. Prentice Hall, 1994.

[1202] Wang, P.P. (ed.), *Advances in Fuzzy Sets, Possibility Theory, and Applications*, New York, Plenum Press, 1983.

[1203] Wang, P.P. (ed.), *Advances in Fuzzy Theory and Technology, Vol. 1,* Durham, NC, Bookwright Press, 1992.

[1204] Wang, P.Z. et. al. (eds.), *Between Mind and Computer: Fuzzy Science and Engineering*, Singapore, World Scientific Press, 1993.

[1205] Wang, P.Z., "Contactability and Fuzzy Variables," *Fuzzy Sets and Systems*, Vol. 8, #1, 1982, pp. 81-92.

[1206] Wang, S., "Generating Fuzzy Membership Functions: A Monotonic Neural Network Model," *Fuzzy Sets and Systems*, Vol. 61, #1, 1994, pp. 71-82.

[1207] Wang, Z. et. al., *Fuzzy Measure Theory*, New York, Plenum Press, 1992.

[1208] Whalen, T. et. al., "Usuality, Regularity, and Fuzzy Set Logic," *Intern. Jour. of Approximate Reasoning*, Vol. 6, #4, 1992, pp. 481-504.

[1209a] Wierzchon, S.T., "An Algorithm for Identification of Fuzzy Measure," *Fuzzy Sets and Systems*, Vol. 9, #1, 1983, pp. 69-78.

[1209b] Windeknecht, T. G., *General Dynamic Process*, New York, Academic Press, 1971.

[1210] Wong, C.K., "Fuzzy Points and Local Properties of Fuzzy Topology," Jour. Math. Anal. and Appln., Vol. 46, 19874, pp. 316-328.

[1211] Wong, C.K., "Fuzzy Topology," in [1034], 1975, pp. 171-190.

[1212] Wong, C.K., "Categories of Fuzzy Sets and Fuzzy Topological Spaces," Jour. Math. Anal. and Appln., Vol. 53, 1976, pp. 704-714.

[1213] Wygralak, M., "Fuzzy Inclusion and Fuzzy Equality of two Fuzzy Subsets," Fuzzy Sets and Systems, Vol. 10, #2, 1983, pp. 157-168.

[1214] Xu, W-L. et. al., "An Algorithm to Solve the Max-Min. Composite Fuzzy Relational Equation," in [996], 1979, pp. 47-50.

[1215] Yager, R.R., "On the Lack of Inverses in Fuzzy Arithmetic," *Fuzzy Sets and Systems*, Vol. 4, #1, 1980, pp. 73-82.

[1216] Yager, R.R. (Ed.), *Fuzzy Set and Possibility Theory: Recent Development*, New York, Pergamon Press 1992.

[1217] Yager, R.R., "Fuzzy Subsets with Uncertain Membership Grades," *IEEE Transactions on Systems, Man and Cybernetics*, Vol. 14, #2, 1984, pp. 271-275.

[1218] Yager, R.R. et. al., *Essentials of Fuzzy Modeling and Control*, New York, John Wiley, 1994.

[1219] Yager, R.R. et. al. (eds.), An Introduction to Fuzzy Logic Applications in *Intelligent Systems*, Boston, Mass., Kluwer 1992.

[1220] Yager, R.R. et. al., (Eds.), *Fuzzy Sets, Neural Networks, and Soft Computing*, New York, Nostrand Reinhold, 1994.

[1221] Yager, R.R., "On the Theory of Fuzzy Bags," *Intern. Jour. of General Systems*, Vol. 13, #1, 1986, pp. 23-37.

[1222] Yager, R.R., Cardinality of Fuzzy Sets via Bags," *Mathematical Modelling*, Vol. 9, #6, 1987, pp. 441-446.

[1223] Yan, J. et. al., *Using Fuzzy Logic: Towards Intelligent Systems*, Englewood Cliff, N.J., Prentice Hall, 1994.

[1224] Yang, X.M., "A Note on Convex Fuzzy Sets," *Fuzzy Sets and Systems*, Vol. 53, #1, 1993, pp. 117-118.

[1225] Ying, M.S., "Some Notes on Multidimensional Fuzzy Reasoning," *Cybernetics and Systems*, Vol. 19, #4, 1988, pp. 281-293.

[1226] Zadeh, L.A., "A Computational Theory of Decompositions," *Intern. Jour. of Intelligent Systems*, Vol. 2, #1, 1987, pp. 39-63.

[1227] Zadeh, L.A., "The Birth and Evolution of Fuzzy Logic," *Intern. Jour. of General Systems*, Vol. 17, #(2-3) 1990, pp. 95-105.

[1228] Zadeh, L.A. et. al., Fuzzy Logic for *the Management of Uncertainty*, New York, John Wiley, 1992.

[1229] Zimmerman, H.J., *Fuzzy Set Theory and Its Applications,* Boston, Mass, Kluwer, 1985.

On Fuzzy Probability, Possibility and Statistics

[1230] Bandemer, H., “From Fuzzy Data to Functional Relations,” Mathematical Modelling, Vol. 6, 1987, pp. 419-426.

[1231] Bandemer, H. et. al., *Fuzzy Data Analysis*, Boston, Mass, Kluwer, 1992.

[1232] Kruse, R. et. al., *Statistics with Vague Data,* Dordrecht, D. Reidel Pub. Co., 1987.

[1233] Chang, R.L.P., et. al., “Applications of Fuzzy Sets in Curve Fitting,” *Fuzzy Sets and Systems*, Vol. 2, #1, pp. 67-74.

[1234] Chen, S.Q., “Analysis for Multiple Fuzzy Regression,” *Fuzzy Sets and Systems*, Vol. 25, #1, pp. 56-65.

[1235] Celmins, A., “Least Squares Model Fitting to Fuzzy Vector Data,” *Fuzzy Sets* and Systems, Vol. 22, #3, 1987, pp. 245-270.

[1236] Celmins, A., “Multidimensional Least-Squares Fitting of Fuzzy Model,” *Mathematical Modelling*, Vol. 9, #9, pp. 669-690.

[1237] El Rayes, A.B. et. al., “Generalized Possibility Measures,” *Information Sciences,* Vol. 79, 1994, pp. 201-222.

[1238] Dumitrescu, D., “Entropy of a Fuzzy Process,” *Fuzzy Sets and Systems*, Vol. 55, #2, 1993, pp. 169-177.

[1239] Delgado, M. et. al., “On the Concept of Possibility-Probability Consistency,” *Fuzzy Sets and Systems*, Vol. 21, #3, 1987, pp. 311-318.

[1240] Devi, B.B. et. al., “Estimation of Fuzzy Memberships from Histograms,” *Information Sciences*, Vol. 35, #1, 1985, pp. 43-59.

[1241] Diamond, P., “Fuzzy Least Squares”, *Information Sciences*, Vol. 46, #3, 1988, pp. 141-157.

[1242] Dubois, D. et. al., "Fuzzy Sets Probability and Measurement," *European Jour. of Operational Research*, Vol. 40, #2, 1989, pp. 135-154.

[1243] Dubois, D. et al., "A Class of Fuzzy Measures based on Triangular Norms," *Intern. Jour. of Gen. System*, Vol.8, pp.43-61.

[1244] Fruhwirth-Schnatter, S., "On Statistical Inference for Fuzzy Data with Applications to Descriptive Statistics," *Fuzzy Sets and Systems*, Vol. 50, #2, 1992, pp. 143-165.

[1245] Fruhwirth-Schnatter, S., "On Fuzzy Bayesian Inference," *Fuzzy Sets and Systems*, Vol. 60, #1, 1993, pp. 41-58.

[1246] Gaines, B.R., "Fuzzy and Probability Uncertainty logics," *Information and Control*, Vol. 38, #2, 1978, pp. 154-169.

[1247] Geer, J.F. et. al., "Discord in Possibility Theory," *International Jour. of General Systems*, Vol. 19, 1991, pp. 119-132.

[1248] Geer, J.F. et. al., "A Mathematical Analysis of Information-Processing Transformation Between Probabilistic and Possibilistic Formulation of Uncertainty," *International Jour. of General Systems*, Vol. 20, #2, 1992, pp. 14-176.

[1249] Gharpuray, M. et. al., "Fuzzy Linear Regression, Analysis of Cellulose hydrolysis," *Chemical Engineering Commun.*, Vol. 41, 1986, pp. 299-314.

[1250] Gil, M.A., "Fuzziness and Loss of Information in Statistical Problems," *IEEE Transaction on Systems, Man and Cybernetic*, Vol. 17, #6, 1987, pp. 1016-1025.

[1251] Goodman, I.R. et. al., *Uncertainty Models for Knowledge Based Systems*, New York, North-Holland, 1985.

[1252] Grabish, M. et. al., *Fundamentals of Uncertainty Calculi with Application to Fuzzy Systems*, Boston, Mass., Kluwer, 1994.

[1253] Guan, J.W. et. al., *Evidence Theory and Its Applications*, Vol. 1, New York, North-Holland, 1991.

[1254] Guan, J.W. et. al., *Evidence Theory and Its Applications*, Vol. 2, New York, North-Holland, 1992.

[1255] Ishibuchi, H. et. al., "Fuzzy Regression, Analysis Using Neural Networks," *Fuzzy Sets and Systems*, Vol. 50, #3, 1992, pp. 257-265.

[1256] Hisdal, E., Are Grades of Membership Probabilities?," *Fuzzy Sets and Systems*, Vol. 25, #3, 1988, pp. 349-356.

[1257] Klir, G.J., "Where Do we Stand on Measures of Uncertainty, Ambignity, Fuzziness and the like?" *Fuzzy Sets and Systems*, Vol. 24, #2, 1987, pp. 141-160.

[1258] Klir, G.J. et. al., *Fuzzy Sets, Uncertainty and Information*, Englewood Cliff, Prentice Hll, 1988.

[1259] Klir, G. J. et. al. ,"Probability-Possibility Transformations: A Comparison," *Intern. Jour. of General Systems*, Vol. 21, #3, 1992, pp. 291-310.

[1260] Kosko, B., "Fuzziness vs Probability," *Intern. Jour. of General Systems*, Vol. 17, #(1-3) 1990, pp. 211-240.

[1261] Kruse, R.E. et. al., *Uncertainty and Vagueness in Knowledge Based Systems: Numerical Methods*, New York, Springer-Verlag 1991.

[1262] Manton, K.G. et. al., *Statistical Applications Using Fuzzy Sets*, New York, John Wiley, 1994.

[1263] Meier, W., et. al., "Fuzzy Data Analysis: Methods and Indistrial Applications," *Fuzzy Sets and Systems*, Vol. 61, #1, 1994, pp. 19-28.

[1264] Nakamura, A., et. al., "A logic for Fuzzy Data Analysis," *Fuzzy Sets and Systems,* vol. 39, #2, 1991, pp. 127-132.

[1265] Negoita, C.V. et. al., *Simulation, Knowledge-Based Compting and Fuzzy Statistics*, New York, Van Nostrand Reinhold, 1987.

[1266] Nguyen, H.T., "Random Sets and Belief Functions," *Jour. of Math. Analysis and Applications*, Vol. 65, #3, 1978, pp. 531-542.

[1267] Prade, H. et. al., "Representation and Combination of Uncertainty with belief Functions and Possibility Measures," *Comput. Intell.* ,Vol. 4, 1988, pp. 244-264.

[1268] Puri, M.L. et. al., "Fuzzy Random Variables," *Jour. of Mathematical Analysis and Applications*, Vol. 114, #2, 1986, pp. 409-422.

[1269] Sakawa, M. et. al., "Multiobjective Fuzzy linear Regression Analysis for Fuzzy Input-Output Data," *Fuzzy Sets and Systems*, Vol. 47, #2, 1992, pp. 173-182.

[1270] Savic, D.A., et. al., "Evaluation of Fuzzy Linear Regression Models," *Fuzzy Sets and Systems*, Vol. 39, #1, 1991, pp. 51-64.

[1271] Schneider, M. et. al., "Properties of the Fuzzy Expected Values and the Fuzzy Expected Interval," *Fuzzy Sets and Systems*, Vol. 26, #3, 1988, pp. 373-385.

[1272] Sudkamp, T., "On Probability-Possibility Transformations," *Fuzzy Sets and Systems*, Vol. 51, #1, 1992, pp. 73-82.

[1273] Tanaka, H., "Fuzzy Data Analysis by Possibilistic Linear Models," *Fuzzy Sets* and *Systems*, Vol. 24, #3, 1987, pp. 363-375.

[1274] Tanaka, H. et. al., "Possibilistic Linear Regression Analysis for Fuzzy Data," *European Jour. of Operational Research*, Vol. 40, #3, 1989, pp. 389-396.

[1275] Walley, P., *Statistical Reasoning with Imprecise Probabilities*, London Chapman and Hall, 1991.

[1276] Wang, G.Y. et. al., "The Theory of Fuzzy Stochastic Processes," *Fuzzy Sets and Systems*, Vol. 51, #2 1992, pp. 161-178.

[1277] Wang, X. et. al., "Fuzzy Linear Regression Analysis of Fuzzy Valued Variable," *Fuzzy Sets and Systems*, Vol. 36, #1, 1990, pp. 125-136.

Index

A

B

C

D

E

F

G

H

I

L

M

N

O

P

Printing: Strauss GmbH, Mörlenbach
Binding: Schäffer, Grünstadt